JUDAISM, SCIENCE, AND MORAL RESPONSIBILITY

OTHER VOLUMES IN THE ORTHODOX FORUM SERIES

Rabbinic Authority and Personal Autonomy
edited by
Moshe Z. Sokol

Jewish Tradition and the Non-Traditional Jew
edited by
Jacob J. Schacter

Israel as a Religious Reality
edited by
Chaim I. Waxman

Modern Scholarship in the Study of Torah: Contributions and Limitations
edited by
Shalom Carmy

Tikkun Olam: Social Responsibility in Jewish Thought and Law
edited by
David Shatz, Chaim I. Waxman, and Nathan J. Diament

Engaging Modernity: Rabbinic Leaders and the Challenge of the Twentieth Century
edited by
Moshe Z. Sokol

Jewish Perspectives on the Experience of Suffering
edited by
Shalom Carmy

Jewish Business Ethics: The Firm and Its Stakeholders
edited by
Aaron Levine and Moses Pava

Tolerance, Dissent, and Democracy: Philosophical, Historical, and Halakhic Perspectives
edited by
Moshe Z. Sokol

Jewish Spirituality and Divine Law
edited by
Adam Mintz and Lawrence Schiffman

JUDAISM, SCIENCE, AND MORAL RESPONSIBILITY

Edited by
Yitzhak Berger and
David Shatz

ROWMAN & LITTLEFIELD PUBLISHERS, INC.
Lanham • Boulder • New York • Toronto • Oxford

ROWMAN & LITTLEFIELD PUBLISHERS, INC.

Published in the United States of America
by Rowman & Littlefield Publishers, Inc.
A wholly owned subsidiary of The Rowman & Littlefield Publishing Group, Inc.
4501 Forbes Boulevard, Suite 200, Lanham, Maryland 20706
www.rowmanlittlefield.com

PO Box 317
Oxford
OX2 9RU, UK

British Library Cataloguing in Publication Information Available

Library of Congress Cataloging-in-Publication Data

Orthodox Forum (14th : 2002)
Judaism, science, and moral responsibility / edited by Yitzhak Berger and David Shatz.
p. cm. — (The Orthodox Forum)
Proceedings of the 14th Orthodox Forum, held in 2002.
Includes bibliographical references and index.
ISBN 978-0-7425-4595-3 (cloth : alk. paper)
1. Judaism and science—Congresses. 2. Science—Moral and ethical aspects—Congresses. 3. Jewish ethics—Congresses. 4. Repentance—Judaism—Congresses. I. Berger, Yitzhak. II. Shatz, David. III. Title. IV. Series: Orthodox Forum series.
BM538.S3O78 2002
296.3'75—dc22 2005009052

Printed in the United States of America

∞™ The paper used in this publication meets the minimum requirements of American National Standard for Information Sciences—Permanence of Paper for Printed Library Materials, ANSI/NISO Z39.48-1992.

Contents

Orthodox Forum: Fourteenth Conference vii

Series Editor's Introduction xi

Introduction 1
Yitzhak Berger and David Shatz

Science, Free Will, and Jewish Thought

1 A Scientific Perspective on Human Choice 13
Haim Sompolinsky

2 Genetics and Morality 45
Robert Pollack

3 Is Matter All that Matters? Judaism, Free Will, and the Genetic and Neuroscientific Revolutions 54
David Shatz

4 Use It or Lose It: On the Moral Imagination of Free Will 104
Shalom Carmy

Personal Accountability and the Religious Community

5 Choice-Diminished Behavior and Religious-Communal Policy 155
Basil Herring

6 If an Abuser Cannot Control His Impulses, What Is the Responsibility of Other Adults in the Community? A Response to Basil Herring 169
Rivkah Teitz Blau

Psychological Perspectives on the Process of Repentance

7 Psychotherapy and *Teshuvah*: Parallel and Overlapping Systems for Change 175
Michelle Friedman and Rachel Yehuda

8 To Whom, to Where, and to When Does One "Return" in *Teshuvah*? 189
Moshe Halevi Spero

Index 293

About the Series Editor 299

About the Contributors 301

Orthodox Forum: Fourteenth Conference

ORTHODOX FORUM
FOURTEENTH CONFERENCE
Sunday & Monday, March 17–18, 2002

Park East Synagogue
164 East 68th Street
New York City

*LIST OF PARTICIPANTS

Rabbi Elchanan Adler	Mazer Yeshiva Program/RIETS, New York, NY
Dean Norman Adler	Yeshiva College, New York, NY
Dr. Howard Apfel	New York Presbyterian Hospital/Columbia University, New York, NY
Dr. Stuart Apfel	Montefiore Medical Center, Bronx, NY
Mrs. Leora Bednarsh	Israel
Prof. David Berger	Brooklyn College, New York, NY
Prof. Ditza Berger	Hebrew Academy of the Five Towns & Rockaway, Cedarhurst, NY
Rabbi Gedalyah Berger	Fleetwood Synagogue, Mt. Vernon, NY
Rabbi Yitzhak Berger	Hunter College, New York, NY
Rabbi Ari Berman	The Jewish Center, New York, NY
Dr. Miriam Berman	Biologist, Teaneck, NJ
Dr. Rivkah Blau	Author and Lecturer, New York, NY

Rabbi Yosef Blau	RIETS/Yeshiva University, New York, NY
Dr. Judith Bleich	Touro College, New York, NY
Dr. Abba Borowich	Psychiatrist, Mt. Sinai Medical Center, New York, NY
Dr. Alan Brill	Yeshiva University, New York, NY
Rabbi Shalom Carmy	Yeshiva University, New York, NY
Rabbi Mark Dratch	Congregation Agudath Sholom, Stamford, CT
Dr. Yaakov Elman	Yeshiva University, New York, NY
Rabbi Daniel Feldman	Yeshiva University, New York, NY
Rabbi Joshua Flug	RIETS, Wexner Kollel Elyon, New York, NY
Dr. Michelle Friedman	Mt. Sinai Medical Center, New York, NY
Rabbi Mark Gottlieb	Maimonides School, Brookline, MA
Prof. Jonathan Helfand	Brooklyn College, Brooklyn, NY
Rabbi Nathaniel Helfgot	Yeshivat Chovevei Torah, New York, NY
Rabbi Basil Herring	Rabbinical Council of America, New York, NY
Rabbi Robert S. Hirt	RIETS/Yeshiva University, New York, NY
Dr. Arthur Hyman	Yeshiva University, New York, NY
Rabbi David A. Israel	RIETS/Yeshiva University, New York, NY
Rabbi Moshe Kahn	Yeshiva University, New York, NY
Dr. Feige Kaplan	McGill University, Montreal, CA
Rabbi Zvi Kaplan	Yeshiva University, New York, NY
Dr. Chaim Kranzler	Albert Einstein College of Medicine/Yeshiva University, Bronx, NY
Dr. Norman Lamm	Yeshiva University, New York, NY
Miss Yael Levi	Bruriah High School for Girls, Elizabeth, NJ
Mr. Nathan Lewin	Attorney, Washington, DC
Rabbi Dov Linzer	Yeshivat Chovevei Torah, New York, NY
Rabbi Adam Mintz	New York Board of Rabbis, New York, NY
Mrs. Sharon Mintz	Jewish Theological Seminary, New York, NY
Dr. Robert Pollack	Columbia University, New York, NY
Rabbi Avraham Rabinowich	Rosedale Jewish Center/Congregation Rodef Shalom, Rosedale, NY
Rabbi Yonah Reiss	Beth Din of America, New York, NY
Ms. Jennie Rosenfeld	CUNY Graduate Center/Visiting Student Harvard University
Rabbi Gidon Rothstein	HAFTR Community Kollel, Cedarhurst, NY
Rabbi Yonason Sacks	RIETS/Yeshiva University, New York, NY
Mrs. Yocheved Schacter	Clinical Social Worker, Brookline, MA
Dr. Yitzchak Schechter	Clinical Psychologist, New York, NY
Dr. Lawrence Schiffman	New York University, New York, NY
Rabbi Kenneth Schiowitz	The Ramaz School, New York, NY

Mr. Carmi Schwartz	Organization Consultant, New York, NY
Rabbi Ezra Schwartz	RIETS/Yeshiva University, New York, NY
Dr. David Shatz	Yeshiva University, New York, NY
Dr. Moshe Z. Sokol	Touro College, New York, NY
Dr. Moshe Sokolow	Yeshiva University, New York, NY
Dr. Haim Sompolinsky	Hebrew University, Jerusalem, Israel
Dr. Moshe Halevi Spero	Bar Ilan University, Jerusalem, Israel
Mr. Eliyahu Stern	University of California at Berkley, CA
Mr. Marc D. Stern	American Jewish Congress, New York, NY
Prof. Suzanne Last Stone	Cardozo School of Law/Yeshiva University, New York, NY
Rabbi Moshe D. Tendler	RIETS/Yeshiva University, New York, NY
Dr. Ziporah Tokayer	Ohel Children's Home & Family Services, Brooklyn, NY
Dr. Jeremiah Unterman	RIETS/Yeshiva University, New York, NY
Rabbi Tzvi Weinreb	Orthodox Union, New York, NY
Dr. Joel B. Wolowelsky	Yeshivah of Flatbush High School, Brooklyn, NY
Rabbi Alan Yuter	Touro College, New York, NY
Dr. Jerry Zeitchik	The Ramaz School, New York, NY

*Current Affiliation

THE ORTHODOX FORUM

The Orthodox Forum, convened by Dr. Norman Lamm, Chancellor of Yeshiva University, meets each year to consider major issues of concern to the Jewish community. Forum participants from throughout the world, including academicians in both Jewish and secular fields, rabbis, *rashei yeshiva*, Jewish educators, and Jewish communal professionals, gather in conference as a think tank to discuss and critique each other's original papers, examining different aspects of a central theme. The purpose of the Forum is to create and disseminate a new and vibrant Torah literature addressing the critical issues facing Jewry today.

The Orthodox Forum
gratefully acknowledges the support
of the Joseph J. and Bertha K. Green
Memorial Fund
at the Rabbi Isaac Elchanan Theological Seminary

The Orthodox Forum Series
is a project of the Rabbi Isaac Elchanan Theological Seminary,
an affiliate of Yeshiva University

Series Editor's Introduction

Judaism, Science, and Moral Responsibility is the fourteenth conference volume in the Orthodox Forum Series. Yitzhak Berger and David Shatz, the volume's editors, have presented us with a broad array of articles on Judaism, science, and moral responsibility.

Current scientific and moral trends stress the need for greater sensitivity to human dignity, but at the same time challenge the very structure and sanctity of traditional Jewish norms.

The contributors in this work explore the issues of Judaism, science, and Jewish moral principles in a manner that should be of interest to the layman and scholar alike.

The Forum Series provides a valuable and relevant resource, bringing the insights of Jewish thinkers to the fore in a rapidly changing society.

We hope that this work will enrich the knowledge of our readership.

Robert S. Hirt
January 2005

Introduction

Yitzhak Berger and David Shatz

Do human beings have free will? Are they genuinely responsible for their actions? These questions have persisted all through the history of philosophy, but in the twenty-first century they arguably have become defined more sharply and clearly than ever. Indeed, today an especially vivid and mighty tension underlies intellectual struggles over free will. On the one hand, the rapid advances of several empirical disciplines, notably neuropsychology and genetics, threaten our instinctive affirmation that human free will and moral responsibility exist. For these disciplines maintain that all events, including those we refer to as human choices, are determined—preordained, as it were—by prior physical states. To be sure, in times past determinism was discussed and often endorsed, but the thesis lacked the massive empirical support it enjoys today. At the same time, the depth and force of our instincts—our powerful intuition that there *is* free will, that there *is* moral responsibility—present, for most people, an almost impenetrable barrier against the sweeping denial of free will suggested by empirical research.

The contemporary battles over free will and responsibility, then, are largely between those who come down on the side of empirical research and those who champion our intuitions. But in the process of debate, philosophers and scientists who ponder free will have posed an array of fascinating questions. Is determinism truly total, or can we identify "gaps" in the causal chain of events in the human brain that allow a role for free choice? Might we carve out a middle ground that accepts both free will and determinism—that is, might freedom prove compatible with a deterministic view of the world? In an *in*determinist world (a world in which not all events are determined) could we provide a coherent explanation of how we make moral choices? And so forth.

Recent remarks of two important players in the debate exemplify the intensity of our resistance to the empirical assault on free will. Philosopher Peter van Inwagen, while arguing that free will is compatible with neither determinism nor indeterminism, humbly sums up with the following: "I conclude that free will remains a mystery—that is, that free will *undeniably exists* and that *there is a strong and unanswered prima facie case for its impossibility*" [italics added].[1] So free will, it would appear, *cannot* exist—and yet it must. Somewhat analogously, neuroscientist Benjamin Libet, who has shown that our awareness of decisions arises only *after* the neural states responsible for them, nonetheless proposes that we possess the ability, via a "conscious veto," to alter our apparently determined course of action. Libet concludes:

> The intuitive feelings about the phenomenon of free will form a fundamental basis for views of our human nature, and great care should be taken not to believe allegedly scientific conclusions about them that actually depend upon hidden ad hoc assumptions.... My conclusion about free will, one genuinely free in the nondetermined sense, is then that its existence is at least as good, if not a better, scientific option than its denial by deterministic theory.[2]

If our natural sense of freedom is strong, commitment to religious principles only adds to our conviction of its centrality. The Torah insists that human beings are faced with a choice between good and evil, life and death, and that they are to "choose life"—that is, perform God's commandments (Deut. 30:19). Moses Maimonides (1138–1204) took this stark presentation of alternatives to signify the existence of a free choice between options and the existence of responsibility as well. Some thinkers who believe in determinism might advocate an end to ascriptions of moral responsibility on the grounds that these ascriptions are indefensible; but in the Bible it is *God* who holds people responsible for their actions, and His judgment is presumed to be infallible. From a religious standpoint, then, there is moral responsibility, and from this it follows that there is free will.[3]

The viability of belief in free will and moral responsibility in our age was the subject of the fourteenth annual conference of the Orthodox Forum, held in March 2002. The papers in this volume are based on the proceedings of that conference and represent collectively a blend of scientific, philosophical, psychological, and social insights into this most mystifying of topics.

The essays in the first half of this volume take up the challenge, from a traditional Jewish perspective, of delineating the parameters of human freedom and deepening our understanding of its workings.

In the opening essay, Haim Sompolinsky presents a version of the scientific challenge to the standard religious conception of human freedom, that is, what is known as *incompatibilist* free will.[4] This most common understanding suggests that the individual decides on a course of action not otherwise determined by prior physical states.

Sompolinsky takes us through recent brain research, demonstrating that what we perceive as the process of conscious decision-making relies overwhelmingly upon the physical workings of the brain. Research on patients having undergone split-brain procedures, for example, suggests that "reasoning, will, and the generation of actions can be fractured and compartmentalized within the brain," and that "cognition and volition are rooted in discrete brain structures." If there remains a role for choices that are not neurobiologically determined, this would require an independent will to operate on the quantum level, where one finds physical indeterminacy. And indeed, chaos theory suggests that it cannot be ruled out that small quantum fluctuations affected by the individual's will might tip the balance of a borderline decision in favor of one alternative or another.

Nevertheless, Sompolinsky argues that if "mental forces result in *event probabilities* in the brain that are different from those predicted by the known physical laws, these violations of the laws should be readily detected and verified by ordinary scientific observation." Such violations of probabilistic laws stand against current scientific evidence. Furthermore, incompatibilist free will requires either a mental state that has *emerged* from the brain having a causal effect on the brain itself, or an independent metaphysical entity acting on the brain. Sompolinsky argues that any such entity or state affecting the brain, or for that matter, any discovered violation of known probabilistic laws, should, in any case, be cast in terms of a revision of our conception of the physical laws governing nature. He proceeds, also, to contest the notion that an "emergent" mental state could have *independent* causal power: in allegedly analogous cases—for example, where the "emergent" solidity of a wheel generates rotation—it is really the condition of the molecules themselves that causes the result, nothing more. In this view, then, only *compatibilist* versions of free will "do not run counter to scientific ideas or findings," for in these versions, the will does not choose between alternatives otherwise undetermined by physical laws.

Robert Pollack states straightforwardly that genetic specificity is an incomplete determinant of behavior: "synaptic connections—the basis of all mental activity later in life—cannot have all been specifically encoded by our genomes." Thus, "human genomes do in fact allow free will."

The inadequacy of genetic makeup in determining individual behavior prompts Pollack to warn against utilizing genetic information "as an excuse for setting aside religious values." In this context, he raises the need to work with every individual regardless of mental capacities in an effort to maximize his/her potential. Furthermore, in minimizing the centrality of DNA in our evaluation of individuals and groups, Pollack argues that Jews should not view themselves in terms of genetic uniqueness: while Maimonides speaks of the "Crowns" of royalty and priesthood as inherited, the "Crown of Torah" is "set aside, waiting for every Jew." It is, then, the individual who may strive for elevation, and through teaching and learning—not by appeal to the DNA of the group.

Pollack proceeds to take issue with James Watson's call for utilizing genetic analysis to reduce "the differences in human beings" in order to create "a society in which we can effectively view all individuals as truly equal." "Our future," writes Pollack, "lies not in minimizing our differences, but in cherishing them." He concludes by calling for "religiously informed, sensitive" policies in contemporary medicine that respect the individuality and autonomy of patients, and work toward helping people in need, "one person at a time."

David Shatz's essay responds to the challenge outlined by Sompolinsky. Initially, Shatz offers that even if neuroscience suggests that all that exists are material entities, belief in a non-material soul and in free will could be defended by an appeal to faith. But even independently of such an appeal, the fact that science has no place for an entity or phenomenon of type X (specifically, non-material ones) does not entail that there are no entities or phenomena of type X—only that such entities and phenomena do not enter into scientific theories, due to methodological constraints scientists adopt. Similarly, God could not enter into a scientific theory, because He is a non-material being and science does not utilize non-material explanations; but a religious scientist will believe, legitimately, that He exists. Finally, some philosophers even today (admittedly a small minority) think that belief in a soul can be supported by evidence.

Despite his resistance to materialism and his belief in free will, Shatz does not think that the contemporary scientific picture is totally alien to Judaism. In the main body of the paper, he bids us to consider how far Judaism is willing to go in accepting materialism and in denying free will—can it affirm these theses in part, and if so, at what stop must it get off the train? In response to this question, Shatz points to such precedents as views that attribute sin to the biological "part" of a human being, Hasidic views that deny the existence of free will, views like Nahmanides' that downplay the value of free will, the medieval assertion that astrology determines human behavior, and compatibilist views (prevalent in medieval times) which maintain that an action can be both free and determined. All of these approaches suggest considerable room for either materialistic explanations of actions, a denial of free will, or a reconciliation between free will and determinism. Even so, Shatz points out some flaws in adopting certain of these precedents.

Thereafter, he turns to the strengths and weaknesses of compatibilism. While he thinks the weaknesses are significant, the fact that compatibilism salvages free will and was accepted by medieval Jewish philosophers (in a form different from today's) gives some motivation for seeking to dispel the objections to it. But since compatibilism and incompatibilism both confront significant challenges, in the end Shatz admits to being mystified by what free will is. If we do not know what it is to have free will, we cannot decide whether we have it.

For Shalom Carmy, free will, in our culture, is impossible without a lively imagination of what it may be like to exercise free will. It is therefore desirable to explore the existential dimensions of freedom and determinism. The attractiveness of freedom is rooted in the introspective conviction, in prevalent beliefs about responsibility and in the experience of creativity. Determinism is attractive because it fits human beliefs about scientific system. It is also thought to promote psychologically and, to some people, morally desirable goals such as the avoidance of anxiety, avoidance of guilt and tolerance of one's own, and others' deficiencies: some maintain that if we don't know that an act is free, it is morally safest to view it as determined, though, in Carmy's view, this position does not stand up to inspection. Lastly, determinism captures the introspective conviction of being unfree.

Philosophical questions about the conceptual coherence of free will, writes Carmy, lead to the conclusion that the fluent experience of freedom escapes systematic exposition because of its inherently self-referential nature. This makes it crucially important to provide suitable models and metaphors for freedom; otherwise one would have no idea what it's like. One such model views freedom in terms of an effort of will that overcomes resistance. Another comprehends freedom along the lines of artistic creativity. The latter does better in capturing intuitions about the unpredictability and mystery of freedom, and even succeeds in explaining cases where freedom is associated with effort. Creativity is particularly important in connection with personal relations, where creative response to the other, and not only effort of will, is necessary for wholesome moral action.

Carmy regards R. Dessler's and R. Kook's treatments of free will as the most significant in twentieth-century Jewish thought. Detailed philosophical and literary examination reveals R. Dessler's affinities to Kantian thought, with its emphasis on moral responsibility, and to aspects of the effort model with corresponding difficulties in dealing with moral luck and with the traditional belief that reward and punishment are not completely determined by effort. R. Kook's discussion, by contrast, is oriented toward the creativity model.

While this model is not geared to the judicial problem of responsibility, it can provide insight in this area too.

Recovery of a living sense of freedom is facilitated by appropriate attention to language as illustrated by close reading of passages in Dreiser, Henry James, and psychotherapeutic literature. It likewise requires experiences of competent willing and honestly earned confidence in one's agency. Lastly, if belief in freedom is not to become a ceremonial vacuity, we need a realistic, concrete sense of its limitations, of human vulnerability and creatureliness. Such an outlook sustains a robust religious sense of self, confident enough to withstand the pressure of ideological conformity and to maintain the dignity of private individuality.

The next section of the volume concerns the proper religious-communal response to problematic behaviors where it is suspected that the violator's free will is constrained. Basil Herring catalogues the near-universal affirmation of broad free will in rabbinic and post-rabbinic Jewish thought and emphasizes its centrality in the fundamental religious imperative of overcoming moral failure and achieving personal redemption. Not only "individual self-discipline," but "enlightened communal censure, social rejection, and institutional education" all must play a role in sublimating "forbidden . . . urges, no matter how widespread or condoned they may be within the larger societal sphere." Nevertheless, "the rabbinic tradition also recognized certain limits and constraints on human behavior that, while not ultimately removing freedom of choice and its consequences, greatly complicate and diminish human freedom." For such circumstances, Herring coins the term "choice-diminished behavior" (CDB).

Behavior that is immoral remains so even if choice-diminished, and where it is harmful to others, Herring affirms that "there ought to be no tolerance or leniency." Where there is no direct harm to others, in contemporary societal settings—particularly in cases of CDB—"a carefully constructed case can be made favoring restraint and caution" on the part of the community. For individuals who engage in problematic CDB, Herring advocates allowing participation in synagogue life, provided that "public health, personal dignity, and communal decorum" are maintained, and that there is no public expression in religious-communal settings of behavior that violates religious principles. Even where these criteria might justify communal sanctions, a balanced judgment must be made as to whether or not this would be appropriate and beneficial in the long term. Furthermore, any policy of communal sanctions must avoid arbitrary selectivity in its application. Herring proceeds to observe that in school settings, where "character formation" is a basic objective, there are cases in which "serious sanctions" might be appropriate even where the behavior is not harmful to others, but he delineates a series of sensitive concerns in the development and implementation of any such policy.

Picking up on Herring's call for no tolerance of behavior harmful to others, Rivkah Teitz Blau forcefully urges the religious community to adopt policies toward abusers of children that reflect a no-nonsense approach: "Once a person has abused a youngster, he cannot serve in any position that will present him with another opportunity." Because of the effects of abuse, "the potential further damage to children in the event that even a one-time abuser is allowed to continue working with them is incalculable." The indifference even of bystanders, not to mention the community leadership, is a form of complicity, and the abuser's other accomplishments must be seen as irrelevant.

To confront the problem systematically, Blau advocates the establishment of "a review board consisting of trained professionals" to investigate all accusations. This method provides a range of benefits for parents, children, employers, and the accused. In addition, it should be part of children's education "to recognize abusive behavior, to resist it, and to report it." The elimination of "surprise and secrecy from the abuser's arsenal" will contribute to the containment of this intolerable behavior, however "choice-diminished."

The volume's final section concerns self-improvement and *teshuvah.* The essay by Michelle Friedman and Rachel Yehuda focuses in particular on similarities and differences between psychotherapy and the process of *teshuvah* (repentance). Both involve the alleviation of emotional pain and are systems for change, but there are other similarities as well as contrasts. (1) In both psychotherapy and repentance, we can identify an authority who stimulates change. But whereas in repentance the locus of authority is God, to whom one relates out of love and/or fear, in psychotherapy the locus of authority is the patient—since psychotherapists maintain that a patient should not change just "for the therapist." (2) In both instances, transference takes place. (Transference is the process by which the patient or potential penitent invests the therapist, or rabbi, or even God with a constellation of feelings whose origins are based on earlier relationships.) (3) Psychotherapy aims, in the first instance, at relieving pain. *Teshuvah,* however, aims in the first instance at changing behavior to conform to halakhah, with relief of pain only a secondary benefit. (4) Judaism believes that good behavior provides greater insight, while in classic psychotherapy (as opposed to cognitive behavior therapy), the reverse is the case—greater insight motivates correct behavior.

Friedman and Yehuda illustrate their points with three case studies. Ultimately they argue that familiarity with psychotherapeutic techniques and processes can facilitate the *teshuvah* process, and conversely, an understanding of *teshuvah* can enhance the psychotherapeutic process. Rabbis and mental health professionals must therefore understand each other's principles. However, behavior that is acceptable to a psychotherapist may not be acceptable to a rabbi, because the rabbi relates to halakhah and not only to the patient's suffering.

Moshe Halevi Spero analyzes the process of change inherent in *teshuvah*, informed by a broad spectrum of philosophical and psychological literature as well as by primary and secondary sources within the Jewish tradition. In this "postmodern inquiry," Spero challenges simplistic conceptions of "return" to a prior state, calling attention to the elusiveness of the concept of the self as a rigidly bounded entity, the limitations of memory, the paradoxical nature of time, and the inadequacies of the medium of language. It becomes necessary, in light of this challenge, to formulate a new approach to *teshuvah* that circumscribes the "impossibilities" that emerge from a more crude understanding.

Spero expounds on the theory that there is initially no differentiation between the self and anything it experiences: the self attains identification through a process of "negativization," which enables "discrete perceptions of self" and "other." A gap thus emerges between the self and the other, and any longing for the other as originally experienced will fall short, as it can exist now in the mind only through symbolic representation, itself dependent upon the limited vehicle of language. Yet in linguistic expression there lies desire, "attempting to escape and seek fulfillment." And indeed, in the "linguistically moderated rituals of *teshuvah*," there "comes a searching of the deepest elements of the human mind for that fundamental flaw or moment of severance from which emanates a cry that gradually took the concrete form of alienation, mistrust, apathy, and sin, but also potentially expresses the search for the original union." And while the original union cannot itself be fully retrieved, *teshuvah* entails a process of "refinding" and "reapproximating" this prior state by means of mental, language-based representation. On the basis of his analysis, Spero advances novel understandings of numerous aspects of *teshuvah* delineated in classical sources, including the distinction between *teshuvah mei-ahavah* and *teshuvah mi-yir'ah*, and the efficacy of *teshuvah be-sha`ah ahat*, in the process providing ingenious interpretations of several aggadic tales of individuals engaged in the *teshuvah* process.

It is our pleasure to thank those who facilitated the production of this volume.

The advice and efforts of Rabbi Robert S. Hirt, Senior Advisor to the President of Yeshiva University and editor of the Orthodox Forum Series, have been immensely helpful at every stage. As always, the assistance of Ms. Marcia Schwartz has been invaluable. Drs. Edward Reichman and Yitzchak Schechter assisted in recommending and contacting potential presenters. Finally, the copy-editing skills of Meira Mintz and Shifra Schapiro made our jobs significantly easier.

The publication of Orthodox Forum proceedings adds to the legacy of Yeshiva University's chancellor and former president, Dr. Norman Lamm, who spearheaded the Forum and remains a guiding force. His continued support and involvement, and now that of Yeshiva's current president, Richard M. Joel, inspires and energizes Forum participants.

Our sincere thanks to all.

Notes

1. Peter Van Inwagen, "Free Will Remains a Mystery," in *The Oxford Handbook of Free Will*, ed. Robert Kane (Oxford, 2002), 159.
2. Benjamin Libet, "Do We Have Free Will?," in *Oxford Handbook*, 563.
3. It is interesting to ask whether we can formulate a role for human choice that does not limit God's sovereignty. The bottom line, however, is that with the exception of certain Hasidic thinkers, Jewish philosophers have usually affirmed the existence of free will in some sense. See David Shatz's essay in this volume.
4. Many Jewish philosophers in the Middle Ages, however, were compatibilists.

I

SCIENCE, FREE WILL, AND JEWISH THOUGHT

1

A Scientific Perspective on Human Choice

Haim Sompolinsky

I. The Problem

WE ARE USED TO THINKING ABOUT OURSELVES as having the freedom to make decisions that affect the course of our lives to a significant degree. We regard our ability to make such choices as a key characteristic of the human condition. At the same time, we are part of a universe that appears to operate in fairly fixed and unalterable ways. This gives rise to what philosophers call the problem of free choice (or free-will) and determinism: what does it mean for us humans to be truly free in a world where the future is determined by the present? Do we possess free will, or are all our actions and mental states inexorably determined?

The notion of free will is one of the most controversial issues in philosophy, and determinism is just one of the difficulties surrounding it. Equally problematic is the role of our will or choice in an indeterministic world, since randomness and chance seem to be blind to the power of our will. In this article, I will sketch the broad lines of a scientific perspective of human choice. In Section II, I discuss the notions of determinism and predictability in classical and modern physics. Section III outlines current research on the neuronal processes that underlie human cognition and choice. In Section IV, I discuss the main attempts to reconcile the strong sense of free will—metaphysical freedom of the will—with scientific conceptions and knowledge. Section V is devoted to a critical appraisal of these attempts, emphasizing the problem of will/brain interaction. Summary and concluding remarks are presented in Section VI.

II. Determinism and Probabilities in the Laws of Physics

1. Determinism in Classical Physics

The problem of free will is commonly referred to as the conflict between the concept of free will and the concept of determinism. There are various kinds of determinism, including logical, physical, and theological determinism. In this paper, I focus on physical determinism.

A fundamental consequence of the principles of classical physics is that physical laws fully determine the way events in nature unfold. According to classical theory, all future states of a physical system are determined by its present state and the forces acting on it. Let us consider, for example, an isolated mechanical system. The physical state of the system at any time consists of the values of the positions and velocities of all the particles comprising the system at that time. These particles are acted upon by gravitational, mechanical, or electrostatic forces. According to Newton's Laws of Mechanics, the positions and instantaneous velocities of all the particles at a given time, together with the forces between them, fully determine the future (and the past) movements of all the particles in the system.

The above principle of determinism applies to an isolated system. In reality, physical systems are in continuous contact with their environment and exchange momentum and energy with it. In such a situation, the principle of determinism still holds, but the interaction of the system with its environment should be included among the forces that affect the motion of the particles in the system.

An alternative exposition of the principle of determinism, which avoids this complication, is to apply it to the entire universe. This giant conceptual step was made by the French mathematician and philosopher Pierre Simon de Laplace. In his *Philosophical Essay on Probability* (1819), Laplace argues that all natural events are connected to previous ones by universal causation: "All events, even those which, because of their small scale, do not appear to keep to the great laws of nature, are just as necessary a result of those laws as are the revolutions of the sun."[1] In other words, there should be a set of scientific laws that characterize the physical forces at all scales of matter, and these laws together with the state of the universe at a given time will fully determine the future course of the entire universe.

What do we mean by claiming that the laws of nature "fully determine" the future course of events? Essentially, we mean that the laws of nature are sufficient "causes" of future events; namely that the future state of a physical system is necessitated by its present state and the physical forces operating on it.

The concept of causality has been the subject of controversy for centuries and an exposition of the myriad of attendant metaphysical questions is beyond the scope of this paper. However, in the present context there is a simple operational definition of "full determination" of the state of a system. If we have full knowledge of the initial state of a physical system and the forces that operate on it we *can calculate* its future state. In other words, classical determinism here implies predictability, at least in theory. This relationship between determinism and predictability was emphasized by Laplace himself:

> We must therefore regard the present state of the universe as the effect of its preceding state and as the cause of the one which is to follow. An intelligence which in a singular instant could know all the forces which animate the natural world, and the respective situations of all the beings that made it up, could, provided it was vast enough to make an analysis of all the data so supplied, be able to produce a single formula which specified all the movements in the universe from those of the largest bodies in the universe to those of the lightest atom. For such intelligence, nothing would be "uncertain," and the future, like the past, would be present before its eyes.

Laplacian determinism projects a naïve view of physical systems as clock-like predictable machines. Modern physics provides a much more complex picture of determinism and predictability. In the following section I will briefly review some of the new perspectives provided by modern physics.

2. Chaos and the Limits of Predictability of Physical Systems

Laplace's equating of determinism with predictability raises the question of the feasibility of determining the initial state of a system with the accuracy required to predict its future state. Of course, Laplace's super-intelligent demon that has full knowledge of the initial condition of the entire universe is an outright fiction. The more rigorous issue is this: how precisely must we know the initial state of a physical system in order to accurately predict its future course?

Obviously, we can never know the state of any system with infinite accuracy. Nevertheless, our daily experiences inform us that we can often make reliable predictions about the future course of natural events. Apparently, some systems' future behavior is not very sensitive to mild changes in their initial condition; hence, we do not need to have very precise information about their initial state in order to predict their future state. Are all physical systems like this?

The modern theory of chaos (originally developed by Henri Poincaré around the turn of the twentieth century) has revealed that many physical systems (e.g., air currents in the atmosphere, a dripping faucet, electronic circuits, heart beats) are so exquisitely sensitive that the slightest disturbance in their initial conditions can affect their behavior radically.[2] The degree of unpredictability depends on the degree of chaos in the system and the length of the attempted prediction. Even in a chaotic system, short-term behavior can often be predicted reliably. Hence, chaos does not imply a complete loss of predictability. Rather, it implies that the error in predicting the future course grows extremely fast with the time elapsed since the state of the system was determined to a given degree of accuracy. The further we want to predict into the future, the more accurately we have to determine the initial state. The amount of accuracy required grows so fast with the time of prediction that long-term prediction is practically impossible for chaotic systems. The weather is an example of a chaotic system; because of its chaotic nature, long-term precise forecasting is virtually impossible. Nevertheless, modern observation and modeling techniques allow for relatively reliable short-term weather forecasting.

This analysis makes it clear that in chaotic systems, it is *predictability* that really breaks down, not the cause-effect chain of events. A chaotic system is no less determined by the physical laws than a nonchaotic system—it is simply more difficult to predict its course over long periods. This is why the more accurately we know the initial state of a chaotic system, the better our prediction of its behavior will be. The fact that there is *no theoretical limit* on the accuracy of determination of the *initial state* of a chaotic system (as opposed to practical limits) implies that there is no *theoretical limit* on the ability to *predict* its future.

3. Statistical Physics

The range of phenomena in the real world that are described efficiently by deterministic laws such as the Newtonian laws of mechanics is very limited. Very often the properties of macroscopic systems at ordinary temperatures are not described by these laws but by statistical laws, formulated by a branch of physics called statistical mechanics.[3] At the core of statistical mechanics lies the distinction between the microstate of a large system and its macrostate. The microstate is specified by the positions and velocities of each atom and molecule, as in classical mechanics. The macrostate corresponds to a few macroscopic properties, such as the system's volume, pressure, and temperature. Rather than attempting to predict the detailed microstate of the system, statistical mechanics establishes *the probability that the system will be in each of its different microstates*, and the relation between these probabilities and the macrostate of the system.

It is an important and interesting fact that most of our accounts of macroscopic systems, including biological ones, are far removed from the deterministic microscopic laws of motion, the kind of laws that Laplace had in mind. Basic physical concepts, such as heat and temperature, are grounded in the statistical description of the states of the system. Noise, randomness, fluctuations, and probabilities dominate the terrain of modern physics. One may wonder about the source of this probabilistic description of nature. What is the justification for ignoring the deterministic picture of microscopic classical physics when dealing with macroscopic objects?

Although a full account of how the probabilistic laws of statistical mechanics arise from the deterministic laws of microphysics may still be incomplete, the general picture is clear. Ordinary macroscopic systems are not isolated. They are in continuous interaction with the environment. The microstate of the system is sensitive to the details of the extremely complex microscopic interactions between the system and its environment. Often, the influence of the environment on the surrounded system takes the form of a force whose value fluctuates strongly in space and time. These fluctuations are called *thermal* fluctuations. The higher the temperature of the system (and its environment), the larger these fluctuations become. However, the macroscopic properties of the system are relatively insensitive to the details of its microstate, and are therefore insensitive to the details of the system-environment interactions. Furthermore, because of the interaction with the environment, the microstate of the system changes very rapidly in an erratic manner, so that during any reasonable epoch of observation, the system has visited an enormous number of different microstates. It is therefore appropriate to treat the interaction with the environment as a source of structureless random force acting on the system. This naturally gives rise to a statistical description of the system in terms of the relative frequency of its visiting any of its microstates, and its overall macro-behavior.[4]

From this account, it is clear that although statistics and probabilities are fundamental to the physical description of the properties of macroscopic systems, this does not imply the abandonment of fundamental determinism at the microscopic level. Once again we can invoke the criterion of *predictability in theory*. If we had full knowledge of the microstate of the system and its environment at a given time, for example, the positions and velocities of all of the constituent particles, we could, in principle, use the microscopic deterministic laws to predict the microstate of the system and all its physical properties at future times. In other words, what statistical physics does is to distribute the causal powers that operate on a system between its internal forces and the forces generated by its interaction with the environment. Probability arises when we average out the effect of the environment. However, the fundamental underlying microscopic laws are still fully deterministic.

4. Quantum Indeterminism

The determinist picture of the microscopic physical world was completely overturned by quantum mechanics, a new theory of physics developed at the beginning of the twentieth century. Heisenberg's Uncertainty Principle provides a convenient way of describing what is wrong with classical Laplacian determinism. According to classical mechanics, in order to predict the future position and velocity of a particle, one has to be able to measure its present position and velocity accurately. Heisenberg showed that such a measurement is *in principle impossible.* The more accurately you try to measure the position of the particle, the less accurately you can measure its speed, and vice versa. In fact, the uncertainty in the position of the particle, multiplied by the uncertainty in its velocity (times the mass of the particle) can never be smaller than a certain universal quantity, known as Planck's constant. Moreover, this limit does *not* depend on the way in which one tries to measure the position or velocity of the particle or on the type of particle: Heisenberg's Uncertainty Principle is a fundamental, inescapable property of the world.

This observation and other discoveries led Heisenberg, Bohr, Schrödinger, and Dirac in the 1920s to replace classical mechanics by a new theory of microscopic physics—quantum mechanics. This theory abandons the formulation of physical laws in terms of deterministic predictions about the trajectories of particles, replacing them by inherently probabilistic laws. According to quantum mechanics, the future occurrence of a particular microscopic event (such as an electron being in a certain position or in a certain orbit around the atom's nucleus) is unpredictable even if we know with full precision all that can be known about the current state of the system. All that we can deduce from physical laws is the probability of occurrences of each event.[5]

The fact that according to quantum theory *there are no laws that predict conclusively the actual occurrences of physical events* is a fundamental revolution in our understanding of the causal underpinning of the physical realm. In fact, for many years, Einstein, one of the early founders of the new physics, was reluctant to accept it as a complete theory of physics because he refused to succumb to the idea that the microscopic underpinning of our universe is indeterminate. The ensuing eighty years of vast experimental and theoretical research have upheld with exquisite accuracy the predictions of quantum mechanics. Furthermore, many of the major technological advances of our era including the transistor, lasers, microelectronics, and more, are based on the results of this theory. It is therefore not surprising that, nowadays, quantum theory is accepted as the fundamental theory of physics at the microscopic scale.

The emergence of quantum mechanics signaled an end to Laplace's model of a totally deterministic universe. This does not mean that the world is lawless. Even at microscopic levels, the quantum laws predict the statistical regularity that underlies the frequency of occurrence of events. Most importantly, since the behavior of *macroscopic* bodies depends on the sum total of an enormous number of microscopic quantum events, most macroscopic properties can be predicted with fantastic accuracy, despite the probabilistic nature of these bodies' microscopic constituents. In summary, as in the case of thermal noise, quantum fluctuations in the physical states are apparent at the microscopic scale but, in most cases, they are averaged out at the macro level.

III. Neuroscience of Human Choice

1. Neuroscience of Cognition

Ultimately, cognitive neuroscience seeks to link neural processes with mental processes involved in human cognition. In the last decades we have made considerable progress in understanding the neural basis of cognitive functions such as perception, learning, memory, attention, decision making, language, and motor planning.

An examination of the neural correlates of a specific cognitive function can be broken down into several issues. The most basic issue is that of *localization*: what neural structures in the brain are involved in the function in question, and how does the level of their activation change as the mental process unfolds? Next, there is the issue of *representation*: how is the activity of nerve cells within a brain area related to the sensory, cognitive, or behavioral variables being processed at a given time? Finally, there is the challenge of discovering the underlying *mechanisms*: what are the ways through which these representations come about and what is the causal chain that connects them to earlier and to subsequent brain activity patterns?[6]

In animals, studies geared at identifying the brain structures active during specific cognitive processes have used a variety of electrophysiological techniques to measure changes in electrical activity in a brain region. Foremost among these is the technique of recording by microelectrodes. By inserting a thin wire into an animal's brain, researchers measure changes in the electrical activity of individual cells or small groups of cells. Localization studies in humans use several noninvasive brain-imaging techniques, the dominant one being functional magnetic resonance imaging (fMRI). This technique effectively determines local enhancement in neuronal activity by measuring the increase in oxygen levels in the blood flow in different brain areas.[7]

Localization data by themselves provide little insight into the nature of the signals encoded in a given structure or the mechanisms that underlie this encoding. To gain further insight, we must know the relation between the activity of neurons in a brain area and the sensory data, the memory, or the behavioral pattern with which this activity is correlated. This defines the neural representation of a brain area. Much of the research in systems neuroscience over the past half century has dealt with uncovering the basic features of neural representations in the different brain areas. Most of this research focuses on the electrical activity at the single neuron level. Using microelectrodes, researchers correlate changes in the electrical activity of individual cells with changes in sensory or behavioral conditions. The results are often expressed in terms of neuronal selectivity. Neurons in sensory areas are selective for specific sensory parameters such as the color of an object, the orientation of a line segment, or the frequency of a sound signal. Neurons in motor areas are active selectively prior to a movement of specific limbs, or even a specific type of movement of that limb. For instance, some neurons in the primary motor cortex of an animal fire strongly before the onset of a planned reaching movement of the arm to a target located straight ahead, but are inhibited from firing before a movement of the same arm to a target located 90 degrees away. Other neurons in the same brain area fire strongly before a movement to the latter target, but will be inhibited by a target located straight ahead of the animal.

As the example above illustrates, different neurons in the same brain area will often have different selectivity or preferences. Researchers can reconstruct the neuronal representation of an area by mapping the different neuronal selectivities within it. This mapping of neural representations is also aided by measuring the effect of stimulating a cluster of cells in a given brain location by injecting electrical current through the microelectrode. Activating clusters of direction-selective neurons in a certain visual area through a microelectrode can induce rhesus monkeys to "report" seeing motion in the direction encoded by the activated neurons. Similarly, microstimulation of localized sites in the motor cortex may trigger a certain arm movement.

But even this level of analysis raises fundamental questions. How are the different neuronal selectivities formed? How are these neuronal patterns of activity deciphered by brain structures downstream? Individual neurons in the cortex typically receive input signals from as many as several thousand other neurons. In a typical experiment, however, the neurophysiologist can characterize the responses of only a few neurons at the tip of an electrode. With such a limited data set, it is difficult to determine exactly how the thousands of synaptic inputs to any given cell are transformed to create the cell's pattern of output activity. It is thus an enormously difficult challenge to decipher the dynamic mechanisms that underlie neuronal representations in the cortex.

In formulating a theoretical understanding of brain processing, an interesting question is how many neurons participate in a given perceptual or behavioral decision.[8] The difficulty in answering this question stems from the fact that the *degree* of selectivity of single neurons varies enormously within a given brain area. A small but noticeable minority of cells shows a high degree of selectivity. These cells can be thought of as signaling reliably—by their activity or quiescence—the presence or absence of a particular sensory or motivational variable. Most cells, however, show poor to moderate selectivity, so that the representation of information in any given area most likely resides in the joint pattern of activation and inactivation of a large number of cells, on the order of thousands at least. Furthermore, even the simplest perceptual or behavioral process involves the simultaneous and/or sequential activation of several brain areas, that is, several distinct neuronal representations.

Therefore, although both brain-imaging and single neuron studies often give the impression of a very compartmentalized functional organization of the brain, in fact, even an act as simple as planning a reaching arm movement toward an object involves a complex sequence of activations of large populations of cells in many brain areas: low level visual areas that process features of light patterns on the retina, high level visual areas where objects are detected and recognized, motivational areas where the desirability of an object is stored and assessed, somatosensory areas where the present position of the arm is encoded, and various motor areas where the movement parameters are calculated and executed.

In conclusion, substantial progress has been made in discovering some of the basic anatomical and dynamic components of the neural correlates of cognition. However, neuroscience is still at an early stage of understanding the complex chain of neural events that underlie cognitive functions. Lacking blueprints, discovering function from structure is a vexing task. It is no surprise that modern research aiming at figuring out how the intricate neuronal circuits in the brain generate the appropriate signals is a multidisciplinary enterprise, involving experts from such diverse fields as neurobiology, computer science, physics, and psychology.

It should be stressed, however, that the basic microscopic building blocks of these complex neural circuits are well characterized. Foremost among these are nerve cells: neurons that communicate through a special electro-chemical signaling apparatus—the synapse. Neurons are made of the same "stuff" as other biological cells; their soma contains genetic material embedded in DNA molecules, their exterior consists of ordinary membrane tissue, and the signals that they transmit take a variety of forms from rapid electrical pulses traversing their membrane (the "spikes") to the slow secretion and absorption of various chemicals. The important point is that at the microscopic level, the brain does not harbor mysterious or unfamiliar entities. On the contrary, at this level all observed processes are accounted for by the basic laws of physics such as electrochemistry and thermodynamics. Thus, the mystery that the brain poses to us is not due to some indiscernible microscopic features, but rather to the enormous complexity of the organization of the microscopic building blocks. In fact, this challenge is shared by several other domains of contemporary scientific endeavor—understanding the macroscopic properties of systems that consist of many relatively simple microscopic units woven into a complex pattern of organization.

2. Neuroscientific Studies of Decision-Making

Over the past twenty years, neuroscientists interested in vision and other sensory systems have studied the brain areas involved in various levels of sensory information processing. Meanwhile, other researchers have studied the mechanisms behind body movements. Where and how does one look for neural correlates for reasoning, decision-making, volition, and choice? One current working hypothesis is that these should be found in cortical areas that link sensory and motor processing.[9]

An example of research along this line is the study of the neural correlates of decision-making in monkeys by Platt and Glimcher.[10] They placed a monkey in front of a viewing screen and repeatedly presented two visual stimuli—light-emitting diodes on the screen. The animal was trained to direct its gaze to one of the two targets signaled by an appropriate visual cue, in order to receive a juice reward. Then, the experimenters removed the cue signal and let the monkey perform "free choices" between the two targets. In sequential blocks of roughly 100 trials, the researchers varied the size of the juice reward associated with each target. In this case, the monkey selected the two targets with frequencies that were in proportion to their relative reward sizes. While the monkey performed this task, the neural activity in the lateral intraparietal cortex (LIP) was recorded. It was found that given exactly the same visual display and exactly the same gaze shift, neurons in LIP responded differently depending on the animal's expectation of reward size—an expectation that was evident in the animal's pattern of decisions. LIP receives inputs from several parts of the brain involved in processing visual information, and it projects directly to the areas that control eye movements. Thus, it may be an important intermediate stage in the transformation of visual signals into neural commands to move the eyes. As this experiment suggests, this area contains a population of neurons whose activity represents "high level" (e.g., reward expectation) decision-related variables that are neither purely sensory nor purely motor in nature.

3. Brain Structures Underlying Human Choice

Research on human choice cannot be carried out by invasive methods such as microelectrode recordings except under special circumstances. However, the recent advances in brain-imaging mentioned above have invigorated the neuroscientific research on human volition and decision making. The dorsolateral prefrontal cortex (DLPFC), together with brain regions connected to it, has been found to be of particular importance. This area is known to be important for working memory,[11] which plays a crucial role in many cognitive functions. In addition, several studies show that there is an increase in activity in DLPFC when actions (such as word generation and finger movement tasks) are being selected and initiated. Recent brain-imaging studies by Spence and Frith,[12] among others, indicate that the *subjective* experience of deciding when to act and which action to perform is associated with a characteristic pattern of activity in DLPFC.

It is of interest that patients with symptoms of a "sick will," that is, inactivity, lack of ambition, autistic behavior, etc., show subnormal activity in the prefrontal parts of the brain. Symptoms of this type are encountered in chronic schizophrenia, depression, and organic dementia.[13] Furthermore, damage to DLPFC in humans leads to stereotyped responses to objects in these individuals' environment, such as putting on glasses or eating food whenever these are placed in front of them. Such patients might be said to have a major impairment of free will since they have become slaves to their environment.

A direct relationship between electrical signals in the brain and awareness of intention to perform movements was also found in a recent study conducted by Fried and coworkers.[14] Working with epileptic patients, they noticed that electrical stimulation of sites in the frontal cortex caused the patients to report an urge to move a specific body part, or a feeling that they were about to move. Fried's results are consistent with the view that conscious awareness of intention is conjoined with frontal brain activity.

These and other results indicate that the exercise and experience of choice is correlated with neural mechanisms that are located in the prefrontal cortex and in the brain regions that are linked to it.

4. Lessons from Split-Brain Studies

Important insights into the constraints on human cognitive functions imposed by the anatomy of the brain have been acquired from studies of split-brain patients. In split-brain procedures, done as a last resort in treating certain types of severe epilepsy, the corpus callosum, which connects the two cerebral hemispheres, is surgically cut, and consequently most cerebral communication between the two hemispheres is lost. Split-brain patients superficially function normally in their daily lives, and perform normally in psychometric tests. Nevertheless, careful studies pioneered by the neuropsychologist Roger Sperry have revealed profound changes in these patients' information processing and decision-making capacities.

In those of us with healthy, intact brains, information presented to our right hemisphere is quickly sent to our left hemisphere, and vice-versa. What happens when the right and left hemispheres of the brain can no longer communicate? Cognitive studies of split-brain patients measure their behavioral responses to stimuli that are presented to only one cerebral hemisphere.[15] Numbers, words, and pictures visually presented to the *left* hemisphere (i.e., presented only in the *right* visual field) are verbally repeated or described with no difficulty because the left hemisphere is dominant for language in most humans. Remarkably, such verbal descriptions of sensory input are impossible for the right hemisphere; if an image is shown to the *right* hemisphere (i.e., presented in the *left* visual field), the person will be unable to describe it and will usually say that nothing is there. Further experiments revealed varying degrees of transfer of inter-hemispheric information through subcortical pathways. However, this is largely confined to implicit unconscious processes; the explicit conscious system does not have access to this information.[16]

Thus, it can be said that split-brain patients have two largely separate mental systems, each with its own abilities to grasp the environment. Furthermore, amazingly enough, it appears as though some split-brain patients have two systems of will. The best example of this is patient Paul S, whose right hemisphere had considerable language ability prior to the operation. Consequently, researchers were finally able to interview *both* hemispheres on their views about friendship, love, hate, and aspirations. Interestingly, Paul's right and left hemispheres gave conflicting statements to such questions as what profession he wanted to have, or his attitudes on political events.

The two hemispheres are not only capable of forming their own perceptions and goals but also of generating independent and, at times, conflicting behaviors. One patient found his left hand struggling against his right hand when trying to pull up his pants in the morning. While the right hand tried to pull them up, the left was trying to pull them down.[17]

Interestingly, it was found that the left hemisphere makes major attempts at compensating for its lack of information (in cases where the information was presented to the right hemisphere alone). Michael Gazzaniga showed that when a split-brain patient is subjected to tests in which the left half of his brain does not know the correct answer, it will often make something up based on the information it does have. In one test, each hemisphere was presented with a picture and told to pick the object that relates to that picture. The left hemisphere was shown a chicken claw, while the right viewed a snow scene. The patient pointed to a chicken with his right hand, and to a shovel with his left. After each hemisphere responded, the left hemisphere was asked to explain the choices. When asked what images he saw on the screen, the patient responded, "I saw a claw and I picked the chicken, and you have to clean out the chicken shed with a shovel."

Trial after trial, this kind of response occurred. While observers of the subject knew exactly why the right hemisphere has made its choice, his left hemisphere could merely fabricate a consistent response. Interestingly, however, the left hemisphere did not offer its explanation as a guess but rather as a statement of fact. Gazzaniga hypothesizes that the left hemisphere of the human brain plays the role of the "interpreter." While in normal, non-split brains, information from the right hemisphere is delivered to the interpreting mechanism in the left hemisphere across the corpus callosum, in the case of a split brain the left hemisphere will *confabulate*, inventively integrating the aberrant information into a reasonable story consistent with the left-hemisphere context.

Split-brain research bears several important lessons for the discussion of human reason and will. First, it clearly discredits the notion of a unified holistic entity that is the locus of an individual's mental capabilities. Here we see that reasoning, will, and the generation of actions can be fractured and compartmentalized within the brain. Cognition and volition are rooted in discrete brain structures that can at times work independently.

Many of our conceptions about the nature of human will are based on our subjective awareness of our own deliberations, reasoning, and conscious choice of action. However, analysis of the reactions of split-brain patients to their own actions serves as a clear warning against taking our subjective intuitions at face value. Our sense of conscious choice may sometimes reflect interpretations that our brain produces about our own actions, rather than factual accounts of how our actions have been actually produced.

5. The Causal Relationship between Willing and Acting

The above experiments provide a general scientific context regarding neural correlates of will and decision making and the constraints imposed on these faculties by the anatomy of the brain, but do not directly address the question of the causal link between them and behavior generation. Perhaps the most interesting direct attempt to address this question by scientific methods is the famous experiments by Libet and coworkers.[18]

Previous work has shown that voluntary movements are preceded by a specific wave of electrical activity, which can be picked up by electroencephalogram (EEG) scalp recordings. This precursor brain activity is known as the "readiness potential" (RP). In Libet and coworkers' experiment, subjects were instructed to perform a simple flick or flexion of the wrist spontaneously at any time they felt the urge to do so. These voluntary acts were performed capriciously, free of any external limitations or restrictions. In these acts, the RPs began on average 550 milliseconds before activation of the involved muscles,[19] indicating that the brain had already "decided" to act about half a second before the act itself. The question posed by Libet was: when does the conscious wish or intention to perform the act appear? To address this question, Libet put a rapidly rotating clock in front of the subjects and asked them to mark the position of the clock arm at the time they felt the urge to act. In the traditional view of free will, one would expect conscious will to appear before, or at the onset, of the RP, and thus command the brain to perform the intended act. Instead, Libet found that the subjective awareness of the wish to move appeared some 350ms *after* the onset of the RP, well after the subject's decision to move could be predicted by observing brain activity. This experiment has recently been replicated by Haggard and Eimer.[20]

It would then seem that "the initiation of the freely voluntary act appears to begin in the brain unconsciously, well before the person consciously knows he wants to act!"[21] In other words, the conscious feeling of exerting one's will is an epiphenomenon, "simply a by-product of the brain's activities but with no causal powers of its own."[22]

These findings have generated heated controversy and conflicting interpretations. Libet's proponents noted that "his experiment brought scientific method to a question that had previously been purely philosophical."[23] Other researchers have voiced strong criticism, raising a number of technical and methodological problems. For instance, reports of conscious experience could be delayed relative to the actual experience itself. Additionally, Libet's subjects presumably divided their attention between viewing the external clock and their own decision making; hence identifying the time of their conscious will with the reported clock position is problematic.[24]

The current neuroscientific understanding of human decision-making, choice, and will is still in a very primitive stage, and future progress in noninvasive techniques will surely vastly improve our capability to measure and characterize the underlying neural processes. However, a rough picture has already emerged: even conscious decisions and intentions to act are consequences of electrical and chemical signals produced within the neural circuitry of the brain. Furthermore, the available physiological data confirm what has been known to psychology for a long time, namely that our conscious awareness of our actions is often not a necessary part of the actual real-time causal chain that leads to their execution. Rather, our awareness is part of the process of evaluation and interpretation of our actions and their consequences. This process forms and molds our representation of the world and of ourselves, a representation that will shape our, perhaps unconscious, future behavior.

It should be realized that the most sophisticated decisions studied in current experiments are still a long way from simple everyday decisions, such as what to eat for dinner. Ultimately, it may be impossible to measure the brain signaling that underlies the complex decision process of the scope that we encounter in everyday life in controlled laboratory settings. However, the general working hypothesis in science is that the constrained processes observed in the lab reveal the underlying mechanisms at work in the more complex settings of the real world. There is no reason to suspect that this is not the case for human cognition and behavior.

IV. Attempts to Reconcile Free Will with Science

1. Metaphysical Free Will

In previous sections, we have surveyed the scientific understanding of the nature of natural laws, and current neuroscientific insights into human volition and choice. In this section and in the following one I will analyze some of the attempts to reconcile this scientific picture with the concept of free will. I will confine myself to one pivotal and highly controversial hypothesis, which is known as the Principle of Alternative Possibilities (PAP),[25] or, put simply, the hypothesis that we have free will in the sense that "we could have done otherwise." Actually, this notion of free will consists of two hypotheses:

A. *Indeterminism of the World*

Human choice is a choice between genuinely open alternatives. At the moment of choice, the two options we have in front of us are both possible outcomes. This means that the future is really open. It is undetermined by the present state of the brain and the environment. Given exactly the same state of the world, we could have chosen otherwise, in which case the outcome trajectory of our actions would have been different.

B. Causation of the Will

When a choice is made between two genuine possibilities it is *our will* that makes these choices. Our conscious free will has a genuine causal power to intervene in the physical world and determine which of the available possibilities will be realized.

For the sake of brevity, I will call this notion of free will "metaphysical free will." It is often denoted as an incompatibilist conception of free will. In contrast, various compatibilistic forms of "practical free will" deny that we choose between genuinely open alternatives. These notions of free will in general do not run counter to scientific ideas or findings. It is the strong sense of free will, the metaphysical free will, which stands in apparent conflict with scientific world-view, as will be explained in the following sections.

2. Quantum Indeterminism and Free Will

Since the study of quantum mechanics began, there have been recurrent attempts to explain human freedom by appealing to quantum indeterminism. Sir Arthur Eddington, a prominent British astronomer and philosopher of science at the turn of the twentieth century, was among the first to speculate that Heisenberg's Uncertainty Principle might explain the mind-body interaction. He suggested that quantum fluctuations in the position of synaptic vesicles in the presynaptic nerve fiber, which contain the synaptic neurotransmitter molecules, might influence the discharge of these molecules. This could influence the firing of the post-synaptic cell and trigger "an unstable cascade," leading to a global change in brain state.[26]

A more recent advocate of the link between freedom and quantum mechanics is the Nobel laureate neurophysiologist John Eccles, who proposed a detailed theory of mind/brain interaction based on quantum mechanics. Eccles's starting point is the well-known fact that synaptic release of neurotransmitters does not follow deterministic law but seems to be a stochastic event. The conventional wisdom is that the fluctuations in synaptic events are due to thermal noise. In contrast, Eccles proposed that the fluctuations in synaptic release are quantum in origin, reflecting the indeterministic transitions of the presynaptic membrane between two quantum states. Second, according to this theory, the quantum probability of release of a vesicle can be momentarily changed by mental events without infringing on the law of conservation of energy. But how does the spiritual mind influence the physical processes in the brain? Eccles posited the existence of nonphysical units of mental intention called "psychons," which act through a quantum probability field to affect the probability of synaptic transmission during a brief period.[27]

The idea that quantum indeterminism enables the freedom that underlies our will is also shared by several contemporary philosophers. For instance, Hilary Putnam argues that a necessary condition for freedom of the will is indeterminism of physical laws, and that this indeterminism is provided by quantum physics.[28] John Searle argues in a recent article that "all indeterminism in nature is quantum indeterminism; consciousness is a feature of nature that manifests indeterminism"; hence the conclusion that "consciousness manifests quantum indeterminism."[29] Few physicists today defend the view that quantum mechanics is intimately related to the phenomenon of consciousness or free will. Notable among them is the mathematical physicist Roger Penrose. His theory of consciousness is based on some esoteric and highly dubious speculations on both quantum theory and brain processes.[30]

3. Free Will as an Emergent Property of the Brain

Eccles's theory is an example of a dualist theory, as it relies on the presumed presence of a nonphysical entity (substance or force) that directly influences microscopic processes within the brain. Dualist theories are rare in current scientific and philosophical thinking. Many contemporary attempts to formulate a coherent theory of free will rely on the concept of *emergence*. In contrast to the dualist explanations for the mind-body interaction, emergence theories of mind-body interactions are committed to a physical conception of causation.

Emergentism was originally articulated at the beginning of the twentieth century by C. Lloyd Morgan, Samuel Alexander, C. D. Broad, and others, and has regained considerable attention in the past few years, as a reaction to straightforward reductionist trends in science generally and in the study of the mind in particular.[31] Neuropsychologist and Nobel laureate Roger Sperry, for instance, wrote in favor of emergentism as an alternative to reductionism in the 1980s. "To attempt to explain an entity in terms of its parts and then the parts in terms of their parts and so on, results in an infinite regress in which one is left at the end trying to explain everything in terms of next-to-nothing. At each step of the way critical pattern components of causality are lost and explanation becomes less and less complete at each lower level."[32]

According to Sperry, when a new entity appears, the properties of the entity as a whole overpower the causal forces of its component entities at all the successively lower levels. The new properties of the system (instead of its own original properties) are the ones that determine its eventual fate. Although the causal forces at the microscopic level continue to operate, they are overwhelmed by the new causal properties that emerge in the whole. Thus, writes Sperry, "instead of a universe completely controlled by quantum mechanics and the basic forces of physics, science presents, by this interpretation, a universe controlled by a rich profusion of qualitatively diverse emergent powers that become increasingly complex and competent." In the brain, controls at the physico-chemical and physiological levels are superceded by new forms of causal control that emerge at the level of conscious mental processing, where causal properties include the contents of subjective experience. Causal control is thus shifted in brain dynamics from levels of pure material determinacy to levels of mental or subjective determinacy. The flow of nerve impulse traffic and related physiological events are no longer regulated solely by events in kind, but are moved by the higher mental controls. These conscious subjective properties are interpreted by Sperry to have causal potency in regulating the course of brain events; that is, the mental forces or properties exert a regulative control influence in brain physiology. The subjective forces are thus realities that supercede microscopic laws. "Just as the programming variables of a TV monitor have to be included in order to account for the electron flow pattern of the system, so also in the brain the subjective, mental variables of cerebral function have to be included to give a full account of the flow patterns of neural excitation."

The idea that human choice is an emergent phenomenon and can therefore exert causal influence on neural events has been adopted in one way or another by several thinkers. John Searle, who was quoted earlier, attempts to combine quantum indeterminism in the microscopic domain of neural events with the emergent causal forces of consciousness that fill the "quantum gap" and act to determine the course of events in the brain.[33] Somewhat analogously to the concept of emergence, a current trend in philosophy of mind utilizes the concept of "agent causation." According to the proponents of this view, "agents," such as human beings, have fundamental causal powers well above the familiar microscopic physical causal forces.[34]

The general idea that emergent phenomena are to some degree detached from the underlying microscopic forces and in turn can have causal influence on the microscopic events within the system has recently attracted considerable interest among philosophers attempting to redefine part-whole and inter-level relationships in complex systems. Emergence, in their view, is a conceptual framework for *nonreductive materialism.* They are committed to a scientific-materialistic world-view but argue against the idea that every phenomenon is reducible or predictable even in principle, by the forces acting on its microscopic constituents.

V. Human Choice and the Microscopic Closure of the Physical World

Having presented the main attempts to integrate the concept of free will with a scientific outlook, I present a critical appraisal of them below.

I begin by addressing the question whether, given our current understanding of neural processes, it is plausible that the brain does *not* follow strict deterministic physical laws.

1. Is the Brain an Indeterministic System?

One of the most prominent characteristics of the responses of single neurons to sensory stimuli is the apparent random nature of these responses. Specifically, the number and timing of electrical pulses ("spikes") that are emitted by individual cells in the cortex following a sensory stimulation vary considerably across repeated presentations of the same stimulus.[35] Additionally, as mentioned above, the signal transmission through synapses, which is the main form of communication within the brain, also exhibits considerable randomness. Often, synapses may completely fail to transmit the electrical impulse of the presynaptic cell to the postsynaptic cell.[36] These are just two examples of well-documented apparent randomness in the way neurons in the brain respond to events in the external world. It is therefore not surprising that most current theories describe the mapping between sensory and behavioral events and neuronal electrical activity patterns in probabilistic terms.[37] From this perspective, our current understanding of the brain supports the view that brain processes are appropriately described by indeterministic, statistical laws.

The origins of the ubiquitous stochasticity of neural signals are only partially understood and are still the subject of experimental and theoretical investigation. It is, however, clear that the main sources of neural noise are forces that can be characterized as thermal or chaotic rather than quantum in nature. They range from the thermal fluctuations in the opening and closing of ionic pores in the cell membrane[38] to the fluctuations in the signals that impinge on the sensory organs in ordinary ambient conditions. The complexity of the dynamics of neuronal circuits in cortex may give rise to chaotic patterns of activity,[39] thereby providing additional source of irregularity in neuronal signaling.

As explained in Section II, thermal noise and chaos are all deterministic in the sense that the future trajectory of brain states is completely determined by the present state of the brain and the environment. We may observe that upon repetition of the same stimulus, the response of a nerve cell varies. This is because neither the initial condition of the neuronal networks in the brain nor the environment surrounding the brain remain the same at each stimulus presentation. These differences are responsible for the observed variations in the neuronal responses. If there is nothing more than thermal noise and chaos at play, were we able to recreate exactly the same conditions for the brain and surrounding, the outcome of each trial would not vary. Thus, neither thermal neural noise nor chaotic brain dynamics give rise to an indeterminism of the nature required by the premise of "metaphysical freedom."

Does quantum indeterminism play a significant role in brain function? Our present understanding of synaptic biophysics does not support the specifics of Eccles's theory regarding the quantum nature of synaptic transmission. More generally, the electrical currents in the brain are made of quantum elements and processes, like the rest of the universe. Therefore quantum fluctuations on the atomic or electronic level will leave some minute imprint on the noisy environment of the neuron. Since, in most circumstances, quantum noise is negligible in magnitude compared to thermal noise, quantum effects probably do not play a noticeable role in the *probabilistic laws* underlying brain function. On the other hand, it cannot be ruled out that quantum effects do meaningfully influence the *particular neuronal firing times* of single neurons in *individual events*, as discussed below.

For quantum noise to affect brain processes, the relevant quantum fluctuations must be noticeable in single nerve cells, on length scales of at least several micrometers and time scales of at least several milliseconds. These scales are many orders of magnitude larger than those of individual quantum events. Furthermore, as discussed in Section III, ordinary behavioral decisions are expected to involve not just a few neurons, but rather the summed activity across large neuronal ensembles, putting additional constraints on the plausibility of the thesis that quantum indeterminacy plays an important role in brain function. The gap between the microscopic quantum scales and macroscopic neuronal activity may be surmountable if the neural system is extremely sensitive to minute changes in the forces acting on it. In fact, such sensitivity is expected if neuronal circuits in the brain are highly nonlinear chaotic systems. Chaos within the brain may amplify enormously the small quantum fluctuations at the atomic scale to a degree that will affect the timing of spikes of neurons in the brain circuits. This might be relevant in particular cases where choice is being made between alternatives with nearly equal "weights." In such borderline cases, small fluctuations in firing patterns of a group of cells may tilt the choice in favor of one of the alternatives.

In sum, given the present state of our understanding of brain processes and given the standard interpretation of quantum mechanics, we cannot rule out the possibility that the brain is truly an indeterministic system; that because of quantum indeterminism, there are certain circumstances where choices produced by brain processes are not fully determined by the antecedent brain process and the forces acting on it. If this is so, then the first prerequisite of "metaphysical free will" (Hypothesis A, Section IV) may be consistent with the scientific understanding of the brain. Before I discuss the second prerequisite

(Hypothesis B), I present below a critical analysis of the concept of *emergence* and its application to the brain.

2. Is Human Choice an Emergent Property of the Brain?

The idea that large complex systems have *emergent* macroscopic properties is a cornerstone of our contemporary understanding of physical phenomena. The simplest examples of emergent properties are the macroscopic states of ordinary matter, such as gaseous, liquid, and solid states. The very distinction between gas and liquid, for example, applies only to large systems of atoms or molecules; it is meaningless to speak of a single molecule or even a small group of molecules as being in either a gaseous or liquid state. We thus can say that gaseous and liquid states, viscosity, solidity, and the like, are all emergent features of macroscopic systems.

The concept of emergence is essential not only to physical laws of the macroscopic domain. Current theories of microscopic physics are also based on the notion of emergence. For instance, the present Standard Model of elementary particles is an incomplete theory of matter on the smallest possible spatial scale.[40] For one thing, it does not include gravitational forces between elementary particles, which would play an important role below an extremely short distance known as the Planck length.[41] Furthermore, the mathematical structure of the Standard Model is unintelligible unless this theory is viewed as an effective theory describing physics at the intermediate scales of the subatomic phenomena. We therefore must treat the very notions of electrons and quarks not as describing fundamental units of nature, but as "emergent" features of a more fundamental theory, as yet undiscovered, which describes the appropriate physics at the fantastically small Planck scale. Theories in physics at all levels are cast in terms of entities that are emergent features at the appropriate level.

The relationship between emergent properties and the underlying microscopic laws is an intricate one and the underlying principles have surfaced only in the last thirty years. On the one hand, emergent properties are largely insensitive to the detailed configuration of microscopic constituents. This is, of course, the reason why we can rely on macroscopic laws of matter. The great triumph of theoretical physics, the above-mentioned Standard Model, is exquisitely accurate in describing the interactions between quarks, electrons, and photons, despite the fact that we do not yet have a good idea what physics at the "Planck scale" looks like! On the other hand, emergent properties do depend on some aspects of the microscopic states. For example, the freezing of a liquid depends on the mass of the constituent molecules, and the nature of the forces between them. Thus, emergent properties are independent of many "irrelevant" parameters of the microscopic systems, but do depend on a small set of "relevant" microscopic parameters.[42] Although this principle is nowadays well understood, the actual determination of which parameters are relevant for any given emergent phenomenon may be extremely difficult.

The electrical signaling of nerve cells operates on length and time scales that are large compared to the underlying molecular processes. Furthermore, typical cognitive functions involve complex interactions between many nerve cells distributed across several cortical and subcortical areas. It is natural to consider them as emergent properties of the microscopic brain processes. The view that cognition and other mental phenomena are emergent macroscopic features of brain processes is bolstered by the substantial level of neuronal noise, which, as discussed above, will anyway "average out" the details of the microscopic state of the brain in any given instance. Which aspects of the neuronal firing patterns are relevant for any emergent cognitive faculty is still largely unknown, and remains a fundamental problem in brain theory.

3. Free Will and the Principle of Microscopic Closure

We are now ready to address the question of whether our current scientific understanding is consistent with the notion of metaphysical free will. As described in Section III, we have abundant and decisive scientific evidence that mental faculties such as volition, intention, and decision are rooted in the underlying patterns of neuronal electrical activity in the brain. Yet, can we identify free will as a causal force that works by influencing the neuronal activity in the brain? Let us begin with the "quantum theory" of free will. Above, I have argued that the hypothesis that the brain is affected by quantum fluctuations in a manner that has some consequences for choice and decisions is not inconsistent with our current understanding of the biophysics and dynamics of the underlying neuronal circuitry. However, as described in Section IV, metaphysical free will requires not only an element of indeterminism but also the existence of an entity, our will, that influences the course of events and fills in the indeterministic gap between events. It is this premise, that we have a will with its own causal power to interact with the physical processes in the brain, that stands in apparent contradiction to our scientific understanding.

According to the theory of Eccles and others, our will is efficacious by virtue of some nonphysical entity or force that intervenes at the molecular levels of, say, synaptic transmission and *modifies the quantum probabilities* associated with this process. If mental forces result in *event probabilities* in the brain that are different from those predicted by the known physical laws, these violations of the laws should be readily detected and verified by ordinary scientific observation. As discussed in Section III, there is overwhelming scientific evidence that the *microscopic* processes in the brain are just those predicted by familiar laws of physics and chemistry. Thus, such a hypothesis stands against all current scientific evidence. Furthermore, if violations of the known laws of nature do occur in the brain, then following their detection by scientific methods, we would necessarily change our theory of nature and incorporate these novel forces as part of the scientific description of the natural causal network. We will need to characterize the spatial and temporal properties of the new forces, their causal chains, and their relation to the causal chains of other physical forces, etc. In sum, these new entities, if they exist, will have to be integrated into the framework of science as part of the physical forces acting in the brain. This raises a serious problem with regard to the very notion of a nonphysical, mental entity that intervenes regularly in microscopic brain processes in violation of the regularities imposed by physical forces.

Let us now turn to the idea of free will as an emergent property of brain processes. As we have discussed, the notion that will and choice are emergent features of the brain is a scientifically sound and appealing idea. Furthermore, it would seem that emergence presents a straightforward resolution of the difficulty posed by the interaction between the mental and the physical. This is because according to emergence theory, the mental does not represent some mysterious new microphysical force. In fact, virtually *every* complex system exhibits emergent properties. Construed as such, we might argue that our will is as physical as the solidity of a solid or the fragility of a glass.

The major problem with the emergence explanation of free will is that it is based on the notion of "downward causation." To explain the causal powers of free will, emergentists assume that because the will is part of the emergent macroscopic realm of the brain, it exerts a "regulative control" on its state. However, since action ultimately requires the appropriate neural electrical signals, the controlling power of the will must come about by its capability to influence these microscopic processes. Indeed, the emergentists hold that even in general complex systems other than the brain the emergent properties influence their microscopic state, working against, or at least beyond, the microscopic causal powers.

While the notion of emergence is familiar to contemporary science, the idea that emergent properties have *independent* causal powers is completely foreign to it. Emergent features of a macroscopic system are determined by nothing other than the microscopic constituents of the system, their spatial arrangement and the microscopic forces acting between them. What is special about emergent features is, first, that they represent outcomes of the microscopic physical processes that appear only on the macroscopic scale, and second, that they bear very nonadditive and complex dependencies on the specific details of the microscopic states and force parameters. Nevertheless, they do not assume causal relations that are not rooted in the underlying microscopic laws.

Emergence theories of metaphysical free will often cite analogies from ordinary physical systems to illustrate the concept of downward causation. Earlier, I cited Sperry's example of the programming variables of a TV monitor as being essential in accounting for the electron flow pattern of the system. Another familiar example used by Sperry is a wheel rolling down a hill; the molecules of the wheel rotate by virtue of the wheel's macroscopic motion. However, none of these analogies actually support the concept of downward causation. The rotation of the wheel could also be seen as either a consequence of the motion of the constituent molecules or (depending upon the precise definition of terms) as equivalent to the organized motion of the molecules. Given our understanding of the laws of motion, it does not make sense to consider the motion of the wheel as causing the motion of its molecules. Likewise, the design of a TV monitor does not exert a force on the electrons inside it beyond the usual microscopic forces, but rather utilizes these forces. The device is configured precisely so that the electromagnetic forces between electrons in the TV tube will accelerate them, according to the known laws of motion, to the desired spots on the fluorescent screen. By introducing an element of intelligent design, the example of the TV monitor also confuses the teleological reasoning commonly used in explaining human action with physical mechanisms acting as causes of events.[43]

John Searle acknowledges the flaw in the above analogies. In defining the nature of consciousness he argues:[44] "The trajectory of each molecule is affected by the behavior of the entire solid wheel. But of course there is nothing there but molecules. The wheel consists entirely of molecules. So when we say the solidity functions causally in the behavior of the wheel and in the behavior of the individual molecules that compose the wheel, we are not saying that the solidity is something in addition to the molecules; rather, it is just the condition that the molecules are in." Likewise, "a system feature [e.g., consciousness] can affect micro-level elements in a system composed entirely of the micro-level elements [e.g., neurons], in which *all causal powers are reducible to the causal powers of the micro-level elements*" (my italics). This account, however, does not leave room for real causal powers of the will, if indeed viewed as just a macro-feature of the brain.

As quoted in Section IV, Searle goes on to argue that free will is different from the motion of the wheel because, in contrast to the wheel, consciousness is affected by quantum indeterminism. It is difficult to apprehend how Searle's argument resolves the inconsistency of the macro/micro relationship. After all, if the macrostate of the brain is nothing but the collective state of its micro-elements, how can this macro state influence neuronal electrochemical processes beyond what is determined by the microscopic laws? Similar objections apply to attempts by other philosophers of the Libertarian camp to defend the concept of metaphysical free will by positing some sort of "agent causation."[45]

Finally, we address the hypothesis that an act of free will should not be considered in terms of physical laws acting either on the micro or on the macro levels of the brain. Rather, it should be considered as a nonphysical (mental or spiritual) influence on brain processes that does not violate physical laws but supplements them. For example, perhaps our will intervenes at the quantum level and determines the *actual realization* of quantum events in the brain but *does not change the quantum probabilities* of the different events. This way, the argument might go, free will acts in complete consistency with the regularities predicted by ordinary physical laws. It just fills in the indeterminacy gap left by them. According to this conception, our free will is severely restricted in its scope of influence as it is constrained by the statistical regularities imposed by natural laws. Furthermore, if the mental influences on behavior are fully consistent with physical laws, it follows that all observable features of human cognition and action can, in principle, be explained on the basis of physical processes within the brain. In such a case, from a scientific perspective, invoking putative nonphysical entities, such as free will, in explaining human behavior is redundant.

What emerges from the above objections is a fundamental scientific view of the causal underpinning of all physical phenomena, which stands in contradiction to the incompatibilist conception of "metaphysical free will." I term this view:

> The microscopic closure of the physical world: *The* only *factors that cause or influence physical events are those that result from (or are reducible to) the physical properties of the constituent microscopic elements and the microscopic forces acting between them.*

Note that this hypothesis does not rule out the existence of undetermined or random events in nature. It merely states that whatever is not determined by the physical forces operating at the microscopic level cannot be determined by something else. On the face of it, the hypothesis of the microscopic closure of all physical phenomena seems to be presumptuous. What is the basis of our claim that physical events cannot be determined by other influences, particularly if these influences work within the *statistical regularities* dictated by the physical laws? The basis for this claim is rooted in the foundation of the scientific conception of the physical world. Science provides us with a framework of describing causal chains between physical events mediated by physical laws. According to this framework, any change in the motion or state of particles (such as that which must occur in the brain prior to any action) is by definition a physical force. As such, its effects are subject to the scientific methods of empirical verification and to tests of their consistency with other known natural laws and principles. As explained above, if there are observable deviations[46] from known natural laws, these deviations will be indicative of the presence of hitherto unknown physical forces and will be integrated into an updated physical theory of the brain. If, on the other hand, the nonphysical entities do not cause any observable deviations from known physical laws, then *admittedly* the physical laws are fully adequate in explaining all observables including human action. Thus, the scientific conception does not provide us with any coherent framework for describing nonphysical influences on physical systems.

The above "microscopic closure" hypothesis is related to the more familiar hypothesis of the "causal closure of the physical domain."[47] The version presented here emphasizes the exclusion of emergent scenarios of downward causality. As mentioned above, no example of emergence understood scientifically exhibits downward causation. They are all predictable from, or reducible to, microscopic causal laws. In fact, it is a fundamental character of the scientific conception that complex phenomena ought to be explained in terms of laws that are derivable from the underlying microscopic level. Whether the concept of downward causation is coherent at all is questionable.[48] Even if philosophers were able to explicate it, it is unclear how a scientific conception of reality could incorporate it any more than it could incorporate the notion of nonphysical forces acting at the microscopic level.

VI. Conclusion

I was asked to address the issue: "What measure of non-biologically determined behavior can coincide with current neuroscientific conceptions of the role of physiology in determining behavior?" What is then my answer to this question? Well, it depends crucially on the meaning of the term "non-biologically determined behavior," and, to some extent, also on what we take to be our "current neuroscientific conceptions" of neuronal mechanisms underlying behavior. As I have outlined in this paper, given the current state of understanding of the brain, neuronal activation and the spread of this activity is influenced by a substantial degree of noise. Hence, we expect that there will be many cases where the chains of events linking stimuli to human cognition and behavior are appropriately described by statistical laws rather than deterministic ones. In particular, such laws would allow, and even predict, that upon recurrence of similar external conditions, our actions might take a different course, due to the small variations in the internal neuronal states as well as in the external forces in constant interactions with the nervous system.

On the other hand, I have argued that both external and internal sources of noise are by and large "classical" in nature. This means that although their realization is unpredictable in practice, they are fully determined by the antecedent condition of the system and the forces acting on it. In this sense, these processes are still *biologically determined.* More specifically, in the context of the problem of freedom of choice or free will, given exactly the same internal and external conditions we would not have been able to act differently from the way we did. Still, I have argued that there might be conditions in which the effect of quantum indeterminacy on neural processes is amplified so that it has a noticeable effect on behavior. This would be the case if the relevant neuronal circuits exhibit strongly chaotic dynamics, which would make them extremely sensitive to small quantum fluctuations. In these cases, the particular behavioral outcome of the brain process would be truly undetermined by previous conditions and processes. Assessing the feasibility of this scenario must await more theoretical and experimental research. In any case, we expect this to be the exception, obtaining, for instance, only in cases where the nervous system is near an unstable equilibrium or extremely close to a threshold,[49] rather than the rule.

Finally, I have addressed the incompatibilist conception of *metaphysical free will.* I have discussed two main lines of defense of this notion of free will. One relies on hypothetical mental influences directing quantum microscopic fluctuations. The other assumes that the emergent features at the macrolevel of complex systems have causal powers to influence future macro- and micro-events (downward causation). I have argued that, from a scientific standpoint, even if certain physical events are not fully determined by the underlying physical laws, there remains no good explanation as to how they can be influenced by other factors, operating either at the micro or macro levels. Thus, even if certain behaviors are not fully determined by prior biological processes, we have no explanation for how they might be affected by an *extra-physical* choice or will, namely by a power that would not effectively fall under the scientific character of a physical force.

"Any philosopher"—writes philosopher Daniel Dennett—"ought to feel at least a little embarrassed that with so much work so little progress has been made."[50] Searle agrees. "The persistence of the traditional free will problem in philosophy," he writes, "seems to me something of a scandal." I am not qualified to judge whether these assertions are just. However, it does seem to me something of a scandal that scientists and scientifically literate philosophers who defend metaphysical freedom of the will pay so little attention to the classical problem of interaction between the will and our brain processes. Since the dismal attempts of Descartes to resolve this difficulty there has not been a single scientifically sound hypothesis that explains how our behavior can be determined by an entity not governed by physical laws.

Our objection to incompatibilist metaphysical free will does not imply that human will and choice are unimportant epiphenomena. It simply implies that they are emergent features of the causal chain of physical events that occur within our brain. Delineating the role of reasoning and choice in human behavior remains one of the most important challenges of science as well as philosophy, moral and religious thought.

"Man," writes Einstein, "defends himself from being regarded as an impotent object in the course of the Universe. But should the lawfulness of events, such as unveils itself more or less clearly in organic nature, cease to function in front of the activities in our brain? . . . Determinism does not stop before the majesty of our human will."[51]

"If physical determinism is true," writes Eccles, "then that is the end of all discussion or argument; everything is finished. There is no philosophy. All human beings are caught up in this inexorable web of circumstances and cannot break out of it. Everything that we think we are doing is an illusion and that is that. Will anybody live up to this situation?"[52]

Here are two divergent reactions of distinguished scientists to the prospect that human will and choice, reason and action, all succumb to the deterministic laws of nature. It is time to adopt the modesty preached by Einstein and to give up Eccles's fatalistic worries. We, human beings, are the magnificent product of God's nature. This is not "the end of all discussions." It is merely the end to the wrong discussions, mistakenly putting man in the place of God. This should be the beginning of the right discussions: how to sustain humanity and human values within the natural order decreed by Him.

Notes

I am grateful to Dr. Ruty Lev Bar-Or for her help in writing this article, to Dr. Yuval Dolev for useful discussions on the subject matter, and to Olivia White for helpful and illuminating comments on the manuscript. I also thank Dr. Yitzhak Berger for useful suggestions. I acknowledge the support of the Center for Systems Neuroscience, Harvard University, where the final version of the paper was written.

1. Pierre Simon, Marquis de Laplace, *Essai philosophique sur les probabilitiés, Oeuvres de Laplace* (Paris, 1847), vol. VII, vi–vii. Translated excerpts in John Cottingham, ed., *Western Philosophy: An Anthology* (Malden, MA: Blackwell, 2000), 226–228.

2. For a general introduction to chaos, see, for example, Lui Lam, *Nonlinear Physics for Beginners* (Singapore: World Scientific, 1998), 17–21, and the popular book by James Gleik, *Chaos* (Harrisonburg, VA: Viking Penguin, 1987).

3. For a general discussion of the history and principles of statistical mechanics, see P. W. Atkins, *The Second Law* (New York: Scientific American Books and W. H. Freeman and Co., 1984), and Lawrence Sklar, *Philosophy of Physics* (Boulder, CO: Westview Press, 1992), ch. 3.

4. I have presented an interpretation of statistical physics in terms of interaction with the environment. This interpretation is the most natural in the context of biological systems. Under certain conditions, statistical physics can also be justified for *isolated* physical systems. In either case, classical statistical physics is fully consistent with classical micro-determinism. See for example, Sklar, *Philosophy of Physics*, ch. 3.

5. Actually, in the "standard" interpretation of quantum mechanics, also known as the "Copenhagen interpretation," the quantum laws do not even predict the probabilities of the occurrences of events but only the probabilities of the outcomes of their measurement. This raises difficult fundamental problems about the role of the observers in physical laws and the definition of measurement. In recent years, there has been a resurgence of attempts to rid quantum theory of the notion of measurement and even to reconcile quantum theory with classical determinism. See Sklar, *Philosophy of Physics*, ch. 4; Sheldon Goldestein, "Quantum Theory without Observers," *Physics Today*, March 1998: 42–46, and April 1998: 38–42.

6. See James Nichols and William T. Newsome, "The Neurobiology of Cognition," *Nature* 402 (1999): C35–C38.

7. Nikos K. Logothetis, Jon Pauls, Mark Augath, Torsten Trinath, and Axel Oeltermann, "Neurophysiological Investigation of the Basis of the fMRI Signal," *Nature* 412 (2001): 150–57.

8. For an extensive discussion of this issue, see A. J. Parker and W. T. Newsome, "Sense and the Single Neuron: Probing the Physiology of Perception," *Annual Reviews in Neuroscience* 21 (1998): 227–77.

9. Bas Kast, "Decisions, Decisions," *Nature* 411 (2001): 126–28.

10. Michael L. Platt and Paul W. Glimcher, "Neural Correlates of Decision Variables in Parietal Cortex," *Nature* 400 (1999): 233–38.

11. Patricia S. Goldman-Rakic, "Topography of Cognition. Parallel Distributed Networks in Primate Association Cortex," *Annual Reviews in Neuroscience* 11 (1988): 137–56.

12. Sean A. Spence and Chris D. Frith, "Towards a Functional Anatomy of Volition," in *The Volitional Brain. Towards a Neuroscience of Free Will*, ed. Benjamin Libet, Anthony Freeman, and Keith Sutherland (Imprint Academic, 1999), 11–29.

13. David H. Ingvar, "On Volition. A Neurophysiologically Oriented Essay," in *The Volitional Brain*, 1–10.

14. I. Fried, A. Katz, G. McCarthy et al., "Functional-Organization of Human Supplementary Motor Cortex Studied by Electrical-Stimulation," *Journal of Neuroscience* 11 (1991): 3656–66.

15. Mark Bear, Barry Connors, and Michael Paradiso eds., *Neuroscience* (Baltimore: Lippincott, Williams & Wilkins, 2001), 591.

16. www.macalstr.edu/~psych/whathap/UBNRP/Split_Brain/Split_Brain_Consciousness.html, excerpted from Roger W. Sperry, "A Modified Concept of Consciousness," *Psychological Review* 76 (1969): 532–36.

17. Kathleen Baynes and Michael S. Gazzaniga, "Consciousness, Introspection, and the Split-Brain: The Two Minds/One Body Problem," in *The New Cognitive Neurosciences*, ed. Michael Gazzanigga (Cambridge, MA: The MIT Press, 2000), 1355–63.

18. B. Libet, C. A. Gleason, E. W. Wright, and D. K. Pearl, "Time of Conscious Intention to Act in Relation to Onset of Cerebral Activity (Readiness Potential): The Unconscious Initiation of a Freely Voluntary Act," *Brain* 102 (1983): 623–42.

19. The time of onset of movement is measured by the onset of electrical activation of the muscles (EMG). In some experiments, RPs precede the onset of movement by as much as one to two seconds. See refs. in notes 21–22 for detailed discussions of this issue.

20. Patrick Haggard and Martin Eimer, "On the Relation between Brain Potentials and Conscious Awareness," *Experimental Brain Research* 126 (1999): 128–33.

21. Benjamin Libet, "Do We Have Free Will," in *The Volitional Brain*, 49.

22. Libet, "Free Will," 55.

23. Patrick Haggard, in Patrick Haggard and Benjamin Libet, "Conscious Intention and Brain Activity," *Journal of Consciousness Studies* 8 (2001): 47–63 (quotation from p. 48).

24. See commentaries in *Behavioral and Brain Sciences* 8 (1985) on the article by Benjamin Libet, "Unconscious Cerebral Initiative and the Role of Conscious Will in Voluntary Action," 529–566. For additional critical analysis of Libet's experiments, see Daniel C. Dennett, *Freedom Evolved* (New York: Penguin, 2003), 228–42.

25. PAP figures prominently in debates on moral responsibility. In this article I am referring to the implication of this principle to the very notion of free will. The implications on the concept of moral responsibility are beyond the scope of this article. For discussions of PAP, see David Shatz, "Irresistible Goodness and Alternative Possibilities," in *Freedom and Moral Responsibility: General and Jewish Perspectives* (College Park: University Press of Maryland, 1997), ed. C. H. Manekin et al., and other articles in this volume.

26. Arthur Stanley, *Sir Eddington: The Nature of the Physical World* (New York: MacMillan, 1928).

27. John Eccles, "A Unitary Hypothesis of Mind-Brain Interaction in the Cerebral Cortex," *Proceedings of the Royal Society of London B. Biological Sciences* 240 (1990): 433–51.

28. Hilary Putnam, *Realism with a Human Face* (Cambridge, MA: Harvard University Press, 1990), 151–56.

29. John Searle, "Free Will as a Problem in Neurobiology," http://philosophy.berkeley.edu/searle/.

30. Roger Penrose, *The Emperor's New Mind* (New York: Oxford University Press, 1989). For a technical rebuttal of some aspects of this theory, see Max Tegmark, "The Importance of Quantum Decoherence in Brain Processes," *Physical Review E* 61 (2000): 4194–206.

31. Jaegwon Kim, "Making Sense of Emergence," in the 1997 Oberlin Colloquium in Philosophy, *Mind in a Physical World: An Essay on the Mind-Body Problem and Mental Causation* (Cambridge, MA: MIT Press, 1998); see also articles in *Emergence or Reduction? Essays on the Prospects of Nonreductive Physicalism*, ed. Ansgar Beckermann, Hans Flohr, and Jaegwon Kim (New York: de Gruyter, 1992).

32. Roger Sperry, "Changing Priorities," *Annual Reviews in Neuroscience* 4 (1981): 1–15.

33. See Searle, note 30.

34. Roderick Chisholm, *Person and Object: A Metaphysical Study* (Allen and Unwinn, 1976), chap. 2; John Bishop, "Agent-Causation," *Mind* 92 (1983): 61–79.

35. For review, see F. Rieke, D. Warland, R. R. de Royter van Steveninck, and W. Bialek, *Spikes: Exploring the Neural Codes* (Cambridge, MA: MIT Press, 1997).

36. The origins of unreliability of synaptic transmission are discussed in C. Allen and C. Stevens, "An Evaluation of Causes for Unreliability of Synaptic Transmission," *Proceedings of National Academy of Sciences (USA)* 91 (1994): 10380–83.

37. For the implications of neural noise on neural codes and perception, see ref. 9 and H. S. Seung and H. Sompolinsky, "Simple Models for Reading Neuronal Population Codes," *Proceedings of National Academy of Sciences (USA)* 90 (1993): 10749–53.

38. John W. White, Jay T. Rubinstein, and Alan K. Ray, "Channel Noise in Neurons," *Trends in Neuroscience* 23 (200): 131–37.

39. C. A. van Vreeswijk and H. Sompolinsky, "Chaos in Neuronal Networks with Balanced Excitatory and Inhibitory Activity," *Science* 274 (1996): 1724–26. Recent discussions of this issue can be found in *Chaos in the Brain?*, ed. K Lehnertz, C. E. Elger, J. Arnhold, and P. Grassberger (University of Bonn, Workshop Proceedings, 1999).

40. For a popular exposition of the Standard Model of elementary particles, see Gerard Hooft, *In Search of the Ultimate Building Blocks* (Cambridge University Press, 1997) and Steven Weinberg, *Dreams of Final Theory* (New York: Vintage Books, 1994), chap. 8.

41. As is well known, the gravitational force between two masses falls off at the inverse of the squared distance between them. At large distances, this force is strong only among objects such as the earth and the sun due to their huge masses. With microscopic particles this force is negligible compared with electromagnetic and nuclear forces, except at extremely short distances—at the scale of Planck length, which is many orders of magnitude smaller than the distances probed by present day most powerful accelerators. See refs. in note 41.

42. The underlying physical theory is known as "The Renormalization Group"; see Laurie M. Brown, ed., *Renormalization: From Lorentz to Landau (and Beyond)*, (New York: Springer, 1993), especially chap. 5; Kenneth G. Wilson, "Problems in Physics with Many Scales of Length," *Scientific American* 241 (1979): 158–79.

43. Unless intelligent reasoning is assumed to have real causal powers independent of the neural processes in the designer's brain. This, however, is the very issue in question.

44. See note 30.

45. See the criticism of such "extra factors" by Robert Kane, "Reflections on Free Will, Determinism and Indeterminism," www.ucl.ac.uk/~uctytho/dfwVariousKane.html. Kane, himself a libertarian, argues that free will is exercised only on rare occasions when we make difficult choices between several equally reasonable alternatives; these choices affect our beliefs and motivations and thereby our future deterministic choices as well (hence they are called Self-Forming Alternatives [SFAs]). See Robert Kane, *The Significance of Free Will* (Oxford, 1996). Kane seems to hold that choices in SFAs are undetermined because of an unstable equilibrium in the underlying neuronal dynamics.

However, the actual choices are the outcome of "decisions" made by internal neuronal noises and not by "external" causal powers. Nevertheless, since the equilibrium itself and the ensuing process of deliberation involve the agents' efforts, beliefs, goals, etc., they are justifiably called the choice of the agent's will, as long as the outcomes of the process are, in fact, compatible with his will. Thus, in the present context it seems that Kane is satisfied with a definition of free will that entails indeterminism (Hypothesis A, Section IV) but not the causal powers of the will in the usual sense (Hypothesis B).

46. Observable deviations may take the form of systematic laboratory recordings of microscopic processes, for example, the detection of novel electrical properties of nerve cells, or systematic psychological observations of behavioral patterns with regularities that are inconsistent with the physically based theory.

47. See Jaegwon Kim, note 32.

48. See note 32 and also articles in: *Downward Causation: Minds, Bodies, and Matter*, ed. Peter Bogh Andersen, et al. (Aarhus: Aarhus University Press, 2000).

49. This is somewhat akin to Kane's SFA, see note 46, and David Shatz, ref. in note 26.

50. Daniel C. Dennett, *Elbow Room* (Cambridge: MIT Press, 1997), 2.

51. Albert Einstein, 1931, quoted in D. Home and A. Robinson, "Einstein and Tagore: Man, Nature and Mysticism," *Journal of Consciousness Studies* 2 (1995): 167–79.

52. Karl Popper and John C. Eccles, *The Self and Its Brain* (London: Taylor and Francis, 1977), 546.

2

Genetics and Morality

Robert Pollack

Genes and Human Behavior

I HAVE BEEN ASKED TO ADDRESS the question, "Are there morally objectionable behaviors that appear to be caused entirely by genetic factors without any meaningful volitional component?" The answer to this question is a simple "No."

What, then, is there left to say? To begin with, we can examine the question to discover what is hidden beneath its apparently simple construction. Then we can look at the science.

What is a "morally objectionable behavior," if it is not also a free-will choice? If there were moral judgments attached to any act that took place without free will and the volition to act on it, death itself would be a "morally objectionable behavior." We all agree that suicide is morally objectionable, but death comes to us all in the end "without any meaningful volitional component," as anyone who has stood by a loved one on the threshold of death must surely know. It is undoubtedly the case that our life expectancy is capped at around a century by germ-line "genetic factors" in service to the needs of the species; unless one wants to consider every death to be a "morally objectionable behavior," we would be forced to say that there are no such behaviors that are completely determined by genetic factors.

And what exactly is intended here when we ask about the ramifications of "genetic factors"? The word Mendel himself used for genes was "factors," which conveyed the meaning of their particulate, unvarying nature. Mendelian inheritance is the inheritance of choices among the appearances—or phenotypes—of a population. So, in a Mendelian sense, genetic factors are the genes themselves, stretches of DNA encoding a particular protein, plus adjacent stretches encoding switches that permit the protein to be made solely at the right time and in the right place in the developmental history of the organism. If this is what is meant by "genetic factors," then, as I have said, it is easy to show that the answer is "No": genes construct us with brain and body capable of free will.

But today "factor" carries the notion of a partial contribution, whether genetic or not. Venality was a factor in the Enron collapse, and so was stupidity and weak enforcement of the law. If we take this broader meaning of "factors," then the words "caused entirely by genetic factors" imply that other factors beyond what they encode may also be provided to us by our genes. As we each begin as the construction of genes from sperm and egg, surely much about us thereafter is "caused entirely by genetic factors," if we allow that these factors include all those aspects of ourselves that begin with our genes. So, by either definition of "factors," the answer remains "No."

Put in other words, the question is whether a person who performs an act that is forbidden by halakhah (a "morally objectionable behavior") can ever be taken as having a diminished capacity of free will and therefore not choosing at all. But this presumes too great a role for the human genome in determining behavior. Human genomes do, in fact, allow free will. There are about three billion letters in the human genome. But there are about a millionfold more synaptic connections in a human brain at birth than there are letters in any human cell's canonical text. These synaptic connections—the basis of all mental activity later in life—cannot have all been specifically encoded by our genomes. At birth some are not functional, nor are many stable or specific; synaptic connections harden into circuits only later.

We are born with a tissue that becomes a mind through social interaction. Our DNA encodes, in other words, a Learning Machine. The Learning Machine is very complicated: it requires that most human genes be present in functional versions. Half or more of the genes in the genome are active in the nervous system, and for the most part only in the brain.

What these genes encode is the capacity of synaptic connections to be stabilized by use, through the activation and repression of genes in nerve cells. The Learning Machine starts up at birth at the latest, activated by initial input signals from the organs of perception. This is the mechanism by which the mind slowly emerges from the brain, through imitation of the minds of those people with whom the infant interacts. Experiences of the first two years, before language emerges, lay down much of the stable circuitry of the thinking brain.

But even after these formative years, the mature brain forever retains plasticity in its circuits, and it never loses the capacity to link past with present experience by familiarity of synaptic pattern. Synaptic connections are made and broken throughout life; these are experienced variously as sensation, perception, memory, repression, and—for my argument's sake, most important—ongoing teaching and learning.

The learning machine requires adequate social interaction from birth on; absent that, sociopathic disasters ensue. Many of these will be "morally objectionable behaviors." Racism, for example, is one of these disasters.[1] Whoever is cast as the Other by adults when they interact with their children, will become the Other to those children. When racism is learned in this way, it is a biological event, in that the synaptic wiring of associations in the brain of the child will have mimicked those in the brain of the adult. This form of inheritance is not through DNA, but it can be as stable, and as long lasting, as genetic inheritance. But we must be clear: it is social, not genetic. That is why there is no contradiction between thinking of race as a social construct—the product of racism—and thinking about it in the language of inheritance.

A crucial further implication of the incompleteness of genetic specificity in determining our individual behavior is that genetic information may not be used as an excuse for setting aside religious values. The search for a causative rather than correlative link between versions of genes and propensities for various behaviors, even if it bore fruit, would not remove the obligation to work with every person regardless of his or her genetic gifts, because the moral context that gives meaning to science through medicine requires the attention of both science and medicine to a person in all his or her complexity and variability.

Perhaps the best way to see this is to understand that although in social terms people tend to aggregate into groups of majority and minority populations—often separated by religion—by the data of our genomes, we are all members of genetic minorities that range in size for the millions of a founder population, to the dozens of an immediate family, to the irreducible minority of one that is at the heart and soul of medicine. It would do us well to acknowledge that nothing in the legacy of human DNA blocks the choice to value the differences among us above the resemblance any of us might have to our idea of an ideal person.

Judaism and the Limits of Genetics

From the Jewish perspective, in particular, genes cannot propel us toward religious pursuits and experiences. Consider the following quote from the Rambam, *Mishneh Torah, Hilkhot Talmud Torah*, chapter 3:

> Three crowns were conferred upon Israel: the crown of Torah, the crown of priesthood, and the crown of royalty.

Aaron merited the crown of priesthood, as [Numbers] states: "And it will be an eternal covenant of priesthood for him and his descendants after him."

David merited the crown of royalty, as [Psalms] states: "His seed will continue forever, and his throne be as the sun before Me."

The crown of Torah is set aside, waiting, and ready for each Jew, as [implied by Deuteronomy]: "The Torah which Moses commanded us is the inheritance of the congregation of Jacob." Whoever desires may come and take it.

Rambam then goes on to make the point in his own special way. He says that when hiring a teacher of Torah for yourself or your child, you should give a learned *mamzer* precedence over an ignorant man, even if that ignorant man happens to be the *Kohen Gadol* himself.

Rambam is pushing a point, to make a critical differentiation. By distinguishing the availability of the "inheritance" of Torah, from the father-to-son inheritance of priestly obligations and privileges, and from the royal prerogative to pass a Jewish Kingdom to one's child, he is telling us that no aspect of Jewish life available to *him* was biologically inherited. The inherited priesthood needed a context of Temple sacrifice, and the Davidic reign needed a Jewish nation. At his time, all that Jews had left was Torah.

With this as background, let's look again at what we may yet inherit in a different way: "The crown of Torah is set aside, waiting, and ready for each Jew." Take these characteristics of the Crown of Torah one at a time.

It is "set aside." In the dichotomy of *kodesh ve-hol*, the study of Torah is intrinsically *kodesh*. The crown of understanding cannot be taken, any more than it can be inherited. It can only be given, as a reward for study, for understanding, and for the enactment of what it teaches.

It is "waiting." The understanding of Torah does not require that one be born into the right family, or into the right moment. Rather, any moment is as good as any other to begin studying, and any Jew is as authentic as any other, in terms of the right and the obligation to begin studying.

It is "ready for each Jew." The depth of understanding of Torah necessary for the gaining of the Crown cannot be the same for each Jew. Instead, for each one of us, it is the saturation of our individual capacity to learn, rather than our particular level of innate, inherited capacity, that yields one of us the Crown, and another not. No wonder we say in the *shema*, *ve-shinnantam le-vanekha*—"teach them thoroughly to your children." We have no other choice; we have no genetic, inherited shortcut for our children.

In summary, Rambam is not only saying that the Crown of Torah is not inherited, but that also, no aspect of one's wish to receive the Crown of Torah can be inherited either. Both are wholly a matter of choice and the will to follow Torah at each generation, neither is at all a matter of biological inheritance.

Moreover, there is no fourth crown to mention at all. There is no crown of sensitivity, no crown of spirituality, no crown of religiosity. Now step back and ask what this has to do with our sense of ourselves as Jews. Whatever *pintele yid*, whatever special Jewish soul we may wish to think we have, I think Rambam is saying that it must finally be what we were taught and what we learned, not what came to us through our ancestors' DNA, that makes us Jews. So, while Ruth might have inherited a genetic propensity toward loyalty from her Moabite ancestors, I rather doubt that we should draw upon that genetic legacy when trying to understand her behavior with Naomi or Boaz. Rather, Ruth teaches us that even the Davidic line—with all it implies for our future—emerges by teachable and learnable action, not through DNA.

DNA is a chemical of great informational density, a text of great importance. But any person's genome—his or her complement of two copies of each of about 60,000 genes, one copy of each from each parent—is no more the complete statement of that person's life and character, than a canonical text is the complete statement of a living religion. Everything interesting in both cases is the product of interpretation and interaction. Appearances begin with the information encoded in DNA, but anyone who knows one identical twin also knows that person to be unique, despite the presence of another person with the same canonical text in each cell.

Genetic differences among us are nevertheless great enough to account for some—but not all!—of the differences between one person and another. From any one person to another, unrelated person, the differences in base-pair sequence—letters in the text—for the coding region of any gene studied come to between one in a thousand and one in a hundred. Imagine a canonical text with that many variations from copy to copy. There would be no chance of any version possibly being the "real" one. There can be no biological data in support of the notion of oneself as the member of a genetically identical group.

All that makes our genomes human—and all that makes us human in a biological sense—is that these six billion different genomes, and only they, are all capable of coming together with each other through sperm and egg to make another generation of people. So the biology is clear: there is no chance of some human genomes being Jewish and others not; the biology of us makes us truly all equal.

No rational explanation of what it is to be Jewish, then, can possibly begin with the claim that one set of this exceedingly large number of genetic variants is Jewish, and another not. Better, the history of our species tells us that we are all the descendants of Africans. The evidence for this comes from many quarters, but in our terms the DNA evidence is most interesting: because Africa is the home of us all, people who are the descendants of the original people—Africans—have the greatest genetic diversity of all human subpopulations.

The rest of us are, in a sense, tribal offshoots, each carrying away only a fraction of the genetic richness of our species, which still remains where it began, in Africa. The irony of universal African patrimony only makes the insult of American racism more stupid, but not less dangerous, than any other dehumanization.

Genetics and Modern Medicine

By each of us exerting our free will to decide whether it is wise for us to know more or to know less about our own DNA at any given moment, the scientific data of DNA-based medicine may be returned to a proper medical context. In light of the DNA evidence we already have, this means stretching the definition of normal variation to include the greatest possible diversity of inherited appearances and behaviors. Our obligation here is as clear in its own way, as the countervailing trend is in current medical science.

The straightforward agenda of scientists and the short-term acquiescence of physicians fifty years ago led to the creation of the National Institutes of Health, each institute named for a disease of the middle-aged white men in Congress who gave out federal money in those days and still do today. These institutes have provided the country and the world with much knowledge of great value, both medical and monetary. But with the creation of cheap, easy scans for mutations in genes like BRCA1—the first human gene to be associated with an inherited propensity to develop ovarian and BReast CAncer (hence its name)—knowledge contributed by NIH-supported science has begun to change medicine in ways that deny the meaning medicine provides to science.

In "The Missing Moment," I drew the following quote from my mentor and teacher, James D. Watson, discoverer of the structure of DNA and founding director of the Human Genome Project. Writing in the Annual Report of his laboratory, he had said:

> If we could use genetic analysis to help work out the biochemical pathways underlying memory and clear thinking, for example, we might be able to find pharmaceutical compounds to improve these most needed human attributes. Thus, those who want to protect the mentally ill or the slow learner may not get what they strive for if they portray them exclusively as victims of their environment. We might like to think otherwise, but only by reducing the differences in human beings will we ever have a society in which we can effectively view all individuals as truly equal.

I admire Jim Watson for his unmatched taste in picking the right question to ask of nature as much today as I did when I first met him in the late 1960s, but I know that here he was deeply wrong. We know from a century and a half of research in ecology and evolution that as a species our future lies not in minimizing our differences, but in cherishing them.

Finally, a further remark on the matter of DNA and the objectives of modern medicine. We have the choice to stop short of cooperating in the stigmatization of members of our own community by a false hope that DNA will relieve us of the difficulties that our gift of genetic diversity presents to us. In civil law, a claim of diminished capacity does not rest on any knowledge of the genetic makeup of a person, nor should it here. If I had been asked not about the reality of free will in the light of genetic science, but about the utility of genetic science in ameliorating a halakhic response to non-halakhic behaviors, what I would then say is that I would hope that halakhic responses to non-halakhic behaviors would always be as generous and gentle as possible, and that my hope was free of all scientific content.

In his essay "Catharsis," Rabbi Joseph Soloveitchik describes this paradox in a way familiar to secular scholars of the absurd: the essential element of heroism in Jewish terms is retreat. The paradigm of the hero who retreats is the patriarch Jacob, who wrestles an angel to the ground and then, instead of consummating his victory, lets him go. From such a withholding of final victory Jacob's descendants—the Jews of today—draw their continued existence:

> The Torah wants man, who is bold and adventurous in his quest for opportunities, to act heroically, and at the final moment, when it appears to him that victory is within reach, to stop short, turn around, and retreat. At the most exalted moment of triumph and fulfillment man must forego the ecstasy of victory and take defeat at his own hands. . . . By freeing the defeated enemy Jacob defeated himself. He withdrew from a position he had won through courage and fortitude. He engaged in the movement of recoil.

To a person guided by Torah as the revelation of an unknowable but caring God, successful medical intervention need not confer any moral grandeur, nor need medical failure imply moral decay. The use of DNA analysis to provide a prognosis of certainty that another person will suffer the ultimate loss of free will that is a mortal illness—without also providing the means to ameliorate this certainty—is a form of medicine so unwittingly cruel as to be its own worst enemy. From this Jewish tradition, medicine can perhaps learn to recoil at the moment of victory, to pull back from the opportunity to take on the inappropriate role of judge of another person's fate.

It will be difficult to change the habits of the day, because for anyone—Jewish or not—who doubts there is an unknowable deity concerned for the moral content of their individual actions, medicine does have a way of filling the gap, sapping a patient's freedom to choose how to live his or her life. The enormous capacity of science to create tools for the manipulation of the natural world has helped confer moral authority on medicine by default. The lesson to be drawn from the Jewish tradition is that doctors and scientists have a moral obligation not to fill this gap with their own certainty, if for no other reason than to avoid acting as if they had somehow lost their own God-given free will in turn.

How might this single lesson from Judaism play out in more general, operational terms in today's medicine? In the most general terms, it would redefine medical practice in the following ways. The profession—from basic researcher to primary-care physician—would accept the full autonomy of the patient at all times; the profession would see this autonomy of the patient as a critical aspect of the patient's identity as a unique and complete human being regardless of physical or mental condition; and these obligations would not be less compelling in the last moments before a patient's inevitable death. As my friend Ed Reichman reminds me to say, this might not bring secular medicine more close to the expectations of halakhah, but it would make medical practice more humane.

For example, while criteria for screening for BRCA1 mutations are in flux, we already can see the outlines of a religiously informed, sensitive policy, one informed by the experiences of the Ashkenazic community, but not limited to that community. Here, for instance, are the guidelines used by Dr. Freya Schnabel, a Columbia University colleague whose practice includes many Ashkenazic families. First, the screen is not to be made available to everyone: in the general population, the grounds for a BRCA1 test are either three cases of breast cancer in the family, two in women under sixty. In Ashkenazic families, the criteria are slightly less restrictive, but still stringent: at least two cases of breast cancer, at least one in a woman younger than sixty.

Second—not a criterion but a boundary—prenatal screening for this adult-onset disease is not to be performed even in affected families. As there is no way to predict the age of onset nor whether the child will indeed develop a tumor at all, there are no grounds to put forward the choice of abortion, as there might be with inherited diseases that occur with complete certainty at birth or at a very early age.

The third, fourth, and fifth criteria speak to the essential inseparability of mind and body and the central importance of psychosomatic and emotional events. Third, anyone entering the process of DNA diagnosis for BRCA1 status must first be counseled and a judgment made of their ability to understand and to accept either a positive or a negative result. Fourth, only people who are able to accept the lack of clarity of either result, and are willing to make decisions for themselves about the consequences nevertheless, should be assayed. And fifth, counseling, both psychological and genetic, should continue for at least as long after the result is in, as before the decision to be tested.

We may not all be members of high-risk families, but we are all at risk. It remains uncertain whether guidelines like these—which go far toward helping people to choose how to approach the future, while not unnecessarily shadowing anyone's free will with useless genomic determinism—will be applied widely, or whether genomic data stripped of any medical meaning will continue to be imposed on us all.

Conclusion

All social constructs—good and bad—are expressions of our biological capacity to transmit brain states vertically by a second, non-DNA channel of teaching and learning. They are made possible by our genomes, but they do not depend on our genetic differences. This allows us to make the assumptions—precious to us as Jews—that one's free will is undiminished when one chooses a forbidden act. Absent free will, there is nothing to forgive, nothing to teach, no chance for reinterpretation at a later date and, most important of all, no chance for *teshuvah* on the part of anyone.

We know as well from millennia of religious insight that there is no possible way to justify any ranking of one person over another on grounds of any aspect of their physical being. From those two insights we have the chance of working toward a properly informed medicine, capable of using any and all insights from science in a context derived from the insights of many religions, and thereby capable of reducing all data to one purpose: to help people in need, one person at a time.

Note

1. The racist thinks of everyone in the Other category as if they were genetically identical clones: "all you people look alike to me." The irony of thinking of the Other this way is more perfect in the American case than any other. Here, the Other is likely to be a descendant of Africans, who are today genetically the most genetically diverse of all people. It is a group of like-minded racists who—thinking alike despite all facts—form a clone; not a genetic clone, but a social one.

3

Is Matter All That Matters? Judaism, Free Will, and the Genetic and Neuroscientific Revolutions

David Shatz

> There are three partners in a person: the Holy One, Blessed Be He, His mother and his father. . . . His father [provides certain biological materials]. His mother [provides other biological materials]. The Holy One Blessed Be He places in him a spirit and a soul.
>
> —*Niddah* 31a

> We should not immediately refute any idea which comes to contradict anything in the Torah, but rather we should build the palace of Torah above it.
>
> —Rabbi Abraham Isaac Kook, *Iggerot* #134

JUDAISM, ALONG WITH CHRISTIANITY, HAS LIVED through its share of scientific revolutions, and it has proven to be in key respects resilient and adaptive. Copernicus's claim that the sun, not the earth, is at the center of the planetary system initially seemed to dethrone humanity and to wreak havoc with biblical texts such as "shemesh be-Giv'on dom," Joshua's command that the sun stand still (Josh. 10:12–13). But we adjusted, figuring out ways to cope theologically with a demoted physical locus and exegetically with seemingly embarrassing texts.[1] The adjustment took a long time, but not many Jews in today's centrist orthodox community (and maybe not even very many on the right) would feel impelled to worry about the belief that the sun is at the center of the universe, let alone to construct a model of the universe with the earth at its center on the grounds that "it's a *be-ferush Rambam* (explicit statement by Maimonides)."[2]

Likewise, Darwin's theory of evolution posed all sorts of problems for Orthodox Jews, from its claim that the earth is billions of years old to the thesis that human beings evolve out of lower animals to the implication that nature was "red in tooth and claw" and that a good God allowed horrid savagery and carnage in the animal kingdom over so long a span of time. Using methods like figurative interpretation of *Sefer Bereishit* (book of Genesis) and an appeal to the hiddenness of *ma'aseh Bereishit* (the work of creation), and aided by midrashic and Zoharic texts that suggested Darwinian themes, the centrist community (though not necessarily those on the right) came to a rapprochement with evolution. Certain ideas were non-negotiable, principally the idea that intelligent design guided the whole process. But affirming that proposition left enough leeway to allow in much of the core story told by scientists. Rav Abraham Isaac Kook went so far as to say that evolution gives more eloquent expression to Kabbalistic notions—of gradual progress and the interconnectedness of being—than any other theory.[3] A nemesis became—for some if not many—an ally.

Again in addressing evolution, Rav Kook writes that "we should not immediately refute any idea which comes to contradict anything in the Torah, but rather we should build the palace of Torah above it."[4] I take that to entail that if we believe Torah is true, then other things we accept as true can only deepen our understanding of it. Looking back, we may affirm that if no Darwinist had ever lived, our religious and intellectual life would now be easier; but it would not be richer, nor closer to the truth. Likewise, if the world had never discovered the complex causes of disease and natural disasters, theology would be much simpler—but we would have a skewed picture of how *Ha-Kadosh Barukh Hu*, the Holy One, Blessed Be He, operates in the world, not to mention a severely reduced capacity for handling adversities. By teaching truth, *madda* (secular knowledge) has enabled our understanding of Torah's teachings to grow.

In this essay, I will be examining the religious implications of a view regnant today known as materialism or physicalism—the thesis that all things are made exclusively of matter and that all facts can be explained by reference to material or physical causes. For many people, both religious and secular, this thesis yields a frightening view of the human being, just as Copernicus's and Darwin's theories were frightening when first propounded. In fact, materialist theories of the person represent a further development of the Darwinian paradigm and interlock with evolutionary theory. At the moment, rapprochement between religion and materialist views of humanity seems impossible, just as rapprochement between religion and those other theories seemed impossible at first. The fact that previous "upstarts" came to be accepted in the long run does not mean that this one should or will; in any case, we have to judge by our current religious standards—"a judge has only what his eyes see."

Not being a scientist, I will studiously avoid trying to defuse the threat by the presumptuous tactic of trying to debunk evidence for current theories in genetics and neuroscience.[5] I am simply not qualified, not a "bar hakhi," to write that chapter. Instead, my aim is to ask a simple question: suppose "the scientists" are right. This is not to endorse the scientific view as the whole truth—we are religiously bound not to, I maintain—but rather to use a worst-case scenario as a heuristic device for sorting out what we can accept and what we must reject. What would follow from the truth of materialism? Can we make our peace with some of these implications? Are there positive spiritual messages we can extract? Even supposing there are, which religious ideas are threatened? How bad is the damage, and can it be contained? Are there precedents in Jewish tradition, or ideas in the general philosophical world, that might help us here? If we go along with materialists for part of the ride, at what stop must we get off the train?

In accordance with this volume's emphasis, my main topic will be how materialism impacts on free will. But prior to turning to that subject, I offer some perhaps loosely related observations about materialism in general. What binds the two parts together—materialism and free will—is the overall aim of trying to take some of the fright out of the present scientific paradigm, while acknowledging that some parts remain deeply upsetting to the traditional outlook and seem irreconciliable with traditional Judaism. By demonstrating throughout that our *baalei mahashavah*, our great thinkers, generally made room, in various and conflicting ways, for biological causes of behavior and for at least partial determinism, by showing that their conceptions of free will at times saw determinism and free will as compatible, and by showing that some questioned the very value of free will, I hope to give us a measure of comfort with the new theories of personality. At the same time it would be foolish to pretend that the news is all good; the comfort is decidedly limited.

The Physicalist Paradigm

I begin by describing the revolution.

Long expected and long nurtured, the revolution we are living through is about how we conceive of mind and personality. Talk of souls, which came so easily to the lips of Plato, Descartes, and others who believed in human beings' dual nature as body and soul, no longer has the scientific respectability it once did, and its philosophical respectability is in precipitous decline.[6] Neuroscientists seem to be successfully mapping everything—from higher level cognitive functions, to emotions and personality traits, to raw experiences of color, taste, and sound[7]—on to states of the brain. Moral judgment, religious experience, and religious belief are linked to neural excitations.[8] Simultaneously, geneticists explain how neural pathways are set and identify genetic causes of character traits, personality disorders, and the like. As gaps in our understanding of the mind get filled in, we become like those who gradually learned the mechanisms behind diseases and earthquakes. Just as the "God of the gaps" invoked in centuries past to fill gaps in scientific explanations could live only as long as such gaps existed for Him to fill, so, too, the soul-of-the-gaps could live only so long as matter was delayed from entering as the underlying explanation, if not essence, of mental phenomena and as the solution to the hoary problem of mind and body.

The assertion that *ve-anokhi afar va-efer*, "I am dust and ashes" (as Abraham puts it in Gen. 18:27), which for Jews always was but one pole of a dialectic between majesty and humility (as Rabbi Joseph B. Soloveitchik, "The Rav" *zt'l* would put it), now boasts—in the literal sense of "dust and ashes"—an almost frightening monopoly. No matter how often presentations of materialism are leavened with caveats about its being a merely a research program or working hypothesis, no matter how often we are reminded that science changes, no matter how recalcitrant the pockets of dissent and resistance are even within the scientific community right now,[9] the sense that something true and important has been discovered about human beings is as intimidating as it is credible and widespread. We are talking about a central image, a model, a sovereign approach—in Thomas Kuhn's famous phrase, a paradigm, or something close to it.[10]

A sufficiently evolved and intricate brain, endowed with an exceedingly complex set of neural pathways, can, we are told, perform and experience anything and everything that once promised to mark the human soul as distinctive. Francis Crick, co-discoverer of the double-helix formation of DNA, expressed the challenge by means of what he called "The Astonishing Hypothesis": to wit, "that 'you,' your joys and sorrows, your memories and your ambitions, your sense of identity and free will, are in fact no more than the behavior of a vast assembly of nerve cells and their associated molecules."[11] Likewise, philosopher Daniel Dennett asserts:

> there is only one sort of stuff, namely matter—the physical stuff of physics, chemistry, and physiology—and the mind is somehow nothing but a physical phenomenon. In short, the mind is the brain . . . we can (in principle!) account for every mental phenomenon using the same physical principles, laws and raw materials that suffice to explain radioactivity, continental drift, photosynthesis, reproduction, nutrition, and growth.[12]

If one accepts Crick's and Dennett's hypothesis, the differences between human and beast lie only in the complexity of their physical hardware, not in the existence of a human soul.

No sooner do we utter the "astonishing hypothesis" than we see the multiple challenges it poses to religious people. What becomes of spirituality? Whither selfhood—is the "I," so long sought out, so long elusive, so long thought to underlie choice, a coded DNA sequence? Must we jettison belief in free will and the language of moral accountability? Must we relinquish the notion that diseases are punishments for poor spiritual conduct, since in fact they are genetically inherited and inevitable? Is it the case that biology is, to borrow from the *Rav's Kol Dodi Dofek*, both fate (*goral*) and destiny (*yiud*)—both what happens to us and what values and actions we embrace? What about the soul and the afterlife: *ha-neshamah mah tehe aleha, olam ha-bamah tehe alav*—what shall become of the soul, what shall become of the world to come? These questions of course are not totally new. The thesis of materialism or physicalism has a long pedigree, extending back to pre-Socratic philosophy. But materialism has now matured from an assertion that was heretofore essentially speculative and that cropped up from time to time in the course of history, to a well-developed, well-grounded scientific theory that musters impressive and ever-growing empirical support.

What spurs the view of the human being as material are not only discoveries on the level of scientific theory, but dramatic applications of science in daily life. The notion that a person is purely biological is experienced by millions of people existentially and viscerally. To begin with, in medical ethics, you are your brain—alive if it lives, dead if it's dead. Furthermore, in his best-selling book, *Listening to Prozac*, psychiatrist Peter D. Kramer thoughtfully and eloquently shows how society's extensive use of mood-altering drugs has altered people's conceptions of their selves.[13]

Traditionally, both psychotherapy and *musar* viewed character change as the result of *heshbon ha-nefesh*: a deep look into the causes of one's behavior (resulting in self-knowledge),[14] matched with a desire to change those traits, and long, hard labor. But today, Kramer contends, such causes of "social awkwardness" as low self-worth, competitiveness, jealousy, poor interpersonal skills, shyness, perfectionism, fear of intimacy, lack of industry, and the like—all of which, let me add, can affect our *Avodat ha-Shem*, service of God, including how many *blatt* (pages of Talmud) we learn and how well we *daven*—can be relieved or removed with a quick fix of medication. "A pill could do in a matter of days what psychiatrists hope, and often fail, to accomplish by other means over a course of years." If medicine can change how we think, feel, and value, it would seem that the traits themselves were not essential to the self, but rather, in Kramer's words, "biological tics." If a patient's self-destructiveness and fragility disappear in response to a biological treatment, "they must have been biologically encoded."[15] The patient's biological constitution seems to have determined his or her social failures. Here now is the issue:

> When one pill at breakfast makes you a new person, or makes your patient, or relative, or neighbor a new person, it is difficult to resist the suggestion, the visceral certainty, that who people are is largely biologically determined. I don't mean that it is impossible to escape simplistic biological materialism, but the drama, the rapidity, the thoroughness of drug-induced transformation makes simplicity tempting. Drug responses provide hard-to-ignore beliefs . . . that society has heretofore resisted. I came to believe that even if we tried to understand these matters complexly, new medications would redraw our map of those parts of the self that are biologically responsive, so that we would arrive, as a culture, at a new consensus about the human condition.

Kramer is not discovering America. The effects of older antidepressants as well as of food, drink, and exercise can likewise stimulate the biological concept of a person. Furthermore, we have long known the effect of "bad drugs" (crack, etc.) on the mind, and this knowledge could have given rise to the biological picture of a person long before Prozac. But Kramer asserts a difference in rapidity and effectiveness between Prozac and earlier antidepressant interventions that suggests something new occurs with recent "good" drugs.[16]

Does Kramer exaggerate? That question must be answered by mental health professionals, not me. I would expect to hear from them that the notion of a quick cure by pills is a cartoon of the psychiatric reality. Kramer himself, for all the materialism of many passages in the book, volunteers that patients benefit most when psychotherapeutic conversation is added to a pharmacological regimen. He also engages the objection that mood brightening drugs interfere with a patient's relationship to reality by disconnecting the individual from unpleasant truths, impeding affect tolerance (i.e., "standing up to what you feel"), reinforcing cultural norms, and encouraging conformity. Pills, say critics, "free the taker from struggling with reality . . . they interfere with adaptive mechanisms . . . they encourage people to understand as illness aspects of the self that are normal."[17] But the point is not so much whether the full dress version of materialism is correct or ethically proper. The point, rather, is that patients have this (perhaps exaggerated) perspective, and that this tells us that it is not the theoreticians alone who are changing our notion of humanity, but practitioners and patients as well.[18]

In sum, we face a clash between two pictures of a human being. Our everyday, "common-sense" picture, the one we use to describe, explain, and even predict behavior, views people as deliberating between alternatives. It sees actions as the result of weighing reasons and alternatives, of hearing insights, gaining insight, etc. The scientific picture sees us as genes and neurons. When we work to improve ourselves, sans medicines at least, when we introspect, when we exert willpower, when we make career decisions, when we reflect on existential questions, when we hold back from venting or decide to indulge ourselves, we experience ourselves as deliberators. But in the scientific picture, deliberating, hoping, regretting, and so forth is just how neural activity feels on the inside. The "we" that effects change is the brain and nervous system. Some writers, notably philosopher Paul Churchland, believe that ultimately neuroscience will develop to the point where we can discard our everyday talk of beliefs, desires, emotions, motives, and the like, a discourse that is highly unregimented, unscientific, and "inaccurate," and speak instead in purely biological terms, explaining human action exclusively in neurological discourse.[19] I do not recall if Churchland considers what history, literature, and cinema would look like if ascriptions of motives, beliefs, aspirations and so forth were replaced by discourse about neurons.[20] Yet, even if folk psychology will always have a place in our language, our everyday lives and our self-perception may be profoundly affected by the scientist's picture.[21] The celebration of the body in the art, literature, and philosophy of the wider culture (coming ironically at a time when we are supposed to overlook physical differences among people), both feeds and is fed by the scientific culture around us. (George Bush, Sr.'s declaring the 1990s "the decade of the brain" in recognition of the importance of neurological research contributed still more PR, admittedly in this respect deserved, for materialism.) In terms of the larger problems, it doesn't help at all to protest that environment plays a role in explaining conduct too. For in a materialist picture, even environmental influences just represent a specific kind of physical determinant. A rebbi teaching a *talmid* (pupil) is for a materialist just another physical interaction, implicating, *inter alia*, the rebbi's brain, his vocal mechanisms, sound waves, the pupil's auditory mechanisms, and the pupil's brain. Furthermore, it might be said, what biology in a narrow sense does not do, environment does; and environmental influences themselves fit a materialist picture. In any case, while my subject is physical causes, the notion of external environmental causes, physical or not, is in the big picture troublesome in its own right.

Unfortunately, while the world has been discussing the implications of neuroscience and genetics for our understanding of self and personality, and has been radically altering its conventional concepts in response, Orthodox Judaism has remained, for the most part, silent. Even factoring in the general paucity of philosophy in the community (a lacuna often bemoaned by the Rav[22]), it is striking how little attention Orthodoxy has given to potentially shattering developments in the field of the mind, as opposed to the relatively lavish attention it has showered on evolution. As an educated guess, I would venture that many observant neuroscientists shove to the side thoughts about the potential implications of their research, and that many observant psychiatrists do not assess in religious terms the assumptions that lie behind prescribing drugs to control states of mind. Finally, I don't know how many Orthodox patients ponder, as Kramer's did, the implications of their treatment regimens.[23]

Well, someone might ask, is it so bad not to give thought to neuroscience and genetics? Granted that the revolution in the understanding of humans is all around us in a world of psychotropic drugs, is it important to see philosophically how the current state of scientific thinking affects our understanding of ourselves? I claim that it is. Understanding the human being is a task deemed important by *Tanakh*, *Hazal*, and all our *hakhmei ha-masorah*; how much more so is it relevant to the Torah u-Madda community. To borrow a thought from Rav Aharon Lichtenstein, applied by him to study of the humanities (he might not agree with my application of it), it would be an odd view of *Torah ve-hokhmah* ("Torah and general wisdom") that would vigorously endorse the religious value of studying atoms and electrons on the grounds that these are part of the world *Hakadosh Barukh Hu* created, but then turns a blind eye to the pinnacle of that creation, the human being. If our discussions of the human being continue to draw exclusively on medieval ideas of passive and acquired intellect, or throw around talk of *neshamah* and *nefesh*—the soul—with no real reflection, we will have failed dismally to confront the fruits of science today and for that reason risk working with an inadequate understanding of what it is to be a human who possesses the *tzelem Elokim*, the image of God.

Faith is a powerful—and needless to say, legitimate—source of our beliefs. Moreover, even if we stay within the arena of reason, the materialist assumption that the only entities that can enter into explanations of things are material ones—that physical systems are closed—is in my opinion unproven. The assumption perhaps sets limits on *scientific* explanations, but not on explanations *tout court*. There are methodologies other than science that may yield truth and correct explanations. To think that science is the only path to truth is what is called scientism, which is an epistemological, philosophical thesis rather than a scientific one. That said, it must be conceded that science yields important truths, and that we ignore them at the cost of an inadequate theology and anthropology.

Religious Messages in Materialism

I spoke earlier of some religious claims being non-negotiable. It goes without saying that a truly cosmic materialism, a metaphysical viewpoint that denies the existence of any entities that are not material and of any causes that are not material, is flatly unacceptable to Judaism or for that matter any of the monotheistic religions. God, after all, is a disembodied being, and, furthermore, though disembodied, He acts on the physical world. As I said a few sentences ago, a religious person could accept cosmic materialism as a methodological constraint on scientific inquiry ("in science, qua scientist, do not form hypotheses about the non-physical"), but, on pain of embracing Mordecai Kaplan's naturalism, he or she must resist it as a true description of the cosmos.[24] To the extent that materialism about the mind draws any support from larger materialist assumptions about what sorts of things can exist or can cause events, to that extent it immediately loses credibility in the eyes of theists.[25] One can separate cosmic materialism from materialism about human beings, and materialism about human beings has impressive support independent of any cosmic materialist reasoning, but in the end a religious person may view the paradigm as too confining, barring as it would knowledge of ethical and aesthetic truths (assuming those exist independently of the mind) and perhaps mathematical truths.[26]

But to what extent can Judaism accept the biological picture? Notice first that, as the Rav never tired of stressing, Judaism accepts the physical aspect of the human being. The first chapter of Genesis describes man, in the Rav's words, as "an integral part of nature . . . a child of mother Nature, as is the brute and the beast." In this regard the Bible accepts "the unity and continuity of organic life, considered an indispensable postulate in all biological sciences. . . . Man and beast share equally in the same biology, physiology and pathology."[27] Needless to say, this acceptance of physicality, of "man-natura," must not eradicate what the Rav calls "man-persona," the spiritual being in Genesis 2 who stands apart from and above the rest of nature and upon whose behavior the redemption of physicality depends. In contemporary science, this man-persona turns out to be man-natura in disguise. That having been said, we cannot and should not hope for a theory that does not see salient strands of continuity between us and the physical world. The question is how much of the science of the brain and the genes we can accept, not whether we can accept any. To reject all physicalizing of humanity is contrary to tradition.

Let me be more specific. Kramer's patients, as we noted, were concerned about the prospect that previous "bad behaviors" must have been due to biology, along with current "good behaviors." Yet for many cases the biological view of one's "bad past" is consonant with Jewish sources. For we frequently encounter the idea that wrong and sinful acts are traced to the body. Rambam writes, for instance:

> All man's acts of disobedience and sins are consequent upon his matter and not upon his form [viz. the intellect], whereas all his virtues are consequent upon his form. . . . The commandments and prohibitions of the Law [Torah] are only intended to quell all the impulses of matter. It behooves him who prefers to be a human being in truth, not a beast having the shape and configuration of a human being, to endeavor to diminish all the impulses of matter.[28]

Likewise, Ralbag (Gersonides), who believes that character traits are astrologically determined but that a person's reason can override astrological determination, concedes by the very statement of this thesis that an act not based on reason is determined by the person's physical makeup.[29] The Rav, as we saw, often sounds as if Torah discipline is the harnessing of matter and physicality, of "man-natura," by "man-persona." Often in Jewish literature, the *yetzer ha-ra* (evil inclination) is represented as a biological urge.[30] It is true that the spirit (or, for Rambam, form) is said to be able to control matter, suggesting that spirit's delinquency is complicit in matter's sin. But the causal antecedents of wrong are still biological. The "spirit's" failure to control the body is like a person's failure to douse a fire with water—the failure isn't a cause of the fire, though the person is responsible for not dousing it. These outlooks, then, view evil as emanating from the biological part of the person. I realize that this idea rings simplistically when asserted with thorough literalness, but if it is religiously acceptable to put it forward as a figure of speech, how objectionable could it be to take it literally?

This is a crucial point in evaluating the biological revolution. In essence, half the materialist thesis, the idea that "the bad" is biological, is consistent with a prominent theme in *mahashavah*. Accepting the other half, however, that "the good" is material—this is new, and frames our problem. Yet I would push further here and say that even when it comes to the good, the Rav follows a model in *Hazal* (our Sages) that specifies creative, healthy outlets for destructive drives of the *yetzer ha-ra* (evil inclination)[31] rather than eradicating them. Biology is redirected, not exterminated.[32] I certainly would grant that, from the Rav's perspective, spirit does the directing, while Kramer's patients' model is one of matter conquering matter. But points of commonality should not be ignored.

In addition, it is worth considering one way to make materialism yield a salutary religious message. The thought goes something like this. Three of the major revolutions in science in the past six centuries—the Copernican, the Darwinian, and now the neuroscientific—have, in one way or another, seemed to dethrone human beings from a special place in the universe conferred upon them by God. Make that four revolutions: Freud made the irrational, instinctual side of human beings a central motif in the explanation of human behavior. Likewise in philosophy, David Hume's stress on the instinctual, non-rational or even irrational side of human cognitive achievements and Immanuel Kant's highlighting of the limitations of human knowledge point to a humbling of human beings. Today's enthronement of biology and the denial of free will by many materialists seems to further reduce human power and standing. What is significant religiously is the irony that the Enlightenment's confidence in the autonomous powers of the human mind, which undergirds all scientific inquiry and has led to the abandonment of religion, has brought us to a day in which humans, the builders of those sweeping theories, the same human beings who dared to replace God, seem reduced to glorified beasts.

Rav Kook, who had a penchant for viewing history as a dialectical drama and assessing scientific theories by their spiritual impact rather than the evidence they can marshal,[33] might well have seen the development of contemporary neuroscience as the point at which the world encounters a needed corrective, a dose of humility, a counter to Enlightenment hubris, after which the pendulum will swing back and the physicalist paradigm will be trashed. In this narrative of changing trends in science, humanity can extract from materialism a healthy message for our time and place. Specifically, the moral import of materialism is *yir'ah*, awe, in the Rambam's sense: humanity's sense of its insignificance (*Hilkhot Yesodei ha-Torah* 2:2) and humanity's shame (*Guide* 3:52).[34] By the same token, attempts to place humans above nature, and in particular to preserve for humans a freedom of the will that escapes physical causation, would in the present context reflect an attempt to assert human superiority and distinctiveness. It is therefore inappropriate, it may be argued, that religious people are the ones holding out for free will—for the attempt to do this could even be viewed as a religiously objectionable attempt to shift power away from an almighty deity into our own hands, and to confer upon ourselves supra-natural control in the form of free will. "[T]hat many believe that we stand above nature in some essential respect suggests that the Enlightenment was more successful in its glorification of the individual than in its challenge to the supernatural."[35] Whereas Rambam treated free will as a type of natural cause, nowadays, with nature identified as the material, free will seems supernatural. It should be stressed that many explications of free will view the essence of free will to be control over the world—the ability to make the world conform to what we want.[36] That religious people argue for free will could be construed as arrogance vis-à-vis God; it is therefore healthy to have a world-view that encourages humility, shame, and insignificance. As we will see later, this idea accords with certain Hasidic teachings.

This is not to say, obviously, that free will, dignity, and *ahavat Hashem* (love of God) have no place! To begin with, *ahavah*, in the classic Maimonidean sense of seeing God's wisdom in nature,[37] might be aroused by the intricacies of the human brain no less than by other marvels in nature (that the brain is said to have evolved does not diminish this reaction). More importantly, there is, for Rav Kook, a dialectic between what the Rav later called humility and majesty.[38] As opposed to the Rav's dominant stress on dialectic within an individual's experience, Rav Kook emphasized history, and not only individual experience, as the arena of dialectical combat. If he lived today, Rav Kook might have argued that although there are times when the right message is "majesty," ours isn't one of them. The recent surge of quests for spirituality in the thick of a materialist scientific paradigm and, indeed, on the psychiatrist's couch might have suggested to Rav Kook, were he alive today, that the road back to *ahavah* and uplifted spirit is to experience the confinement of the *yir'ah*-permeated materialist picture fully, to choke on it before expelling it, so as to appreciate the more liberated life. And he would have paved that road for us, perhaps.

It may be countered that materialism is not humbling; it actually expands human power. If spiritual beings exist, humans would not be able to control them, because scientific mastery would not extend that far. But if everything is matter, then humans can control everything.[39] If scientific method is the only way of knowing about the world, then science projects "the apparent sense of intellectual dominion."[40] Determinism also can lead to fatalism, something that concerned Rambam greatly in his letter on astrology. In addition, we need to look at whether materialism has also led to a devaluation of human life and is connected to noxious trends in end-of-life issues. And, in general, while the irony of brilliant scientists declaring themselves to be clumps of matter may amuse some, none of that provides cause for rejecting the scientist's picture and celebrating human beings in the here and now.[41] *Ein lo le-dayyan ela mah she-einav ro'ot*: we have only what is in front of us, and what we see is distressing to our self-conception and a challenge to the traditional view of the soul.

I appreciate this reply to the attempt to extract a positive message of humility in materialism; but the humility theme remains an aspect we might appropriate even while we shrink from buying the materialist package as a totality. If we take seriously the link between culture and the divine presence, we can extract moral values even in a bad situation. That being said, we must still recognize that the situation is bad, as a materialist perspective seems to deny the existence of souls and free will entirely.

Free Will Challenged

Of all the capacities of humans that are affirmed in ordinary discourse but are jeopardized by neuroscience and genetics, the most significant and difficult is free choice. With regard to other phenomena that could be labeled, in a less than literal sense, "achievements of the spirit," physicalism does not gainsay the fact of "spirituality"; it only questions a particular explanation of that fact. For example, if we identify spirituality with certain capacities—for abstract reasoning, moral intuition or spiritual insight—it is blatantly question-begging to assert that these capacities cannot be realized within a physicalist model. Looking over the Rav's list of man-persona's qualities—the longing for vastness and boundlessness, the sense of loneliness, the ability to discipline the body, the ability to think in moral terms[42]—these, too, all lie within the scope of the materialist's account. Taking it as a datum that materialism precludes achievements of the spirit is therefore facile. Needless to say, if you take sticks and stones as your model of a physical system, you will have a hard time imagining that a physical system could think, feel, or experience. But this is a patently silly way to conceive of physicalism. Physical systems vary in complexity, and according to physicalists, a sufficiently complex physical system can do an awful lot. There is no a priori way of knowing that a sufficiently evolved and complex brain, possessed of sufficiently complex neural pathways, could not reason abstractly, form moral judgment, or feel, any more than there is an a priori reason that a collection of pieces of metal and wire assembled in the form of a computer could never be smart enough to defeat a chess grand master. This is not to say that, at the end of the day, the materialist's explanation of human thought and deed will be successful. But at least materialism, in the cases I mentioned, does not deny the very occurrence of phenomena we deem crucial to human spiritual achievement.

Not so in the case of free will. Here the materialist challenge is more threatening. Here, the materialist denies the very phenomenon. This is because free will generally is thought to involve (whether it does or not will be discussed later) a non-material cause—"agent" or "will" or "self"—that controls the physical body. Agents and wills do not fit into a materialist metaphysics. As a result, scientists who have sought to locate a space for agents, wills, and selves, have met with the charge that, while these entities cannot be ruled out, one who introduces them

> wants to find a non-neural, non-physical basis for free will (some sort of mental conscious control over the brain itself) and he wants to find it doing research predicated on the assumption of neural cause and effect. Such a research agenda, wedded to the a priori goal of defeating mechanism yet rooted in physicalist science, is surely doomed from the outset.[43]

This criticism is ambiguous. Is the criticism that as a neuroscientist, bound by certain procedures of the discipline, one must not posit non-physical entities? Or that even in metaphysics as opposed to neuroscience, we can't do such positing? Would the critic concede that the best scientific explanation of a phenomenon need not be the best overall explanation? Suppose it became absolutely clear that God created the universe and separate species (imagine cosmologists retrieved distant sounds, "yehi or" "totze ha-aretz nefesh hayyah leminah," etc.; and tiny particles revealed the words, "I am the Lord your

God").[44] Would this truth have no place in science, which deals only with material entities? What of Newton's invocation of God in explanations? However that problem is resolved, the critic of free will quoted above certainly brings home that unlike other human achievements, free will is not something a materialist accepts but explains differently from a non-materialist. Rather, free will, at least in the sense of a will that is non-material, is something the materialist, at least qua scientist, denies altogether. The defender of free will either has to refute materialism or else must articulate a conception of free will that a materialist can accept.

In truth, the challenge is twofold. One part pertains to *sakhar va-onesh*, reward and punishment—the fundamental belief, or *ikkar ha-Yahadut*, that what happens to a person is a function of his or her spiritual deeds. "If you walk in my ways" good things will happen to you; "and if you do not walk in my ways" you will be beset by enemies and illnesses, so teaches Leviticus 26, the *tokhahah*. But if your genes determine your liability to cancer or hard arteries, where do your deeds enter into this picture? Contrary to traditional thinking, how you exercise your free will would not determine your fate, unless you interpret "proper exercise of free will" to mean proper physical exercise to ward off health problems lurking in your genes or brain cells.

But the second part of the challenge is more pressing: whether there is space for free will, responsibility, and "reactive attitudes" in the contemporary scientific framework. Notice that the challenge of contemporary science can be formulated in two different ways.

A. The strong form of the challenge: We might wonder whether there is any free will at all if everything is biologically determined. We might also wonder whether there is any responsibility at all for human acts and omissions if everything is biologically determined. Finally, we might wonder whether the gamut of "reactive attitudes" like blame, resentment, condemnation, guilt, gratitude, forgiveness, approbation, indignation, in short, the whole "complicated web of attitudes and feelings that form an essential part of the moral life as we know it," can survive in a deterministic framework.[45]
B. The moderate form of the challenge: Alternatively, we might deny universal determinism and yet still accept determinism at the level of certain specific behaviors. Certain specific types of behavior or tendencies that traditionally are thought of as behaviors that the agent is thought to do/omit freely and for which the agent is responsible, and to which "reactive attitudes" are directed—these behaviors are actually biologically determined and, it is inferred, not free. Homosexuality might be a case in point. Another example is a tendency to violence, as in the form

> of spousal abuse.[46] To worry about behaviors like these is not to fear some universal determinism; it is rather to view those particular behaviors as determined.

We are really talking here about both problems—universal determinism of human actions, and determinism of specific actions. I want to deal with A, the threat posed by the prospect of universal determinism. I hope to discuss B on another occasion.

Let us, as a heuristic device, assume determinism of human behavior.[47] It is understandable to seek precedents for this deterministic challenge, instances in which Jewish philosophers confronted a deterministic theory and responded either by accepting it or finding a way to salvage free will in the face of the challenge. My argument in what follows is that four precedents, namely Hasidic thought, medieval astrological determinism, "devaluation" (associated with Ramban and Rav Dessler), and "restrictivism" (Rav Dessler's restricting the scope of free will) are partial precedents that help reduce the scientific challenge. None is a full precedent, but collectively they may make the glass look half full rather than half empty.

Four Potential Precedents

Jewish literature contains an immense amount of material on conflicts between free will and types of determinism other than the scientific, such as forms of theological determinism. First is the seeming incompatibility between *ha-kol tzafuy* and *ha-reshut netunah* (*Avot* 3:15)—if God knows what I will do before I do it, then I am not free to do it, because there is no possibility of my confuting God's knowledge, and hence no possibility of doing otherwise. Second, there is the seeming conflict between God's having a plan for history and human beings making choices that shape history. For example, Joseph's brothers, and the Egyptians during the time of *shibud mitzrayim* (the captivity in Egypt), made certain choices that one might worry were necessitated by a divine plan and hence were not free.

While these problems present formidable brain teasers, they have little existential impact. The Orthodox community does not evince real angst about the divine foreknowledge-free will problem. It makes for fun conversation even among non-philosophers, but, as a rule at least, people do not bring to these discussions any emotional investment. In contrast, the impact of neurophysiological and genetic determinants of personality is potent—these developments genuinely worry people. I have met many people who, while themselves denying or remaining agnostic about scientific determinism, worry that if all events are determined by physical or environmental causes, no one is free, and therefore become alarmed by the present situation in the sciences. In fact, the theory of scientific determinism was often presented by philosophers merely as a "what if" possibility—the very possibility of it was thought to be scary.[48] In sharp contrast, no one thinks of the very possibility that God exists as scary because of the problems it poses for free will. Agnostics often endorse the view that divine foreknowledge and human free will are incompatible, but I have yet to find one of them worrying "*oy vay*, what if God exists—then He knows everything I'll do and my choices aren't free!"[49]

The discrepancy between calm, composed attitudes to theological determinism and angst-ridden attitudes to scientific determinism bears a further look. One piece of an explanation may be that neither Tanakh nor *Hazal* evince much interest in reconciling free will with divine activity or knowledge.[50] On the contrary, biblical narratives are often written in a way that exemplifies what Yehezkel Kaufman calls dual causality—God and humans bringing about the same events—with no hint of a contradiction.[51] Again, divine providence and human responsibility coexist in the Joseph story, without a sense that there is a logical problem here.[52] God deprives Pharaoh of his free will, but only the exegetes and philosophers, not Pharaoh himself, get exercised over this occurrence. Basically, the problem of foreknowledge–free will or even preordination–free will is not a biblical one.

Hazal were similarly untroubled by the contradiction. In fact, the mishnaic expression I quoted earlier (*ha-kol tzafuy ve-ha-reshut netunah*) may not belong in this discussion. R. Ovadyah mi-Bartenura—as well as the late academic scholar Ephraim Urbach—maintains that *tzafuy* means "known" not "foreknown," so that R. Akiva is not giving voice to a contradiction but to an incontestable truism: since God knows what you are doing (as opposed to what you will do) you should exercise your free will responsibly.[53] This point is of some significance, because by neutralizing the relevance of a famous text it underscores the paucity of early rabbinic literature defining the problem. I concur with Charles Manekin that *Hazal* "simply were not exercised by philosophical problems of this sort."[54]

The biblical and rabbinic silence may explain the indifference to theological determinism among some Jews. But it does not explain the present concern over scientific determinism, which, after all, is not rooted in Tanakh or *Hazal*. Only the medieval philosophers, beginning with Saadyah, raise the problem of necessitated choices, and in Jewish philosophy's early stages, they did do so only with regard to divine determinism of behavior, not with regard to physical causation of behavior by natural causes.[55] So the discrepancy in attitude toward divine knowledge and scientific determinism calls for a second, different explanation beyond the one that appeals to biblical and rabbinic silence.

Here is my proposal: in the worst case scenario, if we are deprived of free will because we affirm that God has a plan for history, or that God knows what I will do, we are sacrificing free will to affirm another religious principle, namely, divine providence. To give up one religious principle in the face of pressure from another religious principle—and in particular, divine providence—is understandable from a religious perspective and reflects pious motives. Surrendering to God what is rightfully His is admirable. By contrast, to embrace scientific determinism and to surrender free will because of that embrace is to surrender a religious doctrine because of science. That step is one that Orthodox Jews are less prepared to take.

The Precedent of Hasidic Thought

Hasidic thought provides a useful context for exploring the notion that free will might be surrendered if it clashes with another religious idea. Hasidic thinkers like R. Shnayer Zalman of Lyadi (*Baal ha-Tanya*), the Izbicer (R. Mordecai Joseph Leiner) and his disciples—his son Jacob and R. Tzadok ha-Kohen of Lublin—denied human freedom because they held to an extreme form of divine providence.[56] Indeed, for Hasidic deniers of free will, the point is not so much that there is no free action or will but that there is no human agency or action at all—the human will is nothing more than God's will.[57]

In these views, asserting free will's existence is a mark of arrogance; denying it is humble and self-effacing. Free will entails control and a sense of ontological separateness. In truth, there is no value in our being the author of our acts and maybe even not of our motives; the value of free will gives way to a competing value, divine providence and an ethically important awareness of His will and/or our own helplessness and indeed nothingness. For this reason, it may be objected that the humble, self-effacing denial of free will found in *Hasidut* is not a full precedent for denying free will in the modern context: there appears to be no positive theological valence in contemporary determinism.

Despite this disanalogy between the Hasidic denial of free will and the scientific denial of free will in the modern context, I would insist that the Hasidic position is not a total washout as a precedent. First of all, although the motive for rejection of free will may be pietistic in *Hasidut*, and the motive for rejecting free will in the face of scientific determinism is not, the fact remains that, at the bottom line, the principle of free will has been given up. This gives some degree of precedent for relenting on free will in the face of scientific challenge. Second, and more crucially, if scientific determinism were accepted, the realm of nature could be taken as an expression of the divine will—nature is the *Yad*

Hashem, the hand of God. Determinism in nature would give expression to the truth that God controls all, and generate an ethic of humility. In other words, by giving a theological reading to scientific determinism, a reading whereby recent discoveries in nature are taken to reflect the sovereignty of the divine will, we might try "reconciling" (or non-reconciling, really) freedom with determinism in a manner that parallels Hasidic attitudes to freedom.[58]

Admittedly, in a kabbalistic framework, direct divine activity is always taking place in the form of emanations; but in the case of the contemporary scientific world, nature has independent existence and runs on its own. One could say, nonetheless, that this is precisely the point—nature isn't really running on its own, since the laws of nature that science discovers are patterns of divine agency. I suspect, however, that most will find this way of speaking forced and not sufficiently assertive of divine immanence to ground a surrender of free will. We should, therefore, look for something better. Even so, the idea that determinism reflects divine sovereignty is not to be taken lightly, especially in light of its potential to prevent human arrogance.

The Precedent of Medieval Astrology

With the large exception of Rambam and minor exceptions like Rabbi Isaac Pulgar, medieval Jews, philosophers and non-philosophers alike, believed in the truth of astrology.[59] Astrology as embraced by medievals was the functional equivalent of contemporary genetics. The stars set your character and natural tendencies (dispositions), just as in modern times genes do.

However, medievals also believed that a person can overcome natural tendencies. Charles Manekin suggested the following analogy in correspondence. It is raining. Yesterday you saw the forecast and resolved to buy an umbrella. But your natural laziness got the better of you. The consequence: you got wet. You get home and your wife says, "You could have brought an umbrella." The last thing you will do is excuse yourself by appeal to your natural laziness. Since wise choice could have altered your drenched fate, you are responsible if, in fact, you put up no resistance and did not try to conquer the *mazzal*. Responsibility thus flows from your ability to battle and conquer naturally determined inclinations. We may call this the "override" theory: freedom and responsibility comes from the possibility of overriding natural tendencies.

The Rav refers to this view in explicating Ravad to *Hilkhot Teshuvah* 5:5. Ravad asserts "And it is well known that all of man's deeds, great or small, were given by God to the power of the Fortunes (*be-koah ha-mazzal*). God also gave man the intelligence (*sekhel*) to elude the hand of Fortune, and this is the power given Man for good or evil." The Rav adds his own point here, that man can set up his reactions in advance and in that way not to have to struggle with powerful drives each time. Here the Rav seems to equate the degree of astrological flow with drives and to say that the flow can be dammed up in advance.[60] Similarly, Charles Manekin tells us about Ralbag: "Nowhere does Gersonides say that one freely chooses to be drawn after one's temperament. Rather he says that one's choice itself is 'drawn' after temperament."[61] In other words, a "choice" made by temperament is determined, and free will is located only in the power of reason to override temperament.

Rambam, who asserts that astrology is nonsense, nevertheless holds to a type of non-astrological "override" theory, according to which the influence of matter is overridden.[62] On the one hand, he states:

> [M]any people who have received from their first natural disposition a complexion of temperament with which perfection is in no way compatible. Such is the case of one whose heart is naturally exceedingly hot; for one cannot refrain from anger, even by subjecting one's soul to very stringent training.... Similarly one can find among people rash and reckless folks whose movements ... indicate a corruption of the complexion and a poor quality of temperament.... Perfection can never be perceived in such people. (Guide I:34)

Yet, in Guide III:8 he states that "it is easy ... to control suitable matter.... [I]f it is unsuitable, it is not impossible for someone trained to quell it." However one resolves the contradiction between the quotations from III:8 and the one from I:34, the key point is that both Rambam and the Jewish believers in astrology view virtuous action, where it does take place, as the overriding of matter/nature through the discipline of Torah.[63]

So, in a certain respect, we might feel about genetics and neuroscience "been there, done that." We had a deterministic theory previously, and the clash with free will was "solved" through the concept of override.[64]

Medieval thinkers assumed that whether or not a person would try to quell matter would be up to him or her, and not predetermined by anything else. To be truly analogous to astrology, the genetic/neuroscientific view would have to say that the choice to override biological dispositions itself is not determined by biology. Self-evidently, in contemporary determinism even the agent's override is determined. Hence, the override theory seems to provide no full precedent. Even so, the fact that many actions are determined is significant. This is a step, if a small one, toward greater comfort with contemporary science.

The Precedent of Devaluation: Nahmanides and Rav Dessler

Another way to cushion the blow of determinism by appeal to precedent is to diminish the value of free will. For Ramban, when Adam and Eve became "*yode'ei tov va-ra*" upon eating of the fruit of the tree (Gen. 3:22), they acquired the capacity to choose either of two alternatives in a situation, the good and the evil (commentary to Gen. 2:9). Prior to sin, they automatically did what was right, just as they automatically breathed and digested. But after the sin, they lost this automatic connection to the right and the good; now doing good became a matter of choice. The loss of innocence, the acquisition of free choice, is a step backward. In messianic times, when the Jewish people repent, this return will be brought about by God's "circumcising your heart and that of your seed" (Deut. 30:6). For Ramban, then, free choice, moral struggle, is not a good; conformity to the good is better, even without choice. Ramban obviously would not deny that people do act with free will—but his approach allows us to adjust, should this denial be the outcome of the conflict with physicalism (at least if our good deeds turn out to be determined).

The Nahmanidean approach does not sit well with certain other theological ideas. In particular, many explanations of evil argue that free will is so great a good that preserving it provides adequate justification for God's allowing evil. Such explanations accord a higher value to free will than Ramban does. But a Jew is not obliged to accept these theodicies. Moreover, there are considerations that support Ramban's devaluation of free choice. For example, the fact that we coerce certain types of behavior, the fact that the Torah seems untroubled by hardening of hearts, and indeed the very fact that God uses promises of rewards and threats of punishments to influence behavior, suggest that conformity to the right counts more than free choice. As we will see later, our educational techniques are strongly deterministic; we have no ethical scruples about educating children through indoctrination and trying to obstruct alternative choices. In addition, we pray "ve-khof et yitzrenu le-hishtabed lakh"—asking God to subjugate our wills to Him. Ramban's devaluation, then, has something to support it.

Rabbi Eliyahu Dessler, in *Mikhtav me-Eliyyahu*, takes a position superficially resembling Ramban's. Rav Dessler believes that to be unfree in the sense that one automatically does the good is greater than being free in the sense that one chooses to do the good. "Tafkido shel ha-adam la-hafokh et atzmo mi-boher le-mukhrah"—"a person's task is to transform himself from one who chooses to one [whose actions are] necessitated."[65] I say that this opinion "superficially resembles Ramban's" because one could argue that in Rav Dessler's thought, it is not being a *mukhrah* per se that is good; the good is turning oneself into a *mukhrah* by using free choice to create that state which is noble. That said, the basic thought that, abstracted from its connection to the good, free will is not something of value might lead us to make our peace with the loss of free will preached by modern science. Certainly for Ramban and Rav Dessler the value of a free evil choice and that of an unfree evil choice are the same, whereas they would not be equivalent on a view that sees free will per se as a value. Hence, to learn that evil choices are unfree is not to lose something of value, nor is the loss of good free choices bad if the good free choices are replaced by irresistible good choices.[66]

Ramban probably would not welcome with equanimity the loss of human responsibility and grounds for *sakhar va-onesh*. Physicalism could thus perturb him just as much as the next person. But does he think *sakhar va-onesh* is only here for a less ideal situation, and in the ideal there is none? The notion that the loss of free choice is not the loss of something of independent value could help cushion the force of determinism. Biological factors that predispose one to altruism or appreciation of the good could be viewed as "automatically" instilling the good in us. To that extent the situation might be viewed as good, even if we intuitively resist calling it Eden-like.

The Precedent of Restricted Free Will: Rav Dessler

Isaac Bashevis Singer once said in an interview: "The greatest gift which humanity has received is free choice. It is true that we are limited in our use of free choice. But the little free choice we have is such a great gift and is potentially worth so much that for this itself life is worthwhile living."[67] Singer is willing to say, then, that we have choice, but states that our choice is limited. Let us use the term "restrictivism" to denote the thesis that there is little free will, although a window for free will exists within modern science.[68]

This is a position with a surprising degree of support. In an instructive article published over a quarter century ago, Solomon Schimmel catalogued evidence that "Although Maimonides and most other Jewish philosophers vigorously defend the doctrine of free will . . . in practical terms the doctrine is never carried to its logical conclusions and is considerably circumscribed in several ways."[69] Schimmel mentions, among other evidence, the educational system, which works by reducing choice.

The most famous champion of restrictivism in Jewish sources is Rav Dessler. The latter argues that the vast majority of a person's actions are determined either by environment or by his or her already shaped character, where the character is shaped by previous choices. A person retains only a *nekuddat ha-behirah*, a tiny province of free will, at which moral struggle takes place.[70] Shalom Carmy's paper in this volume explores Rav Dessler's position in insightful detail, sparing me from elaborating, and I would stress but two points. First, the view that past choices determine present ones prima facie is out of sync with the possibilities of *teshuvah* (repentance) and backsliding. Second, it is not clear to me that one can always retain a morally and religiously significant *nekuddat ha-behirah*. If a person reaches a high level of *hitlahavut* (religious ardor) and his only choices are whether to travel to the yeshiva by train or bus, this is hardly an important *nekuddat ha-behirah*, and decidedly not what Rav Dessler means by the term.

Obviously Rav Dessler's view is a far cry from denying that there is any free will. Indeed, as we saw earlier, a person's task is to transform himself from a chooser to one who is compelled—so there are choosers. Furthermore, there is a vast gap between stressing the role of environment in molding character, as Rav Dessler does, and stressing the role of biology. Still and all, Rav Dessler's willingness to restrict free will's scope helps cushion the blow of determinism.[71]

Intermediate Conclusion

In this section we looked at four views that might afford precedents for the challenge of neuroscientific determinism. One was the determinism asserted by Hasidic views, which led Hasidic masters to deny or extremely curtail the reality of free will. Another was medieval astrological theories, which allowed for the override of the constellations' influence and, alternatively, Maimonidean theories of matter that located free will in the ability to override matter. The Hasidic precedent was found wanting because to invoke it we must be prepared to view neuroscience as expressing the truth that God controls all phenomena, and to see brain events as the *Yad Hashem*, the hand of God. While admissible in principle, this perspective may not hold much attraction. The astrological precedent (and Maimonides's non-astrological override-of-matter theory) seemed at first to be no precedent at all because in the neuroscientific context any "override" would be determined. However, the fact is that bad actions are determined according to this theory. Indeed, we will see later that this medieval override view can be resuscitated for contemporary purposes.

Next, Ramban and Rav Dessler's devaluation of freedom as compared to an automatic doing of the good afforded modest help with our issue. Finally, Rav Dessler's willingness to restrict free will suggests, especially when combined with the sources and points Schimmel cites, that free will is more limited in Judaism than the ordinary person supposes.

The Push toward Compatibilism

Despite the partial success of the preceding quest for precedents, we lack as yet an exact, full-blown precedent in Jewish philosophy through which we can defuse the challenge posed by determinism. It is time, therefore, to pursue a different strategy for addressing that challenge. We need to take the bull by the horns and try to show a flaw in the move from "determinism" to "no free will."

Until now we have been assuming that whether there is free will and moral responsibility, and whether reactive attitudes are defensible, is determined purely by the scientific facts. Implicitly, we have reasoned as follows: if science shows us that thought and behavior are determined, whether by environmental influences or biological ones, then we don't have free will or moral responsibility; if science is unable to do this, then there is at least a possibility of free will and responsibility, and the scientific challenge evaporates. So, either way, goes this thought, scientists settle the question by figuring out whether thought and behavior are determined.

These simple equations, "determinism true = no free will, no responsibility, no reactive attitudes[72]; determinism false = yes free will, yes responsibility, yes reactive attitudes" are naive; indeed, they do not survive more than a few moments of reading in the vast philosophical literature. To figure out from the "facts" whether we do or do not have free will, we need to know what we are trying to find. What is free will? This is a philosophical question, not a scientific one; and the popular assumption that, if behavior and thought are determined, there is no free will, no responsibility, and no justified reactive attitudes, reflects merely a particular view known as incompatibilism (the theory that free will, responsibility, and reactive attitudes are incompatible with determinism). Our discussion so far has proceeded on the tacit assumption that either there is free will or there is determinism; each excludes the other. However, for an extremely long time, the front and center issue to philosophers has been precisely whether determinism actually is compatible with free will.

The strategy to be explained in this section is that determinism is compatible with free will, moral responsibility, and the appropriateness of reactive attitudes. This position is known in philosophical literature as compatibilism. If a version of compatibilism can be motivated and defended, Judaism might exploit that as a response to the challenge of contemporary science, particularly if it can be shown that some Jewish philosophers were compatibilists.[73] Instead of searching for precedents for limitations on free will, therefore, we will study precedents for reconciliation between free will and determinism.

A caveat: I'm in a somewhat awkward position. I have devoted several papers in the general philosophical literature to explaining why, in my opinion, compatibilism ultimately fails by turning into incompatibilism when it gets boxed into a corner.[74] Hence, my attempt in this context to explain compatibilism is the product of a hope that I was wrong and that compatibilism will help us as Jews to cope with a deterministic threat. This is reason enough to present the theory and articulate problems I see with it.

First Compatibilist Thesis: Indeterminism Excludes Free Will

One objection to Libertarianism (the version of incompatibilism that accepts free will) is that the libertarian lacks a plausible model for how free action occurs. Often, libertarians posit a metaphysical "self," "will," or "agent" that guides or makes decisions. This notion of "agent causation" is frequently charged with metaphysical murkiness.

Libertarians sometimes assert that quantum physics reveals the existence of indeterminism at the level of elementary particles—for example, electrons whose path is not determined. Champions of human free will often welcome quantum physics as having shown that determinism is false, thereby salvaging free will. But let us look at the situation more closely. Suppose that science tells us human actions are not determined, as quantum or other stochastic events occur in neural processes. Does that mean automatically that we have free will?

No, say compatibilists. Suppose Yankel robs a bank, and it turns out that he did so because a neuron randomly fired in Yankel's brain. (Obviously scientific precision about such an event isn't called for in this context.) There is no fully deterministic pedigree for Yankel's act; but this is hardly Yankel exercising his free will, and anyone who held Yankel responsible for the act is just wrong. Free will is not just a matter of indeterminism; it's a matter also of control, of agency. Yankel's behavior is not something he does; it's something that happens to him. So even were we to find neurological events that are not fully determined, this would not confer free will upon us. Undetermined neural events do not imply that there is free will, but rather that there isn't. If there is indeterminism, one may argue, undetermined choices are "arbitrary," "capricious," "random," "uncontrolled," "irrational," "inexplicable," or "matters of pure luck or chance." "Undetermined events in the brain or body . . . would occur spontaneously and would be more of a nuisance . . . than an enhancement of freedom or responsibility."[75]

This point quickly leads to a sharper formulation of what the free will problem is. Think of the problem as a dilemma, a "*mi-mah nafshakh*" argument. On the one hand, determinism seems incompatible with free will because free will requires being able to do otherwise. On the other hand, indeterminism or chance at the level of brain events would seem to be incompatible with free will as well because it results in chance actions and deprives us of agency. So, whether we are determinists or indeterminists, free will is a chimera, and so, apparently, is moral responsibility.[76] Since indeterminism cannot salvage free will, compatibilism seems to be the way to go.

Second Compatibilist Thesis: Utilitarian Considerations Suffice for Responsibility and Appropriate Reactive Attitudes

Incompatibilists might argue that if actions are determined, punishment is impossible to justify. Compatabilists respond that punishment is, in fact, not based on retribution for failure to exercise moral responsibility, but instead on social utility, society's need for protection. If this is correct, people can properly be punished for actions even in a deterministic world.

A utilitarian theory of punishment faces problems, since, for one thing, we would not justify "punishing" an innocent person just because people think he or she is guilty and incarceration would prevent rioting and thus prevent bad consequences. For another thing, by invoking utility we could punish minor offenses like double-parking with harsh punishments like flogging.[77] These are real objections, but then again, it is clear that some punishments we do administer cannot easily be justified if we stay within a retributivist straitjacket.

Take, for example, cases where the law assigns "strict liability," meaning that for certain types of acts or omissions we do not accept any excuses, even if there was no *mens rea* (criminal intent). Legislators might oppose allowing mental illness as an excuse for an action on the grounds that it can be too easily feigned. People who say this do not deny that, morally speaking, a person who commits a crime due to mental illness should not be punished, but they argue that legally and socially we cannot accept this excuse.

Bear in mind, as well, that many people justify or oppose capital punishment strictly on the grounds of deterrence, which is a purely utilitarian consideration. Furthermore, Judaism is not short of utilitarian approaches to punishment.[78] Clearly, a distinction could be made between holding a person responsible only because it is utilitarian to do so, and administering a harsh punishment to a wrongdoer only because it is utilitarian to do so. But several philosophers have worked out utilitarian theories of punishment and reactive attitudes that overcome standard objections to such approaches.[79] My point is not that these utilitarian rationales are correct—only that the issue of whether responsibility, punishment, and reactive attitudes make sense in a deterministic world is not simple and most definitely is not purely scientific, indeed scientific at all.

Reactive attitudes may likewise be justified by their utility. Praise and blame function in a utilitarian way to encourage or discourage certain sorts of behavior. Moreover, reactive attitudes need not presuppose indeterminism. We praise people for their intelligence, good looks, and native athletic ability, even though these qualities are not due to any agency of their own, and the bearers of those properties often deserve no praise for these qualities.[80] And we often praise people as "good people" without having the foggiest idea of how much of their goodness is self-initiated and how much is due to upbringing alone. The same applies to wrong actions: how many of us, when blaming perpetrators of Nazi atrocities, really think it relevant whether an individual butcher of Jews was the product of his environment, lacking all exposure to other value systems? How many of us really think it relevant to learn how much an individual Palestinian suicide bomber was brainwashed into being evil, receiving beliefs he was in no position to question, before we figure out whether to blame him? Our judgments are not dependent on whether the actions are determined, nor do we feel they must be.[81]

Third Compatibilist Thesis: Free Will Does Not Require Alternative Possibilities

Like the divine foreknowledge–human free will argument,[82] the argument that determinism excludes free will, moral responsibility, and reactive attitudes supposes that in order for a person to act freely in a given situation and/or be responsible and/or deserve the reactive attitudes, the person must possess a two-way power. He or she must be able to do, in the very same circumstances, something other than what he or she in fact does. The agent must possess what was once called *liberum arbitrium*. This thesis is often called the Principle of Alternative Possibilities (PAP), or the Forking Paths principle (borrowing from Jose Luis Borges's *The Garden of Forking Paths*). If the Forking Paths Principle is false, if freedom does not require alternative possibilities, then, barring further problems, neither God's foreknowledge nor determination by physical causes will deprive an agent of freedom or responsibility.[83] Freedom and foreknowledge would be compatible, freedom and scientific determinism likewise. The issue at hand, then, is whether the Forking Paths Principle is true; if not, compatibilism is viable.[84]

Problems with the Forking Paths Principle I: Locke and Frankfurt

A counterexample to the Forking Paths Principle may be developed using a case developed by John Locke (which I will adapt in my own way). Suppose a man awakens in a room and deliberates whether to leave the room. He decides to stay in the room in order to continue an enjoyable conversation with others there. Unbeknownst to him, the room is locked, so he could not leave even if he wanted to. Does he freely remain in the room? If we accept the Forking Paths Principle, the answer would be no: after all, there is only one viable choice, since he cannot leave the room. But intuitively we say he freely remains in the room. Hence, one could argue, the Forking Paths Principle is false.[85]

Another example of this type: while talking to a relative on the phone, you witness a car theft in progress. You callously decide it's not worth leaving your present conversation to hang up and call the police. Unbeknownst to you, the police phone line is down. You could not have reached the police even if you had tried. Intuitively, we hold you responsible for not contacting the police by phone, and we say that you freely refrained from contacting them by phone.

But here, you do not have the power to contact them, so we have a case of free will in which the Forking Paths Principle is not satisfied.[86]

These cases do not seem that powerful as objections. In each instance there is something you did freely, something over which you had a two-way power: namely, deciding to remain in the room, or, in the other case, deciding not to call the police. The fact the room is locked doesn't mean you could not decide to leave it, and the fact the lines are down doesn't mean you could not decide to call. So you are responsible for your deciding not to call the police, even if you are not responsible for not reaching them. An attack on the Forking Paths Principle by means of examples will require not an example in which only one act is possible yet the person is free, but rather, an example in which only one decision is possible. Can such an example be produced?

Thirty-five or so years ago, the philosopher Harry Frankfurt became a household name—in philosophical households, that is—by producing a family of cases that many, if not most, philosophers view as a decisive refutation of the Forking Paths Principle.[87] Basically, his case involves a person who deliberates carefully and decides to do action A, but who, if he had shown any signs of possibly deciding not to do A, would have been coerced, mind-controlled, or knocked out of commission before coming to that other decision. He does A freely, yet could not have decided otherwise.

Here's one variant of a Frankfurt case: Moishy steps into a Florida voting booth to cast his vote for President. George Bush's and John Kerry's names are on the ballot. Moishy deliberates as well as anyone could: he thinks about their domestic and foreign policies, their personalities, their values, their comportment as leaders, and the like. As a result, he decides to vote for Bush. So far, no one will deny, I presume, that Moishy freely decided to vote for Bush and also freely voted for Bush (again, I treat freely deciding and freely acting separately). Now, if you learned later that some nefarious mind controller had wired Moishy's brain, unbeknownst to Moishy, and had "zapped" him into deciding to vote for Bush while he was still deliberating, you would change this assessment and say that Moishy did not vote freely for Bush. But as long as Moishy's decision was his own, based on his reasoning and his desires, without such external interference, you would say that Moishy freely decided to vote for Bush and freely voted for Bush, and is responsible for that decision to whatever extent responsibility is relevant here. (I suppose in a close election he could be responsible for Bush getting into office.)

So far, so good. But imagine the following added wrinkle. A nefarious mind controller is watching Moishy on a surveillance camera. Being adept at interpreting Moishy's expressions and movement—or, if you will, having a monitor that shows Moishy's brain and nervous system—he can tell when Moishy is inclining toward voting for Kerry and when he's inclining toward voting for

Bush. The mind controller is a Republican (maybe even a Supreme Court justice). If he thinks Moishy is about to form a decision to vote for Kerry, he will "zap" him with an electric current before the decision is made, thereby making him desire to vote for Bush. In point of fact, however, the mind controller sees no Kerry-leaning indications, so he leaves Moishy alone. Given the presence of the mind controller, it is clear that Moishy could not have decided otherwise than to vote for Bush, and could not have failed to vote for Bush. So, Moishy does not have a Forked Path, alternative possibilities. But his decision to vote for Bush is still free in the revised case, as is his act of voting for Bush. The fact he could neither decide to do otherwise, nor do otherwise, has no bearing whatsoever on Moishy's freedom.

Many philosophers accept Frankfurt's argument, though the matter remains contested.[88] The literature on Frankfurt's example is enormous. Issues about the examples abound: maybe PAP applies to omissions and not acts; maybe it applies to bad actions but not good actions; maybe the guy isn't really responsible, etc.[89] I will not go over this literature here.[90] The point is that a denial of PAP and the Forking Paths Principle would refute the argument for the incompatibility of free will and determinism.

Problems with the Forking Paths Principle (Principle of Alternative Possibilities) II: Acting in Character

Many everyday actions we regard as free, and assign responsibility for, are determined by a person's character and psychological states. Consider this passage from David Hume's essay, "Of Liberty and Necessity":

> Were a man, whom I know to be honest and opulent, and with whom I live in intimate friendship, to come into my house, where I am surrounded with my servants, I rest assured, that he is not to stab me before he leaves it, in order to rob me of my silver standish; and I no more suspect this event, than the falling of the house itself which is new, and solidly built and founded.[91]

Hume's point is that in real life we treat people as if their acts are determined. If someone wrongs you, your decision whether to criticize him to his face will depend on how well, based on your experience, he takes criticism. If someone is self-absorbed and therefore likely to listen to other people's problems with only half an ear, or if someone is a gossip-monger, you won't choose that person as someone in whom to confide. And, as Hume so pointedly puts it, "A prisoner . . . discovers the impossibility of his escape, as well when he considers the obstinacy of the jailer, as the walls and bars, with which he is surrounded." As Hume also points out, history, politics, and literary criticism depend on our ability to identify both general human motivations and motivational patterns in specific people. We may talk like libertarians when we delve into philosophy, but in everyday life we think like determinists.

Rav Soloveitchik made this very same point in a *teshuvah derashah* published under the title "Repentance and Free Choice":

> Free choice does not mean a state of chaotic anarchy, with sudden and frequent changes of mind that have no rational explanation . . . we usually expect a certain display of consistency in a man's thought and actions; we expect him to embody a certain way of life with its own consistency of character. The law of cause and effect, action and consequence, does prevail in a man's life. . . . The question now is: where does "free choice" come in?[92]

Some deeds are psychologically impossible—unthinkable[93]—for us or for people we deal with. If you attend a *shiur* by a renowned Torah scholar, you know very well that he will not pull out a machine gun or tell scatological jokes. If you are offered $5 to torture an innocent child, you will not and cannot comply. You cannot now spread a vicious rumor that your colleague, who you know is morally upright, is a child molester. When we say of someone, "he cannot hurt a fly," we are not saying he lacks free will. When Martin Luther stated, "Here I stand, I can do no other," he did not mean that he is not freely standing up to the Catholic Church and that he bears no responsibility. If George Washington could not tell a lie, this hardly means he lacks free will and gets no credit for confessing to his chopping down the cherry tree.[94] I would even say that in cases where someone has a powerful impulse to do a wrong thing, a certain type of person might not be able to actually do that thing, however much the person imagines he could because he feels a strong *yetzer* (urge), and imagines that he's conquering it by means of free will. People who have suicidal urges may genuinely be incapable of acting on them at a particular time. In none of these cases does the impossibility of doing otherwise signal that the agent is unfree.[95] And here is the moral to draw, therefore: to act freely or decide freely it is not necessary that you be able to do otherwise, decide otherwise, desire otherwise, and so on.

It is, in fact, folly to find alternative possibilities in these cases. (I call them character cases.[96]) After all, what are you looking for? Basically, you want to find a real possibility that the agent would do something that is contrary to all of her values and reasoning. But if this is the desired "alternative," the demand for an alternative possibility amounts to the demand that the person be able to do something insane, like Dostoevsky's Underground Man who believes that to be truly free is to be able to act against all his logic and all his interests. To insist that being able to act insanely is a condition for free will is strange, quite apart from the utter lack of evidence that the people in my examples are able to act insanely.[97]

Can a defender of the Forked Path Principle account for the freedom of the agents in character cases who lack alternative possibilities? One strategy for retaining the principle may be found in Aristotle. Aristotle maintains (*Nicomachean Ethics* 3:5) that although people who act in character seem to have no real choice in the matter, since their characters are fixed, in point of fact people choose their characters and are therefore responsible for actions that flow from those characters of their choosing. Philosopher Robert Kane, the leading advocate today of libertarianism—the idea that free will requires alternative possibilities, but we have free will—puts forth the following claim in explicating the case of Martin Luther, who declared "here I stand, I can do no other":

> If Luther's affirmation did issue inevitably from his character and motives at the time it was made, then his moral accountability for it would depend on whether he was responsible for being the sort of person he had become at that time, whose character and motives would issue in such an act.[98]

Similarly, the Rav asks, "if man's behavior and responses are, so to speak, set and predictable, where does 'free choice' come in?" and suggests that

> it lies within man's power to determine the framework of cause and effect within which he lives and acts. . . . Man can be the architect of his own personality; he has the ability to fashion his own character and map out the path he will follow. Indeed, man is capable of determining in advance what his reactions will be to given phenomena and events in the course of his life.[99]

In evaluating this strategy, the first question we face is whether it is true that we choose our characters. Let us bear in mind that in educational institutions and home environments, rewards and punishments and indoctrination are used extensively to inculcate values in children from a young age. As Solomon Schimmel puts it, "Rather than cultivating and encouraging the exercise of free-will as some internal operating factor, these techniques tend to preempt freedom by substituting determinants of behavior whose locus is the social environment rather than the individual's will or reason."[100] I would add that the notion that teachers should be role models is an instance of trying to determine behavior through means one presumes can be effective. When a teacher is not providing a good enough role model, we seek a better role model. Finally, to Schimmel's list as a whole I would add that in *shiddukhim* we always assume that the home environment helps shape behavior. Both our educational and social practice thus belie the idea that character must be the result of choice. Rabbi Shlomo Wolbe goes so far as to state, as "a fundamental principle in the chinuch of others and of ourselves" that "one should relate to every person as if he had no bechira, and as if he is 'compelled' by his 'nature,' education, habits and emotional needs."[101] It is true that a person could later in life reflect on his or her acquired values and retain or reshape personality as a result of that evaluation. But our intuition that people will and act freely when they will and act in character is not calibrated according to whether this evaluative process took place.

In addition, I do not think that the fact a person once chose a certain character entails that every choice made thereafter as a result of that choice is a free one. Consider an addict who is overwhelmed by irresistible cravings for drugs, but who brought on those cravings by earlier repeated usage of drugs at times when he could have resisted the desire for them. I do not think the addict shoots up freely just because of this causal connection. Rather, a more accurate description of his problem is this: right now he acts unfreely, due to earlier free choices. Similarly, suppose you choose to travel through a dangerous area in which muggings are frequent. While there, you are coerced into handing over valuables. Shall we say that you freely surrendered the valuables just because you made a choice that enabled the later situation to arise? Clearly not.[102] The proposition that all choices are free even if not made in the context of alternative possibilities, provided that they resulted from some earlier free choice that was made in the context of alternative possibilities, is not correct. A person can freely sell himself into bondage, but bondage is bondage.

What seems more plausible is that when a person makes choices in the context of alternative possibilities, and those earlier choices cause him to do certain later actions without entertaining alternative possibilities—as in the example of the addict or the mugging victim—the person bears a degree of responsibility for the later act even though it is not free. If someone chides him, "See, I told you! Once you take one drug, you'll have no control later," the friend is producing an effective reason for assigning responsibility. Even here, however, I would claim that the degree of responsibility for the later act is diminished. An epileptic bus driver whose bus crashes because he had a seizure, when he should have foreseen this possibility, is not as responsible as if he had crashed the bus deliberately.

My own view is that the best strategy for a libertarian to adopt in the face of character cases is Rav Dessler's, that is, to say "in hakhi nami" ("just so")—free will is restricted, and the agents in question do not have it—but to find a theory of responsibility that explains why agents in those cases are responsible for what they do. In another essay, I have tried to develop a form of libertarianism that differentiates the bus driver example from character cases and that seeks, for better or worse, to deal with character cases. I further argue against a variety of compatibilist attempts to deal with the very character cases that they sling against libertarians.[103] My goal to this point has been merely to show that compatibilism has substantial attractions. Under the combined weight of Frankfurt examples and character cases, the standard notion that free will requires alternative possibilities and a Forking Path is under serious attack. A determined choice can be free and responsible, say compatibilists.

I want to say a few words, though, about why developing this alternative to libertarianism will be difficult.

What's the Alternative Theory?

If free will does not entail alternative possibilities, as the foregoing examples suggest, what do compatibilists think it entails? Shall we say that acting or deciding freely is just a matter of doing what you want to do? No, that won't work. If that were all that is involved, we may as well assign responsibility to cats and dogs—they also carry out their desires. Furthermore, intuitively we do not regard compulsive behavior (e.g., of kleptomaniacs), or the behavior of people who have become addicts through no choice of their own, or of people suffering from phobias, as free, responsible behavior. But why not? If free action is a matter of acting on desire, why exclude people in the grips of phobias or addictions who act on their desire? What is the difference between free and unfree action or willing, according to compatibilists? What makes addicts, phobics and compulsives, and animals different from free and responsible human agents?

Many contemporary philosophers who reject the Forking Paths Principle propose the following alternative. Freedom of the will is a matter of conforming the will to the judgments of either (a) reason or intellect, or (b) one's "deep self" (in Hebrew, the *ratzon elyon* as opposed to *ratzon tahton*, the "higher" will as opposed to the "lower").[104] The notion that free will lies in conformity with reason is a conception that has deep roots in Plato, Aquinas, Spinoza, Kant, and others in general Western thought, and is found as well in Rambam, Ravad, Ralbag, and others on the Jewish side. Josef Stern writes that "[Rambam's] views both in the legal works and in the Guide rest . . . on one feature traditionally associated with classical compatibilism, in particular the notion of self-determination." Moshe Sokol also regards Rambam as a compatibilist.[105] On this view, there is in Rambam, an indifference to determinism vs. indeterminism—his view on free will is consonant with either position.[106] This is not atypical of medieval Jewish philosophy. Charles Manekin informs me that no Jewish philosopher before Isaac Arama used the term "free will" or "free choice" (*behirah hofshit*). Those terms, he relates, enter philosophical Hebrew from the Latin. Jewish philosophers tended to see reason, not liberty, as the main component of freedom; only afterwards did they come to require being able to do otherwise. The basic thought here is that a person with free will is someone who conforms his or her desires to his or her value judgments—period. No ability to act evilly or insanely is demanded. Because being free is acting rationally, an agent must be able to act rationally in order to freely act irrationally (evilly). But a person who is acting rationally is free without needing to be able to do the wrong thing.

A classic example of a "deep rational self" view (combining the two elements of intellect and self) is contained in a famous ruling of Rambam (*Hil. Gerushin*, 2:20) about a man who refuses to give a *get* (writ of divorce) when *Hazal* required him to. *Get* requires *retzon ha-ba'al*, the husband's desire to effect the divorce, yet a Jewish court can coerce him until he says "*rotzeh ani.*" "I want." Offhand, there are various ways in which this Halakhah could be explained, but Rambam proposes that insofar as this man is violating a precept of the Halakhah, he has been attacked and coerced by his evil inclination. The beating weakens the impulse and allows the true self, the self that represents what the person most values—and is the objectively right thing—to be expressed.[107]

This conception of freedom runs into the following major difficulties:[108]

Selfhood: Why is the self identified with intellect and desires deemed external to the self? Can this approach be defended in the face of Humean and Freudian views of the self, for example? From a Jewish point of view, however, rationality reflects the *tzelem Elokim*, so this objection does not have much power.

The first-person perspective: In his defense of a conception of free will that does not include a condition that the agent be able to act otherwise, John Martin Fischer makes the following intriguing concession:

> Moral responsibility is surely only one aspect of a complete understanding of agency. Another dimension of agency pertains to our view of ourselves as practical reasoners. That is, we deliberate about the future. . . And in this deliberation and practical reasoning, we typically take ourselves to have genuine alternative possibilities. The view of the future as a garden of forking paths is a central picture of the natural, intuitive picture of ourselves as agents who deliberate about and plan for the future. . . . [I]t is not a straightforward task to give a picture of deliberation and practical reasoning according to which we may not have alternative possibilities.[109]

As I read Fischer, he is acknowledging that the denial of the Forking Paths Principle fits a third-person perspective—we ascribe responsibility and freedom even in the absence of forking paths. The denial does not fit a first person perspective. The end result is that rejection of the Forking Paths Principle can be but temporary; we need to adopt the principle in order to make sense of deliberation, unless we say that the world of deliberation is all an illusion.

It is difficult to argue that determinism precludes deliberation; determinists still speak of self-control, effort, etc.[110] Indeed, as Schimmel points out, just as Jewish thinkers who affirmed free will in practice circumscribed it, deterministic psychologists who espouse determinism, including Skinner, invoke concepts like rational-emotive control of behavior, phenomenological freedom, self-control, delaying gratification.[111]

> *Akrasia vs. compulsion*: The idea that freedom of the will is a matter of conforming will and behavior to value judgments does a nice job of explaining why people who act compulsively, or who have phobias, are unfree: their behavior does not conform to their value judgments. But notoriously human beings are prone to acting against their better judgment even in cases where their behavior is free and they are responsible for it.

I know I have to grade a mountain of exams and return them to students as quickly as possible. My best judgment, what "reason" tells me, is that I should grade the exams. But, averse to such tedious work, I find some other way to fritter away the time, defaulting on my responsibility to my classes. Do I "freely" fail to grade the exams? Of course. Am I "responsible" for not returning them on time? Of course. But if free will is a matter of acting in accordance with one's best judgment, the answer to both these questions should be no. Instances of this sort are called akrasia, or weakness of will. We experience them all the time (anyone who has cheated on a diet knows what's it like to do what you think is wrong). But akrasia is very different from compulsion; my turning away from the exams I must grade is very different from a kleptomaniac's acting against his better judgment that he ought not steal.

Intuitively the difference is this: the akrates is able to act differently, to do the right thing; the compulsive is not. But to say this is to reinstate a version of the very Forking Paths Principle that the "reason view" of free will was meant to supplant.[112]

> *Evildoers*: A particularly difficult corollary of the akrasia problem is that only rightdoers, not wrongdoers, are classified as free, and so no evil acts are free. This is a strange result. To cope with it, some philosophers have suggested that freedom involves being able to act rightly. When people act rightly we declare them free even when they are not able to act otherwise—they are, after all, able to act rightly, since they do act rightly. But when they act wrongly and were not able to act rightly, we do not hold them responsible and free.[113] To require that they be able to act rightly, however, is to re-introduce the Forking Paths Principle in modified form, with all the problems that entails in a deterministic world.

Brainwashing: Another problem with the "reason" view is that it does not properly handle cases of brainwashing, what we call *tinok she-nishbah* ("the captured infant," that is, a person whose belief and value system is the result of external influences to which he has not chosen to be subjected). For free will, it is not enough that a person's actions conform to his values; we must ask how he acquired those values. If they were implanted in him by brainwashing, hypnosis, and the like, the person did not act freely. The "reason" view does not explain why this is so. Once brainwashing is ruled out somehow, we will be back to the question of why acculturation or physical determination should not also serve to excuse all behavior as unfree.

Coercion: If freedom is a matter of acting in accordance with reason, then the absolute paradigm of free action is coerced action. In coercion, you have just one rational choice. I don't mean only coercion by muggers. I mean also the coercion of a strong impulse—a person with an uncontrollable urge to take drugs is doing the only rational thing in taking the drug to relieve pain! Obviously these acts should be classified as unfree. Yet on the reason view, since giving in to the threat or impulse is often the only rational alternative in the immediate situation, such deeds ought to be classified as paradigmatically free, which is absurd.

Clearly the smoothest way to deal with these objections is to reintroduce the Forked Path Principle and thereby surrender to incompatibilists. Any strategy but surrender will seem like deliberately ignoring the obvious thing to say about the cases and points that were just presented. But incompatibilism has its own problems as well, as we saw—Frankfurt cases and character cases. Here, then, is the real problem: we don't know what free will is. No matter where you turn to explain what free will is, eventually you run into a brick wall. Every definition has problems attached to it. If we don't know what free will is, how can we figure out whether we have it in a deterministic world? No wonder the late Harvard philosopher Robert Nozick wrote that the problem of free will is "the most frustrating and unyielding of philosophical problems."[114] It is that frustrating and unyielding character that leaves us unsure of what direction to take in responding to the challenge of genetics and neuropsychology.

Conclusion

This paper expresses certain general theses, beyond the obvious implication that fusing Torah and *madda* is our best way of getting a grip on the problems and potential solutions to the challenge of materialism. I approach the topic of contemporary science by taking seriously Rav Kook's notion that we must seek to extract the positive moral message in all phenomena, while rejecting the rest. In addition, I believe that thinkers like Rambam, Ravad, Ramban, and Rav Dessler adopted views that go against today's simplistic sloganeering about "behirah." There is a surprising amount of biological explanation and deterministic explanation found in *rishonim* (medieval authorities) and in the work of recent *baalei mahashavah.* I further believe that medieval and modern *baalei mahashavah* adumbrated some of the approaches to free will found in philosophical literature today.

I close with a telling anecdote. In an article in *The Jewish Observer* (April 1996), Dr. Benzion Sorotzkin, a clinical psychologist, drew upon Rav Dessler's discussion of free will and *nekuddat ha-behirah* to argue that much of human behavior is caused and that psychotherapy can be beneficial precisely by recognizing this truth.[115] He criticizes those for whom "any suggestion that there are limitations to human freedom is seen as apikorsus [heresy]," who object to "the search for explanations and causes of behavior inherent in psychotherapy," and who would sooner say "he did it because of his yetzer ha-ra" than look for causes. Sorotzkin's thesis was that "hashkafa [our philosophical outlook] acknowledges that past experiences limit bechira" in that they define a person's *nekuddat ha-behirah.* He went on to say that people ought to work on behaviors that are changeable rather than ones that are outside their *nekuddat ha-behirah* at a given time.

Notwithstanding his candor, Dr. Sorotzkin was criticized in a subsequent issue (March 1997) because

> there is widespread consensus amongst Torah leaders and teachers throughout the generations against the practice of psychology of searching into a person's past experiences and environment to determine the decisive cause of his errant behavior. . . . Torah hashkafa does not acknowledge that past experiences limit bechira and that psychology can define a person's nekudas habechira. . . . To focus on psychological causes does not help most people to change their behavior. . . . Ein davar omed lifnei haratzon.

What about Rav Dessler, though? Doesn't he view much of what we do as externally caused?

> It is a grave mistake . . . to assume that these descriptions were intended to serve as precise definitions of the *nefesh* from a theoretical point of view. When one learns *sifrei mussar* with the intended attitude, he will discern between an objective definition of a phenomenon and an illustration intended to motivate *avoda.* Without heeding this principle, there is a danger of accurately quoting sections of *sifrei mussar* while totally misrepresenting their intent.

This little exchange speaks volumes about how sensitive issues of free will are in our community. While the topic of the exchange is environmental influence and I have concentrated on biological influence, the underlying issue is the same. I suggest that a reality check will vindicate not only the view that much of human behavior is caused, but also that our *hakhmei ha-masorah*, "Torah leaders and teachers throughout the generations"—Rav Dessler included—were under no illusions to the contrary, even while they affirmed that some actions are free. The challenge before us is to utilize their insights, combine them with the best that scientists, psychiatrists, psychologists and philosophers have to offer, and "build the palace of Torah above it."

We cannot do otherwise.[116]

Notes

1. On Jewish reactions to the Copernican Revolution, see Andre Neher, *Jewish Thought and The Scientific Revolution of the Sixteenth Century: David Gans (1541–1613) and His Times*, trans. David Maisel (New York: Oxford University Press, 1986); David Berger, "Judaism and General Culture in Medieval and Early Modern Times" (to which I owe other references in this note) in David Berger, Gerald J. Blidstein, Shnayer Z. Leiman, and Aharon Lichtenstein, *Judaism's Encounter with Other Cultures: Rejection or Integration*, ed. Jacob J. Schacter (Northvale, NJ: Jason Aronson, 1997), 133–35; Hillel Levine, "Paradise Not Surrendered: Jewish Reactions to Copernicus and the Growth of Modern Science," in *Epistemology, Methodology, and the Social Sciences*, ed. R. S. Cohen and M. Wartofsky (Dordrecht: Reidel, 1983), 203–25; David Ruderman, *Jewish Thought and Scientific Discovery in Early Modern Europe* (New Haven, CT: Yale University Press, 1995), 266–68. Tobias Cohen, Yosef Shlomo Delmedigo, David Gans, Tovia Katz, Isaac Lampronti, and David Nieto were among Copernicus's rejecters, with Nieto calling the heliocentric theory *piggul* (see Berger, 134–35 and Ruderman, 266–67).

2. One group actually constructed a geocentric model at a Jewish expo, but I have seen texts from "anti-*Torah-u-Madda*" circles that conceal the fact that *rishonim* believed in a geocentric universe. This testifies either to the concealers' acceptance of the heliocentric picture, or their reluctance to appear to have rejected it.

3. *Orot ha-Kodesh* (Jerusalem: Mossad HaRav Kook, 1985), II:537.

4. *Iggerot ha-Reayah* #134, as trans. by Tzvi Feldman in *Rav A. Y. Kook: Selected Letters* (Jerusalem: Maaleh Adumim, 1986), 14. This letter admittedly contains a potpourri of strategies, at times conflicting ones, for responding to evolution. Works that attempt to integrate evolution with Torah include: Nathan Aviezer, *In the Beginning* (Hoboken, NJ: Ktav, 1990) and *Fossils and Faith* (Hoboken, NJ: Ktav, 2001); Gerald Schroeder, *Genesis and the Big Bang* (New York: Bantam, 1992) and *The Science of God* (New York: The Free Press, 1997); Nosson Slifkin, *The Science of Torah* (Southfield, MI: Targum, 2001) and *Mysterious Creatures* (Southfield, MI: Zoo Torah in Association with Targum/Feldheim). Some modern Orthodox Jews, while endorsing the idea that

Torah and evolution can be harmonized, object to the level of detail with which some of these books "find" contemporary science in Genesis, partly because science changes. As I write this Slifkin's books have been placed under a ban by some Torah luminaries, who affirm that the world is 5,765 years old. It remains to be seen how matters will play out down the road.

5. Indeed, even if I were a scientist, I wouldn't try this approach without being a scientist of the right kind. I have taken careful notice that the "scientists" who sign petitions disclaiming evolution are not biologists. Whether, as critics say, this phenomenon is due to specialists' biases toward orthodoxies of their own field, reinforced by peer pressure, or instead, as I speculate, to non-specialists' being in no position to appreciate certain data, is an interesting question.

6. Most so-called dualists today are what are called "property dualists," meaning that they deny there are nonphysical substances but accept that there are nonphysical properties like the redness of a mental image or the sharpness of a pain. This is the position presented in, for example, what is in some circles regarded as the strongest case against materialism yet made, David Chalmers's *The Conscious Mind: In Search of a Physical Theory* (New York: Oxford University Press, 1996). Yet substance dualism—which accepts an immaterial self—is not lacking in support among academic philosophers. See Robert Audi, "Theism and the Mind-Body Problem," *Faith, Freedom and Rationality*, ed. Jeff Jordan and Daniel Howard-Snyder (Lanham, MD: Rowman & Littlefield, 1996), 155–69; J. Foster, *The Immaterial Self* (London: Routledge, 1991); W. D. Hart, *The Engines of the Soul* (Cambridge University Press, 1988); and Richard Swinburne, *The Evolution of the Soul* (New York: Oxford University Press, 1997). See also Shubert Spero, "What Is Self-Theory and Does Judaism Need One?," *The Torah u-Madda Journal* 12 (2004): 130–57.

7. These raw experiences are thought by many to be a thorn in the side of neuroscience, however. At several places in his writings, the Rav takes the inability of science to capture such experiences to signal a need for an explanation of the world beyond the physical. See *Lonely Man of Faith* (New York: Doubleday, 1992), 12–14, 18; *U-Vikkashtem Mi-Sham*, in *Ish ha-Halakhah: Galuy ve-Nistar* (Jerusalem: World Zionist Organization, 1992), 137–41. For an argument to the effect that neuroscience cannot *fully* explain sensory experiences see Robert Merrihew Adams, "Flavors, Colors, and God," in *The Virtue of Faith and Other Essays in Philosophical Theology* (New York: Oxford University Press, 1987), 243–62.

8. Cf. Eugene d'Aquila and Andrew Newberg, "The Neuropsychological Basis of Religion: Why God Won't Go Away," *Zygon* 33, 2 (June 1998): 187–201.

9. Dissenting neuroscientists voice their view in Benjamin Libet, Anthony Freeman, and Keith Sutherland (eds.), *The Volitional Brain* (Thorverton, UK: Imprint Academic, 1999). See also Jeffrey M. Schwartz, M.D., and Sharon Begley, *The Mind and the Brain: Neuroplasticity and the Power of Mental Force* (New York: Regan Books, 2002). Cf., however, the dissenters from the dissenters, notably Thomas W. Clark, "Fear of Mechanism: A Compatibilist Critique of 'The Volitional Brain,'" Libet et al., 279–93.

10. Kuhn, *The Structure of Scientific Revolutions* (Chicago: University of Chicago Press, 1962). I put aside the "subjectivist" view of science that some think emerges from Kuhn, and only employ his term.

11. Crick, *The Astonishing Hypothesis: The Scientific Search for the Soul* (New York: Charles Scribners' Sons, 1994), 3.

12. Dennett, *Consciousness Explained* (New York: Little, Brown, 1991), 33.

13. Peter D. Kramer, *Listening to Prozac* (New York: Viking Press, 1993), 21.

14. See, however, the exchange from *The Jewish Observer* described in the conclusion to this paper.

15. Kramer, *Listening to Prozac*, 18.

16. The culture of pharmacology for the mind breeds other questions. Traditionally, responsibility is tied to the notion that certain behaviors are "owned" by the patient and reflect his or her "true self." Who is the real self when medicine alters mood and behavior? Is the patient accurate in declaring, after a few days on an anti-depressant/anti-anxiety regimen pursuant to thirty horrid years of entirely different behavior and affect, that after those few days on the pill "now I feel like myself?" Or has he or she, during those recent few days, been living as somebody else? Even people who take even blood pressure medications or medications that slow heart rate sometimes wonder, when they exercise, "was that me or the medicine that went 5.1 miles an hour at 4% elevation without losing breath?" Athletes who take steroids often are thought to be not "really themselves" when they are capable of hitting baseballs 500 feet; hence bans on steroid use in competitive sports. *A fortiori* when the changes are psychological. Should a person get credit for the better behaviors even when there has been no "internal" labor? Is a pharmacological enhancement "external"—like someone pushing a long-since trailing runner over a finish line—or "internal" to the self?

17. Ibid., 258. chapter 9 deals with this and the other objections.

18. Some critics of psychopharmacology claim that the biological perspective that Kramer articulates can only gain strength from economic realities. Taking their marching orders from H. M. O.'s, critics say, psychiatrists are encouraged to utilize the least expensive treatment, medicine, at the expense of dialogue that might provide a more humanistic view of the patient. For this and other criticisms of present-day psychiatry, see J. Allan Hobson and Jonathan A. Leonard, *Out of Its Mind: Psychiatry in Crisis: A Call for Reform* (Cambridge, MA: Perseus Publishing, 2001). My knowledge of the book is gleaned entirely second hand from a review by Andrew Solomon in the *New York Times Book Review* of October 7, 2001, p. 20.

19. Churchland's books include *Matter and Consciousness* (Cambridge, MA: MIT Press, 1988) and *The Engine of Reason, The Seat of the Soul* (Cambridge, MA: MIT Press, 1995). See also his wife Patricia Churchland's *Neurophilosophy* (Cambridge, MA: MIT Press, 1986). A brief exposition of Paul Churchland's views is his essay, "Folk Psychology and the Explanation of Human Behavior," in *The Future of Folk Psychology: Intentionality and Cognitive Science*, ed. John D. Greenwood (New York: Cambridge University Press, 1991), 51–69.

20. Rav Lichtenstein sees history's value, in part, as revealing the choices of individuals. See R. Aharon Lichtenstein, "Torah and General Culture: Confluence and Conflict," in *Judaism's Encounter with Other Cultures*, ed. Jacob J. Schacter (Northvale, NJ: Jason Aronson, 1997), 242.

21. Some writers think that the two pictures of humanity are viewing the same phenomenon from two different perspectives, each valid in its own right. A recent ar-

ticulation of this view is Hilary Bok, *Freedom and Responsibility* (Princeton, NJ: Princeton University Press, 1998). But critics complain that this is not a solution to the problem, but only a restatement of it; for the two perspectives seem incompatible. See Linda Ekstrom, "Introduction," *Agency and Responsibility: Essays on the Metaphysics of Freedom* (Toscadero, CA: Ridgeview, 2001), 4–5. The philosopher Elizabeth Anscombe once wrote of such attempts at reconciliation (e.g., Kant's) that "[they] have always seemed to me either so much gobbledegook, or to make the alleged freedom of action quite unreal." Anscombe, "Causality and Determination," in *Metaphysics and the Philosophy of Mind: The Collected Papers of Elizabeth Anscombe* (Oxford, UK: Blackwell, 1987), 146.

22. See, for example, his words in the selection "Religious Immaturity" in Aharon Rakefet's *The Rav* (Hoboken, NJ: Ktav, 1999), 2:238–241 as well as in *Community, Covenant and Commitment: Selected Letters and Communications*, ed. Nathaniel Helfgot (New York: Toras HoRav Foundation, 2005), 93–105.

23. It is not natural to think in philosophical terms that may prove discomfiting. Proof of this is the popularity of computer science as a career for Orthodox Jews who are wary of entering other professions for fear of spiritual attrition. This trend is in one respect amusing. Computers represent a fundamental challenge to any understanding of human intelligence as special, since computers' capacities suggest that human thought could be explained, just like a computer's behavior, purely by means of physical components and processes. In fact, before the breakthroughs in genetics and neuroscience, artificial intelligence was the key challenge to human distinctiveness. Yet rare is the Orthodox Jew today (or even twenty years ago, for that matter) who regards computers as a challenge to *frumkeit*. (What an irony: it takes a philosopher to spot the *apikorsus* [heresy]). The conclusion I draw is that any field could be regarded as religiously safe—once the person studying it gives no thought to the problems it generates.

24. Similarly, a behaviorist might acknowledge that there are inner mental processes, but might claim, as B. F. Skinner did, that since they are unobservable they have no place in science. Likewise, someone could believe that God creates each species by a special act of creation, but acknowledge that God should not be invoked in a *scientific* explanation, because of constraints on what entities "science" can allow in. See the debate over this point between Alvin Plantinga and Howard van Till in *Christian Scholars' Review* 21,1 (1991): 8–41. Times have changed in this regard. To handle planetary motions that did not fit his equations perfectly, Sir Isaac Newton posited that God adjusts orbits every once in a while. No scientist today could do that! (A final note: ironically, Mordecai Kaplan, naturalist though he was, insisted on free will.)

25. I think that the existence and functioning of the soul is a non-negotiable demand. There is a problem with positing a soul if the latter is only the soul-of-the-gaps: namely, that once the gaps are closed by reference to science, "we have no need of that hypothesis" (as the eighteenth-century mathematician-philosopher Pierre Laplace said of God). If one can find functions for the soul to perform, this objection will be met. Strategies for understanding what the soul does are offered by the "substance dualists" cited in note 6. These writers, unfortunately, do not necessarily emerge with views that allow for immortality of the soul.

26. As you can see already, the problems with physicalism are numerous. See the essays in Howard Robinson (ed.), *Objections to Physicalism* (New York: Oxford University Press, 1993) and Steven Wagner and Richard Warner (eds.), *Naturalism: A Critical Appraisal* (Notre Dame, IN: Notre Dame University Press, 1993). In addition to difficulties in construing ethical facts in a materialist way, there are problems with normative concepts of other kinds, such as rationality. (See, e.g., my essay in the Wagner-Warner volume, "Skepticism and Naturalized Epistemology.") Cosmic materialism, in short, has a lot of explaining to do.

27. *Family Redeemed*, ed. D. Shatz and J. B. Wolowelsky (New York: Toras HoRav Foundation, 2000), 6. See also *The Emergence of Ethical Man*, ed. Michael S. Berger (New York: Toras HoRav Foundation, 2005).

28. *Guide of the Perplexed* III:8, trans. Shlomo Pines (Chicago: University of Chicago Press, 1963), 431, 433.

29. See Charles H. Manekin, "Freedom within Reason?: Gersonides on Human Choice," in *Freedom and Moral Responsibility: Jewish and General Perspectives*, ed. Charles H. Manekin, assoc. ed. Menachem M. Kellner (Bethesda: University of Maryland, 1997), 165–204. Note especially Manekin's statement:

> The power possessed by humans to make opposing choices follows, according to Gersonides, from two facts about their condition: they possess intellect, which motivates them to choose according to reason, and they are corporeal creatures whose material faculties (sense, imagination, etc.) motivate them to choose in opposition to intellect (177).

Also see p. 180, a passage I cite later in the paper: "Nowhere does Gersonides say that one freely chooses to be drawn after one's temperament. Rather he says that one's choice itself is 'drawn' after temperament." Manekin's reading of Ralbag leads him to an important revision of the standard view that, according to Ralbag, God cannot know "future contingents." See his "On the Limited-Omniscience Interpretation of Gersonides' Theory of Divine Knowledge" in *Perspectives in Jewish Thought and Mysticism* [memorial volume for Alexander Altmann], ed. Alfred Ivry et al. (Amsterdam: Harwood Educational Publishing, 1998), 135–70.

30. Cf., however, R. Samson Raphael Hirsch's reading of *yetzer mahshevot libbo* in Gen. 6:5 as not instinct or impulse but rather that which the person, qua *yotzer*, forms.

31. See e.g., *Bereshit Rabbah* 9.

32. It is true that Rambam's model, unlike the Rav's, was dissociating as much as possible from matter, not ascribing channeling. But on the whole, mind-body dualism is less sharp in Judaism than in Christianity.

33. For an elaboration on this remark, see Tamar Ross, "The Cognitive Value of Religious Truth Statements: Rabbi A. I. Kook and Postmodernism," in *Hazon Nahum: Studies in Jewish Law, Thought, and History Presented to Dr. Norman Lamm on the Occasion of His Seventieth Birthday*, ed. Yaakov Elman and Jeffrey S. Gurock (New York: Ktav and Yeshiva University Press, 1997), 529–56, as well as my "The Integration of Torah and Culture: Its Scope and Limits in the Thought of Rav Kook" in Elman and Gurock and my "Rav Kook and Modern Orthodoxy," in *Engaging Modernity*, ed. Moshe Sokol (Northvale, NJ: Jason Aronson, 1997), 91–115.

34. For more on the connections in Rambam between matter, shame, and *yir'ah*, see Josef Stern, "Maimonides' Conceptions of Freedom and the Sense of Shame," in *Freedom and Moral Responsibility: General and Jewish Perspectives*, esp. 247–66. More generally, see *Family Redeemed*, 73–85.

35. T. W. Clark, "Fear of Mechanism," in Libet, Freeman, and Sutherland, *The Volitional Brain*, 280.

36. The subtitle of John Martin Fischer's book, *The Metaphysics of Free Will: An Essay on Control* (Cambridge, MA: Blackwell, 1994) is indicative.

37. See Maimonides, *Mishneh Torah, Laws of the Foundations of the Torah* 2:2.

38. "Majesty and Humility," *Tradition* 17, 2 (Spring 1978): 25–37.

39. On these themes see also John Haught, *Science and Religion: From Conflict to Conversation* (Mahwah, NJ: Paulist Press, 1995), 83–84, 93. Cf. however, Freud, *The Future of an Illusion*, trans. James Strachey (New York: W.W. Norton, 1961), which offers a theory of religion predicated on the idea that whereas humans are unable to control nature by natural means, religion claims to give them the ability to control nature by influencing the divine through supplication and other means.

40. See Haught, 83.

41. Thomas Huxley was once asked by an anti-evolutionist, "I understand you are descended from apes. Is that on your mother's side or your father's?" And a cartoon depicts a hostess introducing a person to a monkey with the remark, "the two of you must have a lot in common to talk about. After all, you share 99% of your DNA." With an eye on such humor, it will be said that the spiritual dimension of humanity cannot be dismissed, and that physicalism tries to dismiss it. But while the humor and irony point up how radically thinking was changed by Darwin, physicalists will not be perturbed.

42. The list is from *Family Redeemed, passim.*

43. Clark, "Fear of Mechanism," 281. The criticism is directed at Benjamin Libet.

44. Robert Nozick wrote an amusing essay on how God could go about proving His existence. The particle scenario is one possibility he raised, only to show that skeptics would find some way around it. See Nozick, *The Examined Life* (New York: Simon and Schuster, 1989).

45. The term "reactive attitudes" was coined by British philosopher P. F. Strawson in a classic and frequently reprinted essay, "Freedom and Resentment," *Proceedings of the British Academy* 48 (1962): 1–25. Responsibility for an act or omission does not entail that the act or omission was done freely (we can speak of omissions being "done" when they are deliberate, like deliberately not phoning someone). A person who negligently flicks cigarette ash on a patron in the next row of seats in the stadium, is not "freely" doing that because he does not will it, and may feel badly once it is done, but he is responsible for the negligence.

46. The existence of free will might affect the scope of *kiddushei taut*. See the argument given by Rabbi J. David Bleich, "*Kiddushei Taut*: Annulment as a Solution to the Agunah Problem," *Tradition* 33,1 (Fall 1998): 91–94.

47. In the eighteenth century, mathematician-philosopher Pierre Laplace envisaged a giant, omniscient predictor that would know all the laws of nature and all the conditions

obtaining in the universe at a given time and deduce therefrom the state of the world at all future times. Laplace's predictor is often invoked as a picturesque way of thinking about the problems posed by determinism: if all events are determined, then human actions are too. Laplace associated determinism with predictability.

48. In the case of divine foreknowledge it may be that since God's knowledge doesn't *cause* a choice, it is less worrisome. But the divine plan problem does not admit of this explanation. Despite the discrepancy I note in the text, John Martin Fischer, in his highly regarded book, *The Metaphysics of Free Will* (Oxford: Blackwell, 1994), works throughout with the analogy between the two problems and implies they should be equally worrisome. For more on scary "what if" scenarios, see my "The Metaphysics of Control," *Philosophy and Phenomenological Research* (1997): 955–60.

49. In suggesting that the *problems* of divine foreknowledge and divine planning are irrelevant to this discussion, I am not suggesting that the *literature* on them is irrelevant. On the contrary, the literature contains certain conceptions of free will that make free will compatible with determinism. These conceptions also render free will compatible with determinism by physical causes, so the solution to one proves to be a solution to the other. I shall allude to this point later again.

50. On this point, see Charles Manekin's astute introduction to Manekin (ed.), *Freedom and Moral Responsibility*, 8–11.

51. See for example his commentary to The Book of Joshua (Jerusalem: Kiryat Sefer, 1966).

52. The narrative *seems to* play with the theme when Joseph *seems to* exonerate his brothers from responsibility, raising the argument that "you did not send me here, but rather God did." But close reading of the verses suggests that he is not exonerating them, but merely providing a reason why "*ve-attah al te'atzevu*," "do not be pained over what you have done, for it has turned out for the good." For discussion of this and other texts in which divine planning and human initiative are present, see Shalom Carmy and David Shatz, "The Bible as a Source of Philosophical Reflection," in *The Routledge History of Jewish Philosophy*, ed. Daniel Frank and Oliver Leaman (London: Routledge, 1997), 25–29. For interesting detailed analysis of the Joseph case, see Charlotte Katzoff, "Divine Causality and Moral Responsibility in the Story of Joseph and His Brothers," *Iyyun: The Jerusalem Philosophical Quarterly* 47(1998): 21–40 and her "The Selling of Joseph—A Frankfurtian Intepretation," in *Moral Responsibility and Alternative Possibilities*, ed. David Widerker and Michael McKenna (Burlington, VT: Ashgate, 2003), 327–38.

53. Ephraim Urbach, *The Sages: Their Concepts and Beliefs* (Jerusalem: Magnes Press, 1979), 257–58.

54. This absence of biblical and rabbinic material contrasts with the ample grappling we find with the problem of evil.

55. See Arthur Hyman, "Aspects of the Medieval Jewish and Islamic Discussion of 'Free Choice,'" in Manekin and Kellner, *Freedom and Moral Responsibility*, 133–52; also see Manekin, "Introduction," 10–11.

56. To be more exact, all denied the possibility of free action. In the realm of thought and decision, some of these thinkers restricted freedom to the ability to believe one of two mutually exclusive alternatives: either that God willed the act (which

is the truth) or that He did not. To sin is, in this construct, is to think you have a will of your own beyond the will to affirm or reject the truth that God does all. *Ha-kol bi-yedei shamayim hutz mi-yir'at Shamayim,* "all is in the hands of Heaven save for fear of Heaven" (*Berakhot* 33b) means that *ha-kol mi-yedei Shamayim,* all is *from* the hands of Heaven, except for your attitude about providence. See Jerome I. Gellman, "Denial of Free Will in Hasidic Thought" in Manekin and Kellner, *Freedom and Moral Responsibility,* 111–31, to which the foregoing explanation is indebted.

57. This argument is also found in the Kalam philosophers whom Saadyah sought to refute. R. Shnayer Zalman of Lyadi's view is that even the separate existence of created things is illusory. The doctrine of *tzimtzum* was understood by R. Shnayer Zalman epistemologically rather than ontologically. Which is to say: it *appears* to us that the *Ein Sof* has withdrawn from the world, but the truth is that He is in everything; hence the creaturely consciousness of separatedness is only an appearance—in actuality is only *bittul ha-yesh,* the eradication of any existence but the *Ein Sof.* For R. Shnayer Zalman; free will is no more real than tables and chairs.

58. The position that nature represents divine action is held in *Guide of the Perplexed* 2:48, where Rambam equates prophetic statements of the form "God does X" with "X happens in accordance with natural law and the course of nature." It also dovetails with many other attempts in Jewish thought to see nature as reflective of divine activity and providence. Note that while some Hasidic thinkers give up freedom of action rather than freedom of decision, contemporary determinism stresses that thoughts, too, are determined—and R. Shnayer Zalman, recall, went so far as to say that thought is also not free.

59. See Dov Schwartz, *Astrology and Magic in Jewish Thought of the Middle Ages* (Heb.) (Ramat Gan: Bar Ilan University Press, 1999).

60. Rabbi Soloveitchik in Peli, *Soloveitchik on Repentance* (Ramsey, NJ, 1984), 173.

61. Manekin, "Freedom within Reason?," 180.

62. Rambam writes in *Shemonah Perakim* that "It is not possible for a person to possess virtue or vice by nature. . . . Still, it is possible to be naturally disposed toward a virtue or a vice, so that it is easier to perform the actions that accord with [a particular virtue or a particular vice]." Through training, a coward can become brave—sometimes easily. The translation is from *Maimonides' Ethical Writings,* ed. Raymond L. Weiss and Charles Butterworth (New York: Dover, 1975), 83–84. For references and further points pertaining to Rambam's view, see notes 63, 73, 78, 102, and 105–7.

63. For further discussion of the various Maimonidean passages about matter as they relate to free will, see Moshe Sokol, "Maimonides on Freedom of the Will and Moral Responsibility," *Harvard Theological Review* 91, 1(1998): 25–39; Josef Stern, "Maimonides' Conceptions of Freedom and the Sense of Shame," in Manekin and Kellner, *Freedom and Moral Responsibility,* 217–66.

64. I find override theories in need of fuller exposition. Supposing for the moment that determinism is incompatible with free will, it seems to me that when the agent does not exercise the override, his situation *should not* be described as one of exercising free will. It is true that he/she could have (in the medieval theory) overridden the "natural" (astrological) determinants of action. But it hardly follows that the determined acts were free. If I'm standing at a bus stop and fail to brace myself for a strong

wind, and as a result of my unpreparedness I topple onto someone else and injure him, I do not freely will to topple onto the person just because I could have (freely) braced myself. Likewise, if my soda is spilling as I walk and someone slips on the liquid, I do not freely spill the soda just because I could have (freely) held the can more firmly. What does follow from my being able to do otherwise is that I am responsible for the results of my negligence (if negligence it indeed was). By analogy, in the case where an act of mine is astrologically determined and I do not override the astrological determinant, I do not freely will the act that is determined just because I could have freely done something else, even though I am responsible for the act. Manekin, "Freedom within Reason?," 167–68, notes that in Ralbag's (and others') theories of providence, rewards, and punishments are built into nature. This, indeed, is the idea of being "drawn after" temperament: I do not freely perform the temperament-induced action. Quite apart from this complication, however, the medieval override theory does not seem to furnish an effective response vis-à-vis determinism today.

65. *Mikhtav Me-Eliyyahu* 2:51–63. Actually, Rav Dessler has a still higher stage beyond *mukhrah.* At this stage, there is no subjective feeling of compulsion at all, but rather *avodah me-ahavah,* service out of love—"one cannot speak of compulsion to do something that one loves. At this point the human being becomes truly free, finding no resistance within him to the good which he loves. . . . 'No one is free but he who occupies himself with the Torah.' . . . This man of the spirit is the truly liberated man." (Translation is from *Strive for Truth,* 2:63.)

66. Robert Nozick expresses a similar devaluation. Nozick coins the term "tracking bestness" to refer to the process of doing the right without free choice. "Let us ask which you would choose for yourself: tracking bestness (without the process) or the process of free will (without tracking bestness). Our difficulty simply in choosing the second indicates the value of the first, of tracking bestness." See Nozick, *Philosophical Explanations* (Cambridge, MA: Harvard University Press, 1981), 327. Cf. my "Compatibilism, Values, and 'Could Have Done Otherwise,'" *Philosophical Topics* 16, 1 (1988), esp. 167–75.

67. Interview by H. Flender in *Writers at Work,* ed. George Plimpton (New York: Penguin, 1968), cited in Libet et al., "The Volitional Brain," 57.

68. I take the term from philosopher Peter van Inwagen, "When Is the Will Free?," *Philosophical Perspectives* 3 (Tascadero, CA, 1989), 399–422.

69. Solomon Schimmel, "Free Will, Guilt and Self-Control in Rabbinic Judaism and Contemporary Psychology," *Judaism* (1977): 418–29, at p. 425.

70. For similar views in the secular literature, see C. A. Campbell, "Is Free Will a Pseudo-Problem?" *Mind* 60 (1951): 446–65; Peter van Inwagen, "When Is the Will Free?" A tough-grading philosophy professor I know, who is not Jewish, assigned an 'A' to a student paper that he later learned was a plagiarism of Rav Dessler's works. He was impressed, obviously, with the philosophical quality of the argument.

71. As a group of contributions to *The Jewish Observer* some years ago makes manifest, the Orthodox right has its problems with this view. See the conclusion of this paper.

72. Shorthand for "no *appropriate* or *justified* reactive attitudes."

73. Key players in the debate over whether Maimonides was a compatibilist are Altmann, Pines, Jerome Gellman, Moshe Sokol, and Josef Stern, all already cited.

74. These criticisms of compatibilism are explained most fully in "Free Will and the Structure of Motivation," *Midwest Studies in Philosophy X*, ed. Peter French et al. (Minneapolis: University of Minnesota, 1985), 451–85; "Irresistible Goodness and Alternative Possibilities," in Manekin and Kellner, *Freedom and Moral Responsibility*, 33–73, esp. 51–62; "Compatibilism, Values and 'Could Have Done Otherwise,'" *Philosophical Topics* (1988): 151–200, and "The Metaphysics of Control, " a review essay on Fischer, *The Metaphysics of Free Will, Philosophy and Phenomenological Research* (1997): 955–60. Also see "Hierarchical Theories of Freedom and the Hardening of Hearts," *Midwest Studies in Philosophy* XXI, ed. Peter French et al. (Notre Dame, IN: Notre Dame University Press, 1997), 202–24.

75. The past two sentences are actually borrowed from an eminent *critic* of the argument that indeterminism implies no free will: see Robert Kane, "Responsibility, Luck and Chance: Reflections on Free Will and Indeterminism," *The Journal of Philosophy* 96 (May 1999): 217–40. Kane's *The Significance of Free Will* (New York: Oxford University Press, 1996), as I mention again shortly, tries to address this problem and construct a model of free will based on indeterminism. However, Kane does not believe in "agent causation," a special sort of causation.

76. In addition, even if there are undetermined neural events that result in actions that can plausibly be called free, we need to know how often these undetermined events occur and to how many people. Suppose scientists discovered that on October 18, 1924, a farmer in Fayetteville, Arkanas, arose early and deliberated whether to feed his chickens or go back to sleep. In an undetermined fashion, scientists conclude, he decided to go back to sleep, causing the chickens the pain of hunger. Now, if that were the only case of undetermined behavior in the history of the world, would we say "Yippee! There's free will! *Mazzal tov*!"? Hardly. Moreover, even if we knew there were a large number of undetermined actions in real life, we want to know *what precise choices those involved.* Suppose that due entirely to physical events in his nervous system, a person decides to beat his neighbor to death. However, suppose further that the person's choice whether to use a rock or a baseball bat to carry out the beating was undetermined, and that the perpetrator entertained arguments on both sides before deciding on the bat. In terms of the question, "Is there free will?," we would hardly find this as "consoling" a thought as the discovery that his choice to beat his neighbor was not determined. If your most significant choices in the world were "strawberry or vanilla?" or "should I use a fork or a spoon," or "a machine gun or a bomb?" this would not be, in Daniel Dennett's phrase, "a variety of free will worth wanting." (The term is from Dennett, *Elbow Room: The Varieties of Free Will Worth Wanting* [Cambridge, MA: MIT Press, 1984].) (Some of this note reflects observations I heard made long ago by the late Sidney Morgenbesser.)

77. There is a world of literature on utilitarian vs. retributive theories of punishment. For a brief guide to the issues, see Joel Feinberg, "The Classic Debate," in *Philosophy of Law*, ed. Joel Feinberg and Jules Coleman, 7th ed. (Belmont, CA: Wadsworth/Thompson Learning, 2004), 799–804.

78. Maimonides, for one, includes elements of a utilitarian theory regarding quantity of punishment. See, for example, Guide 3:41 (frequent crimes are punished more severely than rare ones, crimes that are easily incited are punished more severely,

crimes that are done in secret warrant heavier punishments). For a lengthy listing of harsh punishments meted out by rabbinic authorities on utilitarian grounds (e.g., killing of *moserim*), see Aaron Schreiber, *Jewish Law and Decisionmaking* (Philadelphia: Temple University Press, 1979), 375–424. Familiar examples include *mishpat ha-melekh* (the king's law) and *makkin ve-oneshin she-lo min ha-din* (a court may inflict harsher measures than the law prescribes).

79. Examples include Strawson, "Freedom and Resentment"; Derek Pereboom, *Living without Free Will* (Cambridge, UK and New York: Cambridge University Press, 2001), chaps. 5–7.

80. This point is a standard one. See for example Pereboom, *Living without Free Will*, 139–140; Eugene Schlossberger, *Moral Responsibility and Persons* (Philadelphia, 1992), 4–7, 79, 101, 112, 117–18.

81. Now, you might argue that what these examples show is that attributions of moral responsibility are only one type of evaluation; maybe in blaming the perpetrators of atrocities we are evaluating their actions and not *them*, the *hata'im* and not the *hote'im* (in Beruriah's famous phrase—the sins, not the sinners). There is something importantly correct in this response—we must be wary of homogenizing assessments of responsibility with assessments of reprehensibility. Likewise, being *admired* for one's goodness is different from being *credited with* one's goodness. We do want to say that some people are not to be blamed for their heinous acts because they are victims of horrible upbringings. My point, however, is not that moral praise and blame definitely are consistent with determinism; it is rather that whether they are is a philosophical question, not a scientific one. On the distinction between moral and nonmoral evaluation, see Michael J. Zimmerman, "Responsibility Regarding the Unthinkable," in *Midwest Studies in Philosophy XXI: Moral Concepts*, ed. Peter A. French, Theodore Uehling, Jr., and Howard K. Wettstein (Notre Dame, IN: Notre Dame University Press, 1996), 213–14; cf. Robert Merrihew Adams, "Involuntary Sins," *Philosophical Review* 94 (1985): 3–31.

82. In his acclaimed *The Metaphysics of Free Will* (Oxford: Blackwell, 1994), John Martin Fischer treats theological determinism and scientific determinism as parallel arguments.

83. Classical compatibilism, the version of compatibilism developed by David Hume, John Stuart Mill and other British philosophers, actually did not reject the Forking Paths Principle so much as reinterpret it. The British compatibilists insisted that a person acts freely provided that (a) the person does as he desires, because he desires to; (b) if the person were to desire not to do the action, the person would not do it. The problem with this view is that the only person who isn't free is someone who is chained. Kleptomaniacs, drug addicts, phobics, and the like all turn out to be free using this constraint—an incorrect result. Indeed, not only humans but animals could satisfy the conditions of the British compatibilists. For this reason philosophers today opt either for a wholesale rejection of PAP or a different reading than the compatibilists gave it.

84. I am assuming that determinism cancels alternative possibilities. It is surprisingly difficult to show this, as is clear from Fischer, *Metaphysics of Free Will*, part one.

But, in the end, Fischer thinks the entailment holds: determinism rules out alternative possibilities. (Alternatively, one might think that determinism excludes free will, moral responsibility, and reactive attitudes because if our actions are determined, we do not "own" them, they are not ours. That, too, is a controversial philosophical thesis. It blithely implies that we know what ownership means and that determinism deprives us of it, this in the face of immense problems in understanding such concepts. I will stick here to the Forked Paths Principle.)

85. See Locke, *Essay Concerning Human Understanding*, Book 2, chap. 21. Locke's view is that the man who chooses to remain is more fortunate than someone who would want to leave but cannot—but he is not more free.

86. The example is given by Peter van Inwagen, "Ability and Responsibility," *Philosophical Review* 87 (April 1978), 201–24.

87. Harry Frankfurt, "Alternate Possibilities and Moral Responsibility," *The Journal of Philosophy* 66 (1969): 829–39. For excellent collections of essays on the Frankfurt cases and other issues pertaining to moral responsibility, see John Martin Fischer (ed.), *Moral Responsibility* (Ithaca, NY: Cornell University Press, 1986); Fischer and Mark Ravizza (eds.), *Perspectives on Moral Responsibility* (Ithaca, NY: Cornell University Press, 1993; Widerker and Mckenna, *Moral Responsibility and Alternative Possibilities.*

88. Frankfurt does not mention a contrast with Locke's view, but I believe the contrast is between applying the Forked Paths Principle to actions (Locke) and applying it to decisions (Frankfurt).

89. John Martin Fischer and Mark Ravizza, along with Susan Wolf, are key players in evaluating these possibilities. See especially Fischer and Ravizza, *Responsibility and Control* (Cambridge, UK and New York: Cambridge University Press, 1998).

90. David Widerker has shown that Frankfurt's examples cannot be effective in the case of decisions because the conditions those examples presuppose about the scenario are ones that a libertarian defender of PAP would not accept. See, among others of Widerker's papers on the subject, "Libertarianism and Frankfurt's Attack on the Principle of Alternative Possibilities," *Philosophical Review* 104 (1995): 247–61.

91. Hume, *Enquiry Concerning Human Understanding*, VIII.

92. Peli, *Soloveitchik on Repentance*, 172.

93. The phrase is Frankfurt's. See his "Rationality and the Unthinkable," in *The Importance of What We Care About* (Cambridge, UK: Cambridge University Press, 1988), 177–190. Cf. Zimmerman, "Responsibility Regarding the Unthinkable," *Midwest Studies in Philosophy* XX, ed. Peter French et al.

94. This paragraph is an amalgam of cases used in the literature to illustrate the point I am making. See: Daniel Dennett, *Elbow Room: The Varieties of Free Will Worth Wanting* (Cambridge, MA: MIT Press, 1984), 131–52; Peter van Inwagen, "When Is the Will Free?" *Philosophical Perspectives* 3 (1989), ed. James Tomberlin (Tascadero, CA, 1989), 399–422; Susan Wolf, *Freedom within Reason* (New York: Oxford University Press, 1990). Mark Twain once said: "I am a greater person than George Washington. George Washington could not tell a lie. I can—but I won't."

95. In presenting these examples to people, I have found that they keep responding, "no, you *could do* the other thing, you just *won't*." I do not agree with this. We can

picture the person doing the act, but that's not the same as saying the person is *able* to. If we take psychic determinism seriously, you are not able to do things that go against your settled character. You could over time be exposed to bad people, or find yourself in a desperate situation, or develop some brain disorder that makes you go for the $5 bribe and torture the child. But *baasher hu sham*, as things are right at this moment, the moment you decide and/or are acting, you can neither decide otherwise nor do otherwise. Again, I have heard people say: "well, maybe I wouldn't torture a child for $5, but I would be able to do it for a higher price, $5000." This, I suggest, is irrelevant to what you can do in *these* circumstances of the $5 bribe. I believe that the temptation to think the $5000 bribe is relevant comes from a confusion between being able to do X and having the ability to do X. A concert pianist with her hands in a cast has the ability to play the piano, but she is not able (then) to play the piano; Michael Jordan *has the ability* to hit a lay-up even if his hands are tied, but he is not then *able* to hit the basket. You have the ability to torture the child, but are not able, because certain enabling circumstances (like a *humongous* bribe) are absent. I elaborate on this and the other distinctions here in "Compatibilism, Values and 'Could Have Done Otherwise,'" "Irresistible Goodness," and "The Metaphysics of Control."

96. See Wolf, *Freedom within Reason, passim.*

97. See again Wolf's attack on the idea that freedom requires the ability to act insanely.

98. Kane, "The Dual Regress of Free Will and the Role of Alternative Possibilities," in *Philosophical Perspectives 14: Action and Freedom*, ed. James E. Tomberlin (Boston and Oxford, UK: Blackwell, 2000).

99. Peli, *Soloveitchik on Repentance*, 173.

100. Solomon Schimmel, "Free Will, Guilt and Self-Control in Rabbinic Judaism and Contemporary Psychology," *Judaism* (1977): 418–29, at p. 425.

101. R. Wolbe, *Alei Shur* 1, p. 156, as quoted by Dr. Baruch Sorotzkin, "*Bechira*: How Free Is Free Will?," *Jewish Observer* (April 1996): 21. For the intriguing fate of the Sorotzkin article, see my "conclusion" section. For more general criticisms of the Aristotelian position that we choose our characters see Joshua Dressler, *Understanding Criminal Law* (New York, 1987), 695–96; Michael Moore, "Choice, Character and Excuse," in *Crime, Culpability and Remedy*, ed. Ellen Frankel Paul et al. (New York: Oxford University Press, 1990), 45.

102. Rambam in *Hilkhot Yesodei ha-Torah* 5:4 states that if a person can escape a hostile ruler and does not, and as a result is coerced into worshipping idols, he is not an *anus* (i.e., he is not acting under duress) but instead an *oved avodah zarah be-mezid* (i.e., he is deliberately committing idolatry). I believe that this ruling is best understood as a ruling about the person's responsibility, not his freedom, and can be understood in terms of an analysis of *ones* I hope to develop on another occasion. For further scrutiny of such situations, see Teshuvot Tashbetz #63. (I thank Rabbi Dov Linzer for the Tashbetz reference.)

103. "Irresistible Goodness and Alternative Possibilities," esp. pp. 63–73.

104. Philosophers who adopt (a) often rationalize (a) in terms of (b), invoking the Platonic-Kantian idea that the true self is the rational self. While I question the thesis, I will work with a version of (a) that accepts this rationale.

105. In particular, he reads Rambam as endorsing the type of "sane deep self view" expounded by contemporary philosopher Susan Wolf. Charles Manekin sees an affinity between Ralbag's and Wolf's view. See Stern, "Maimonides' Conceptions of Freedom"; Sokol, "Maimonides on Freedom of the Will and Moral Responsibility"; and Manekin, "Freedom within Reason?" (The latter title is a play on Wolf's). The use of these examples is inexact since I do not draw their distinction between theoretical and practical intellect, interested as I am only in the general flavor of the theories.

106. See also Stern, "Maimonides' Conceptions of Freedom," 232, note 25.

107. I won't here enter into the question of whether all Jews, according to Rambam, have the desire to obey the law as their "deepest self." On one reading of Rambam's approach to the hardening of Pharaoh's heart (especially the version in *Shemonah Perakim* 8), Rambam holds that there too Pharaoh's deep self is restored, in which case the key to Pharaoh's freedom is not God's restoring the choice that reason establishes but merely God's allowing the true self to have its way. In truth I think Rambam has a different view of the Pharaoh case. See also Nechama Leibowitz's chapter on this topic in her *Studies in Exodus* (Heb.) (3rd ed., Jerusalem: World Zionist Organization, 1973), 110–17.

108. I discuss these in much more detail in the articles mentioned in note 74; especially relevant is "Compatibilism, Values and 'Could Have Done Otherwise.'" There I propose, *inter alia*, a further subdivision of "reason views" into "objectivist" and "subjectivist" ones.

109. Fischer, *The Metaphysics of Free Will*, 206.

110. Schimmel, "Free Will, Guilt and Self-Control," 427.

111. I once read an interview with Skinner in which, stepping out of his professional persona, he spoke freely of the inner motivations, desires, and beliefs of Americans. I was reminded of Bertrand Russell's quip in reviewing a book by Skinner. Russell said that a psychologist denying inner mental life is like a tailor running through the streets yelling, "There are no pants! There are no pants!" Worthy of being cited, too, is the question that Sidney Morgenbesser posed to Skinner: "Are you saying that we should not anthropomorphize *people*?"

112. See Gary Watson's strangely neglected paper, "Skepticism about Weakness of Will," *The Philosophical Review* 86 (1977): 319–39.

113. See Wolf, *Freedom within Reason*; Allen Wood, "Kant's Compatibilism," in *Self and Nature in Kant's Philosophy* (Ithaca, NY: Cornell University Press, 1984), 73–101.

114. Robert Nozick, *Philosophical Explanations* (Cambridge, MA: Harvard University Press, 1981), 293.

115. Sorotzkin, "Bechira."

116. I thank Yitzhak Berger, Yitzchak Blau, Michael Broyde, Shalom Carmy, Dov Linzer, Charles Manekin, Meira Mintz, and Shifra Schapiro for valuable communications relevant to this paper.

4

Use It or Lose It: On the Moral Imagination of Free Will

Shalom Carmy

Dr Johnson shunned to-night any discussion of the perplexed question of fate and free will, which I attempted to agitate: "Sir (said he), we *know* our will is free, and *there's* an end on't."

—James Boswell[1]

For it is not only the two antagonists who are there, but he himself as well, and who really knows his intentions? His dream, though, is that some time in an unguarded moment—and this would require a night darker than any night has ever been yet—he will jump out of the fighting line and be promoted, on account of his experience in fighting, to the position of umpire over his antagonists in their fight with each other.

—Franz Kafka[2]

For the [eschatological] removal of choice that is an intensification of the power of choice is nothing but a manifestation of the singularity of choice. . . . But because, in the final analysis, choice is suspended, the heresy most suited to the end of days is denial of man's singular capacity for choice, from which flows all the contempt for the human image in our age.

—R. Yitzhak Hutner[3]

A third cause for self-deception in the matter of freedom is that many people experience it very seldom themselves. We should not forget that free will is not a cheap guest; it stays in very expensive hotels [*akhsanyes*], in people who are located on a high moral stage.

—R. Reuven Agushewitz[4]

FISHMONGERS KNOW THE HADDOCK from the hake by the color of his eye. There is no corresponding standard by which experts on the human condition can infallibly distinguish a freely willed action from one that is determined. We can all think of cases when an individual seems to act freely, though in fact his or her behavior is being manipulated. The determinist claims that all behavior is caused by previous states of affairs conjoined with whatever psychological laws can be invoked. The causes may be external, identifiable with some factor in the environment, or they may be desires or traits of character internal to the agent. The individual may identify himself with such desires or dispositions, but these too, according to the determinist, are caused by what has come before, insofar as the person's desires and character are not his own creation, but derive from some combination of independent genetic and environmental factors. It is true that any particular action may be brought about this way, and it is not impossible that all actions are determined in this manner.

Determinists are generally divided into two camps. Soft determinists believe that it is meaningful to speak of free actions when the act is not externally caused. That one's actions are ultimately caused by factors beyond one's control does not, in their opinion, diminish the individual's responsibility for them. (Such a position must distinguish between external and internal causation: is a possible chemical imbalance an external interference with one's free choice, or is it part and parcel of who the person is?) Soft determinists are called compatibilists because, on their view, freedom of choice (as they conceive of it) is compatible with determinism. Hard determinists do not adopt the external/internal distinction. The hard determinist must therefore reject or substantially revise ordinary intuitions about human freedom and its connection to conventional notions of responsibility. In theory, all that is required to refute determinism is to establish one exception to the network of mechanical causation. By the same logic, it would be sufficient to make belief in free will tenable if such an exception were logically possible, without offering evidence that such exceptions actually exist.

This defense of freedom, of course, would not be an argument for belief in free will. The philosopher, or the human being, who chooses to believe in free will, cannot be satisfied with the minimal assertion that it is possible. Further intellectual commitments, explicit or implicit, must be added to arrive at a reasonable justification of the belief. We shall discuss some of the reasons supporting belief in free will later on. The strongest kind of argument would claim that free will is true, or probably true. A weaker claim would merely assert that an individual already committed to the belief has insufficient reason to forsake it. If, in fact, an incompatibilist treatment of free will were the only account of the subject compatible with Judaism, then whatever grounds a philosopher, or a human being, possesses for belief in Judaism would also support belief in free will. In that case, we could heartily endorse Samuel Johnson's blunt avowal of free will, and be done with it.

Unfortunately, however, we cannot dismiss the matter so easily. For many, determinism is an attractive doctrine, for a combination of philosophical and existential reasons. Perhaps the most important obstacle to belief in free will is our difficulty in imagining what freedom is really about. The threat, from a purely philosophical perspective, is that free agency may be impeached as an incoherent concept. An incoherent concept, one containing a contradiction, is an impossible one. If this is the case, one may still counter as follows: free will is only incoherent from the viewpoint of logic (that is, human logic); nonetheless, we must affirm it, despite the absurdity, because it is one of those truths of heaven and earth that do not make sense to our limited minds. Apart from the intellectual discomfort this move engenders, it also presents a dismal challenge to the imagination: how can one go about acting incoherently; how does one conceive of oneself as a metaphysical impossibility? How can you exercise freedom if the whole idea of freedom makes no sense to you? I will argue below that inadequate psychological models of free will endanger the existence of freedom as a dimension of the human person.

Determinism, we just said, is refuted by even one free act, and the necessity of determinism is refuted by the mere possibility of such an act. But the refutation of necessary determinism is not tantamount to the demonstration of freedom, or even the possibility of freedom, in any significant sense. For free will to be a meaningful aspect of human life, it cannot be a rare, miraculous occurrence. Freedom must be exercised frequently enough to be part of the human landscape. By the same token, it cannot be limited to trivial decisions; at least some momentous decisions must come under the category of free will. Many would deem it important that these actions not only be free actions, but also that they must be perceived, or recognized to be free actions. While I disagree with the last claim, it expresses a deep-seated intuition that deserves, and will receive attention.

Some maintain that our age confronts radically new challenges to the doctrine of free will. The argument is that contemporary science, in the form of genetics and biochemistry, has stamped out the possibility of free will by demonstrating the truth of determinism. I believe that this claim is false; science has not established a rigid, indubitable network of cause and effect, though, as a non-scientist, I must stand ready to be corrected, and cannot judge whether the ambitious deterministic program is likely to be realized in the future. The present crisis regarding free will and responsibility did not appear out of nowhere in the past couple of decades. To be sure, by asserting more clearly and insistently the claims of genetic and biochemical factors, the new science highlights the limits of freedom. Insofar as free will proponents believe in freedom as a meaningful factor in human existence, how much free will is a significant question for the existing individual, if not for the analytic metaphysician. However, the major intellectual, social and personal issues have been around for a long time, the crisis has been evident for over half a century, and much of the urgency of our debate is a function of prolonged neglect.

Our modest effort to promote that long-overdue discussion will begin with an examination of the reasons and impulses that make libertarianism and determinism attractive. We will move on to consider the paradoxical nature of free will that leads so many philosophers to regard it as a hopelessly muddled idea. Throughout, we will concentrate on those arguments and ideas that express and mold our thinking and our living. The student of free will and determinism confronts a huge literature dealing with the subject from a strictly logical vantage point. In everyday life, however, I suspect that the crucial considerations are those that arise from our deepest experience of reality. The logic comes afterward, and though a purely logical chain of reasoning may lead us to abandon a position to which we were initially attracted, it is always possible to cobble together some sort of account that yields and shields the convictions we really care about. Therefore, we will have more to say about how Don Newcombe, the Brooklyn Dodger pitcher, conducts himself than about the celebrated problem about foreknowledge and fatalism, invented by the physicist William Newcomb, that has launched so many philosophical fishing expeditions of the past generation. If my reasons for employing this strategy seem obscure at this stage, by the end of the essay I hope it will all become clear.

The introductory sections set the stage for a study of two profound accounts of free will created by twentieth-century Jewish thought, those of R. Eliyahu Dessler and R. Abraham Isaac Kook. Their teachings recapitulate, respectively, the two major approaches that emerge from our observations on the general philosophical culture. Finally, at the risk of trespassing on the territory of the psychologists and educators, we will not avoid looking at some of the practical implications of our analysis.

Why Free Will?

Introspection

Catch an agent in the midst of ordinary action. His or her immediate feeling is that the action is free and not determined entirely by prior states of affairs. It is often suggested, therefore, that the phenomenology of action derived from introspection, leads to a perception of oneself, and one's actions, as free.

This subjective sense is not incorrigible. No matter how strong the conviction of freedom, the agent may come to believe that he was deluded and that his action is not free. This would happen, for example, if the agent sees videotape of himself being given hypnotic suggestion, or of a scientist stimulating the appropriate area of his brain. (Not that such evidence would be a final refutation of freedom: a person who, after effective hypnotic suggestion, goes ahead and does exactly what he would have done without the hypnosis, is still acting freely!) Nevertheless, in the absence of evidence to the contrary, we tend to take this subjective experience as probative. It strongly affects the outlook of the ordinary person, and even, *malgré lui*, that of the skeptic.

Current philosophy and psychology are not enamored of introspection as a source of knowledge. Right now I am less interested in assessing its validity in the face of deterministic counter-argument than in uncovering one of its presuppositions and implications. The subjective conviction that our actions are free can arise only on the basis of repeated experiences of successful, meaningful volition. When that happens we may, rightly or wrongly, experience our success as the manifestation of our own free authorship. A weak or oppressed person, especially someone who has experienced helplessness from childhood on, someone for whom such easy agency is an irregular or rare feature of life, would be unlikely to arrive at a firm conviction of free will. That is why Nietzsche once wrote that free will is the invention of the strong.[5] At a practical level, the individual who genuinely adopts free will, not only as nominal dogma, but also as a psychological reality, must have had opportunities for the successful exercise of the will. Otherwise, the feeling of free will may remain undeveloped or atrophy.

Freedom as a Moral/Juridical Postulate

Proponents of free will commonly claim that the fundamental moral practices of praise and blame, and the institutions of reward and punishment, are meaningless unless the individual is the free author of her actions. Under determinism, what the individual does cannot be credited to her moral worth or liability. Rambam is an influential representative of the moral and religious authorities who argued for free will on these grounds.[6] The primary motive for soft determinism is the desire to salvage these practices. Kant, who held that free will, as an exception to the principle of universal causation, can be neither proved nor disproved, persisted in upholding it as a Postulate of Practical Reason, necessary for the idea of moral activity.

Forty years ago, the English philosopher Peter Strawson outlined the basis for a version of the argument from praise and blame derived from social judgments or intuitions. Where the typical introspective claim is that examining my own actions convinces me of my freedom, Strawson suggested that it is contemplation of other people's behavior that compels us to acknowledge their freedom. When another person harms me, I resent it, as I do not resent the harm brought about by an animal or an inanimate object. This implies that I attribute responsibility, and therefore freedom, to the other individual. Strawson posits a reactive expression recognizing free agency, but the awareness is not first person, as when one experiences guilt or pride, but second or third person. If his proposal is true to our experience, and not merely inference from a moral attitude, it may appeal to the kind of modern person who does not vividly experience his own free agency (or responsibility), but retains a robust sense of resentment.[7]

Even at this early stage of our discussion, we should note that the intimate connection between free authorship of one's acts and moral judgment of them is open to question independent of the metaphysical question of free will. Strawson, for one, doesn't care if we are free, so long as the experience he describes governs moral judgment, and will continue to do so. Free will, let us recall, means that a significant part of what we do is a product of free moral choice. Yet even the strongest champions of free will concede that freedom does not operate in a vacuum. Often we intend an action, but the external world does not cooperate. Moreover, the difficult, sometimes impossible, situations we find ourselves in, are not entirely the consequence of our own choices. Nor can we take credit for the formation of our characters, which are largely molded by physical and biological constraints and by the intervention of others. Human beings may create their destinies, but surely not *ex nihilo*. Notwithstanding, common sense morality and our legal institutions continue to hold agents responsible, even while acknowledging the role of "moral luck," as philosophers dub these undeserved factors in the moral landscape.[8] Hence the argument from moral desert, deployed as a weapon against determinism, poses problems for the libertarian as well.

Freedom and Creativity

We frequently characterize our choices and their results, as well as those of other persons, as creative, as realizing something that was previously only a potentiality. This lustrous apprehension of the freshness of the universe, and the creativity of our endeavor, is diminished when it is regarded as nothing but the discovery of an inevitable outcome that was already on the books, so to speak, from the beginning of time, or as the final outcome of causal chains independent of any personal intervention. From a purely logical point of view, there is no contradiction between a deterministic world and our enjoyment of its unpredictability and freshness, as we, in our unscientific ignorance, perceive it.

Soft determinism, which takes freedom seriously in the here and now while cutting off its metaphysical roots, does a more plausible job of accounting for our intuitions about creativity than it does in conserving our traditional moral judgments: for our aesthetic delight does not depend on the negation of determinism, but only (in the worst case) on our ignorance of determinism. Zest for the future may indeed, if we are fortunate, survive conversion to determinism; or he who gains knowledge may end up gaining dullness, in which event, the truth will make us less happy. William James, perhaps the most colorful exponent of this philosophy, contrasts the pluralism committed to an open future with the determinist's "block universe," concluding that our choice of free will or determinism does not depend on the weight of scientific evidence, but is rather a matter of temperament.

Two additional remarks, for the time being, before we move on: First, the freedom that makes James's pluralistic universe such an exciting place is much broader than the human free will that is our ostensible topic. James prefers a contingent, rambunctious world that is not rigidly predictable. Such freshness goes well with the idea that human beings are free to make choices that cannot be predicted on the basis of prior states of affairs, but it also goes well with other elements of freedom and unpredictability. In particular, the pluralistic universe is eminently compatible with a consciousness of divine freedom manifested through His personal intervention in our lives.

Second, the kind of freedom connected to the creativity theme seems to be very different from that derivable from standard introspection or from moral-juridical argumentation. The immediate consciousness of freedom fits most easily the situation in which the agent has a specific goal in mind, wills that goal and the means to that goal, and is rewarded by the transparent success of his or her effort. Likewise, the claim that responsibility presupposes freedom implies that the free action proceeds directly from the executive will, and can be identified with the result of the volitional process. By contrast, in speaking of freedom as an attribute of creativity, one implies a surplus of creativity that goes beyond what anyone (including the agent) could have predicted. There is a mysterious quality to such action; the individual, no less than his audience, is surprised by freedom. The tension between these two ways of thinking about free will comes up again later in our discussion.

Why Not Free Will?

The Scientific Imperative

For determinism the prime, often the only ground adduced, is science. If this means that science has produced an exhaustive deterministic system, the claim is false. Based on what science has already achieved, one may believe, hope or dread that such a system is in the offing. But human psychology is blessed with a peculiar dialectical nature, that is to say, our thoughts and beliefs about psychology affect the truth of those beliefs, and this makes it hard to imagine what such a system would look like.

A more cautious formulation would maintain that scientific investigation requires belief in determinism. Science succeeds only because the scientific community is committed to the search for an all-embracing network of explanation. In purely pragmatic terms, the psychiatrist and philosopher George Ainslie identifies his proposals as deterministic, conceding that the causes may not be identified: "Some are probably too hidden or complex or internally fed back ever to be useful for practical prediction; but again, explanations of behavior should never depend on some uncaused or otherwise imponderable factor."[9] One can go beyond this and turn universal causality into a necessary philosophical principle: Accounting for part of the recorded data by summoning the intervention of occult agents, themselves uncaused, existing beyond the range of scientific inquiry, precludes explanation, and renders impossible the development of scientific theory. Make one exception, and you are committed to an endless number of them.[10]

Against this it can be objected that whether belief in free will impedes scientific progress or not is an empirical question. One may adopt a program of universal causality as a methodological principle without believing in its total applicability.[11] It is hard to improve on William James's remarks about science-inspired determinism and the mentality that fosters it (allowing that theists also ascribe freedom to God):

> [M]any persons talk as if the minutest dose of disconnectedness of one part with another, the smallest modicum of independence, the faintest tremor of ambiguity about the future, for example, would ruin everything, and turn this goodly universe into a sort of insane sand-heap or nulliverse, no universe at all. Since future human volitions are as a matter of fact the only ambiguous things we are tempted to believe in, let us stop for a moment to make ourselves sure whether their independent and accidental character need be fraught with such direful conclusions to the universe as these.[12]

But in the absence of any serious intellectual alternative, ad hoc working hypotheses (*she-lo lishmah*) become unchallenged dogmas (*lishmah*).

A sharper objection, derived from quantum mechanics, has flourished for generations, albeit in an overstated popular form. Apologists for free will infer from Heisenberg's theory about indeterminacy of the motion of subatomic particles that determinism is false, and that libertarianism is therefore true. This argument succeeds in putting flesh on James's insight that a little indeterminacy does not topple the scientific enterprise. Quantum uncertainty should cool the deterministic ardor of a person whose imagination has been fired by Laplace's vision of a world like a set of tightly packed billiard balls, each of which moves predictably in conformity with the mechanical momentum imparted to it. It does not, however, provide the scope and range required for meaningful freedom; as we shall see, it offers little to a coherent conception of freedom.[13]

The Morality of Determinism

Contrary to the common wisdom that morality unambiguously favors free will, some of the strongest motives for belief in determinism arise from the moral orientations common among modern people. This surprising thesis can be explained, in part, by the fact that modern society often embraces, as the height of moral sensitivity, attitudes, and judgments that traditional morality cannot but regard as amoral or immoral. But some of the motives we shall consider reveal genuine ambiguities in our conception of freedom and in our moral ideas.

Please note that the positions we shall explore in this section are not always deterministic in the metaphysical sense. People who subscribe to these views may be libertarians at the abstract level. But their insights and intuitions about themselves, about others, or about morality, lead them to restrict the domain of freedom, limiting the persons who are judged free, or the range of actions and feelings that are deemed free. As we have seen, the upholder of free will is overall interested in expanding the domain of freedom: a freedom that is too constricted may actually be meaningless. Moreover, the tendency to narrow the realm of free choice locally can precipitate a way of thinking that spills over naturally into a global determinism. Note also that the popular positions we are looking at are often held without philosophical reflection, so that those who propose them do not always realize where they are headed. Where there is self-consciousness about the question of freedom, it can be allayed by wishful appeal to some notion of soft determinism that hopefully avoids sacrificing the advantages of freedom.

1. The Management of Anxiety

Kierkegaard and Sartre have taught us that freedom engenders anxiety because the free individual must confront an indeterminate future for which he is responsible, whether directly, through his own actions, or indirectly, whenever he is called upon to respond to the free activity of others. Many people, contemporary people not least, dislike and fear anxiety. Partial or total escape from freedom takes care of anxiety, and thus contributes to what many people cherish as the ideal of mental health. Spinoza is a spectacular example of a major philosopher who constructed his entire system on the premise of the "block universe," a world comprehensively defined by the philosopher's vision of rational determinism, totally excluding freedom and deprecating as irrational the anxiety felt by the unenlightened about the future. Among less rationalistic folk, the same results can be obtained by assiduously consulting one's horoscope, confirming the aphorism attributed to Harry Wolfson, that superstition is the heresy of the ignorant, while heresy is the superstition of the intellectuals.

In addition to eliminating anxiety, a thorough determinism about oneself also alleviates guilt, an even greater threat to the "feeling good about me" that is the *summum bonum* of the therapeutic ethic.[14] This point is too obvious to require further elaboration.

Determinism also makes it easier for us to deal with people we abhor. To recognize obnoxious people as free agents entails the possibility that they may change. This forces us to relate to them as fellow human beings, with whom we share a moral universe, and for whose moral destiny we may have to take responsibility. (This would likewise be true for the soft determinist, because his actions can become the cause of change in the odious other.) If hateful people or groups are powerless over their characters, there is nothing we can do to change them. Such perceptions both ease our anxiety about how to treat these people, and also assuage the guilt we may experience in connection with our hatred and contempt. This kind of localized determinism finds appropriate expression in the prominent use of zoological imagery, so noticeable in demotic journalism and crowd-stirring political speeches, when depicting the type of people we are invited to loathe. Two-legged animals (except that, as orators never tire of adding with an air of apologetic triumph, as if they had invented some stunning and original witticism, the very comparison is an insult to the four-legged kind!) and the less popular insect species do not rate the treatment accorded to human beings.

It is good for our equanimity and our all-important self-esteem to believe that those who diverge from our behavior and ideas are not exercising their capacity for free choice, but are instead crazy and therefore irresponsible for their decisions and desires. The closer the deviant is to us, the more incomprehensible his or her deviance is to us, or the more we are fascinated and repelled by the possibility of becoming like him, the more we are tempted to deny the freedom of that person. Communities offering lip service to the doctrine of free will nevertheless employ such techniques of dismissal and denial more than occasionally in order to avoid coming to grips with unwelcome truths about members, or former members. The desperate desire to fix deterministically future affairs that are, in truth, all too vulnerable to the working out of free destinies, is similarly displayed when we engage in superstitiously nitpicking researches into the nooks and crannies of a prospective spouse's family antecedents. The anxiety to avoid a disastrous marriage is understandable; the irrelevant questions and obsessive ruminations are ridiculous and sad.

The motives and examples of the anxious determinist mentioned so far ought to have made a deterministic orientation morally discreditable. It is only fair to examine the other side of the anxiety management problem. Consider the following: Your brother-in-law, or another adult loved and trusted by your children, turns out to be a child molester. The doctrine of free will encourages us (let's assume, for purposes of the argument, correctly) that this individual could have acted differently, that he could have withstood temptation. What this entails, and what proponents of free will have no difficulty accepting, is that the child molester is, in principle, our brother under the skin. To be sure, we do not feel any desire to make sexual use of children, and having reached a certain age, we may be reasonably confident that we will not be tempted in this way. But we experience other temptations: One might be sexually attracted by members of his own sex, or distracted by a roving eye for members of the other, or inordinately fond of liquid consolation, idle gossip, or spiteful remarks. If the child molester could have acted otherwise, his or her situation is not qualitatively different from ours. We may be fortunate in that our temptations, we hope, do not involve very high stakes. Our vices are ordinary: we stumble and rise up again. The pedophile, the homosexual, or the adulterer, are not so fortunate. But for the grace of God, however, we would be in their place. But for the grace of God, in other words, we should look in the mirror and see a potential monster.

So reasons the libertarian, eager to interpret as much of life as possible in categories of freedom and responsibility. Now explain all this to your children! No doubt our children, the children of the sheltered upper middle classes, are overly protected from the realities of this world, which is why so many of us are devoutly convinced that our society must mobilize armies of therapists for the children (and the more "sensitive" adults) whenever reality reveals its harsher aspects. No doubt it would be better if our children understood more about the world than we permit them to. But can one baldly teach them that the person they trusted is not "sick" but is rather a moral agent who has chosen to yield to temptation, just as any other human being is liable to fall? Is such instruction likely to provide the child with a more realistic orientation to the world of others, and to his own task of self-examination? Or is it likely instead to "destabilize" the child's moral universe, to uproot the very core around which the child's moral personality ought to crystallize?[15]

So you tell your children that Uncle Ben is "sick," that he is not a participant in the great moral drama in which people make free choices and have the privilege and onus of repenting those that are evil. Is this a "white lie"? If so, whether to tell it or to present the hardheaded free will outlook is an ethical dilemma, not a philosophical provocation. As we continue our analysis, however, we may conclude that the truth about this case is elusive, and that we do not face a simple choice between truth and falsehood, but rather the difficult task of communicating, as best as we can, a complex and subtle moral-psychological reality.

A different way in which the morally benign aspects of a deterministic orientation come into play can be illustrated by the following anecdote: Having ruined his stellar pitching career with drink, Don Newcombe was reborn as an effective crusader for the rehabilitation of addicted athletes. A fun-loving soul once surreptitiously fortified the punch bowl from which Newcombe was drinking with a liberal dose of the hard stuff. The disease theory of alcoholism, to which the ex-fireballer subscribed, holds that the alcoholic is powerless against his addiction, which is why the only remedy is total abstinence. Imbibing any amount of the spiked beverage would trigger a binge as surely as exposure to pollen would bring on the symptoms of hay fever in victims of that malady. The prankster was presumably looking forward to a merry entertainment. Newcombe, experiencing the warm familiar buzz, instead of drinking himself under the table, proceeded to identify the perpetrator, and with all the sober dignity of his imposing physique, made him sincerely regretful about the whole affair. Although the joke fell flat, the incident is rich in irony. Does not Newcombe's exemplary behavior demonstrate his power over alcohol and thus contradict his own theory?[16]

In the context of the real world, it is not at all clear that Newcombe is refuted. Let us re-describe the situation in the following manner: Newcombe can take a drink and resist the urge to continue drinking. Therefore, he is not totally powerless over the chemical. He does not, however, enjoy a free and easy mastery of liquor. To that extent his freedom is severely limited. A nuanced psychological profile of his situation respecting alcohol would require a thick, detailed account of his past and prospective interaction with the substance, and would probably be incomplete without an exhaustive depiction of other facets of his existence. Under the pressures of the real world such microscopic analysis may be useless. In order to conduct his life reasonably, it is necessary for the agent to portray himself to himself with broader brush strokes. Given the choice between "I can imbibe alcohol within limits like most people" and "I am powerless in relation to alcohol," both of which are inadequate descriptions of the situation, Newcombe chooses the latter formulation as both closer to the truth and more prudent. While this discussion is psychological, and independent of the classic metaphysical free will problem, it highlights the need for complexity in making ascriptions of freedom or lack of freedom.

2. *The Cult of Sentiment*

It goes without saying that the most convenient effect of deterministic belief is to eliminate traditional ideas of moral responsibility. This is an obvious benefit for the individual determinist. It is also useful in exculpating people with whom one identifies, or with whom one has shared interests. The agenda of blamelessness is also furthered systematically when it is mobilized in a general way, as in Madame de Stael's famous bon mot, *tout comprendre est tout pardonner* (to understand all is to forgive all). It is in the interest of those who wish to be forgiven a great deal to nurture a culture in which people habitually dissolve the condemnation of evil in a sentimental bath consisting of deterministic stories where the perpetrators of evil are helpless playthings of determinism.

Again let us not forget that one can invent deterministic explanations of behavior that effectively appeal to one's own Rousseauistic sentiments of self-pity and to the indulgence of others while circumventing hard metaphysical determinism. The difficulty (perhaps impossibility) of internalizing a genuinely consistent determinism makes such selective determinism necessary. Additionally, because the man or woman of sentiment is not only self-forgiving and self-pitying but also resentful and envious, it is useful to have someone to blame for one's frustrations and unhappiness. The non-judgmental morality so beloved of the therapeutic classes means, in effect, the morality of blaming others.

We should also recall that a deterministically oriented outlook was vigorously inculcated by powerful sections of the cultural and intellectual elite for most of the previous century. I have come across would-be molders of post-Orthodox opinion, anxious to be ahead of the zeitgeist, who assert that R. Soloveitchik's emphasis on the individual's personal relationship with God was fine and dandy in his individualistic, self-reliant time, but is outmoded now that we have learned the individuality-destroying truth that we are shaped by outside forces. This thesis insouciantly evades knowing the face of twentieth-century culture, showing unintentionally how those who worship history are condemned to be ignorant of it. For the cultural situation half a century ago, the following icons are emblematic: Remember that long before John Adams's opera *The Death of Klinghoffer* confirmed enlightened sympathy for frustrated, freedom-hungry Palestinians who can't help throwing overboard the American Jew in his bourgeois wheelchair, *West Side Story*'s "Gee, Officer Krupke," showed singing gang members (the Sharks and the Jets) expertly mimicking the cant formulas of psychiatrists and social workers who blamed society for juvenile delinquency. Around the time that the movie *Inherit the Wind* immortalized Clarence Darrow's version of his intellectual victory over a straw man fundamentalist back in the 1920s, his other great blow for enlightenment in that era, the successful defense of the psychotic killers Leopold and Loeb through the invocation of a sentimental cracker barrel determinism, was celebrated in *Compulsion*, encapsulating the formula of numberless politically correct movie and TV scenarios, and inspiring new saccharine pieties to their audiences. A mighty engine of public instruction, crying the good news that we are, at bottom, and when all the self-dignifying euphemisms are unwrapped, irresponsible robots, ran on tracks laid by social scientists, psychoanalysts and their influential, media-savvy acolytes, fueled by the vapors of liberal sentimentalism. The shallowness of the entrenched zeitgeist arrayed against free will does not preclude the possibility that current knowledge supports determinism more seriously than the scientific shibboleths of yesteryear. But it should alert us to the fact that ideological antagonism to free will is an old story, and that the attitudes we confront today are deeply rooted in the moral convulsions of the early twentieth century.

Here, too, with respect to the exculpatory tendency of the modern liberal mentality, we should be wary of oversimplification. Once more, I will try to illustrate the other side of the argument with an anecdote. Some years ago, a young man was diagnosed with a correctable physiological deficit that had, until then, impeded, to some degree, his intellectual progress. His father, a man deservedly known as a *ba'al hesed* and a *mensch* par excellence, confessed privately that he had previously regarded this son as an underachiever and had on occasion indicated a measure of disappointment in him. Now that he had discovered that the opposite was the case, and that in light of his unsuspected handicap, this son had achieved much more than could have been expected, he realized that he had been wrong. Recalling his remarks, I cannot forget the tinge of mortification and regret that accompanied his joy at the happy medical resolution. The lesson seems clear: Judge not, lest you be mistaken.

The very first sentence of this essay implied the asymmetry between determinism and libertarianism. An act that appears to be free may turn out to be determined: we may discover so much about the causal factors operative that we absolve the person of fault (e.g., the story just cited). About an act that appears to be determined, it is possible to be mistaken, and to discover the mistake: One believes that homosexual orientation or schizophrenia, for instance, are caused by family dynamics, and then comes to believe that genetics play a much larger, or exclusive role. But it can never happen that an act for which we have a plausible causal explanation is demonstrated to be a free act. There will always be a plausible deterministic explanation; usually, with respect to individual events, there will be a plethora of such explanations.[17] If an individual indeed acts in a manner so extravagant that it cannot be subsumed under any kind of plausible causal story, we would immediately think of it as a bizarre occurrence, a miraculous event, but by no means as the product of ordinary free agency.

Given this asymmetry, the voice of determinism will have a natural advantage every time. Given the reluctance of decent people to cause others pain, the determinist's skepticism is bound to overcome the libertarian's convictions. Given the risk of blaming someone for actions that may turn out not to have been within her power, and the apparent safety of refraining from assertions that can never be proven in any event, *shev ve-al ta'aseh adif*—the principle of not inflicting harm—will win out. The halfhearted libertarian is reduced to a wan smile, and the determinist is free to spin his mechanical yarns.

The trap here is the idea that not treating other persons, or oneself for that matter, as free agents is refraining from doing them harm. The truth is closer to being the opposite. Not to regard another person as free and responsible is condescending if not downright humiliating. The father who was so pensive upon learning that he had been hard on his son for the wrong reason would truly have been a bad father if he had reasoned all the years that there must be a cause for his son's relative slowness, and that it was not for him to make demands on the boy, but merely to wait patiently for the day when medical science would explain it all, assuming that day would ever come. He would even have failed as a father if, knowing that his son was held back by a physiologically caused impediment (one that, let us assume, could not be rectified), he ceased to expect the best effort from him, saying to himself that the youngster wasn't going to amount to much and that taxing him to the limit of his abilities (or more likely beyond, since the limits of human intellectual capacities are rarely to be calculated with precision) would expose him to accusations of cruelty. In that case, the son would have been cheated out of the education he did complete with great effort, and of the opportunity to strive manfully against his limitations; the father, for his part, would have been spared some regrets, but he would also have missed out on the pride and inspiration of his son's well-earned vindication.

The lesson, therefore, is not that we should refrain from holding others (and ourselves) responsible, but that we must do so with a well-developed sense of humility and fallibility. If Don Newcombe's paradox teaches us how difficult it is to articulate our intuitions about the scope and nature of our freedom in particular situations, the story of the father and the son teaches us that the knowledge on which we base our judgments and articulations is shrouded in uncertainty. Let me suggest that this insight is relevant not only to ethical practice, but alters our conception of what it is to be free and what it means to ascribe freedom to oneself or to another.

Introspection

Determinists rarely, if ever, seek validation for their beliefs in the data of consciousness. It probably strikes them as unscientific; besides, there is nothing to be gained by engaging an area where the libertarian seems to hold the home court advantage, so to speak. As our purpose is to peer into the soul of the determinist rather than to list her arguments, I see no reason to ignore elements of the inner life that enhance the determinist mentality.

Just as individuals report an introspective sense of freedom, they also know of moments in which the overwhelming quality of the inner world is one of paralysis. Such reports have become increasingly familiar with the recent boom in clinical depression. Addicts offer similar narratives of helplessness. Many of us have experienced shorter episodes of uncontrollable mirth or blistering anger, or longer periods of sexual obsession or low spirits. In these moods we experience ourselves possessed, or burdened by forces that seem totally alien to us. I have already alluded to the predicament of the person who has no paradigmatic experience of free will, and whose psychological horizon is therefore entirely occupied by the experience of non-freedom. One may emerge from such an experience with a powerful conviction of *not* being free in any meaningful sense. Within the religious realm, the sense of powerlessness to do what one ought to do is more frequently found in Christian spiritual life (especially where the Pauline elements loom large) than in Judaism, and it may not be irrelevant that twentieth-century psychotherapy traces its descent to the secularization of liberal Protestantism. Nonetheless, Jews too know something of the "dark night of the soul." In a more benign context, moral resolve is often experienced as coercive: "Here I stand, I can do no other." And at the highest level of religious development, the prophet of God experiences himself as passionately consumed by the word of God: "And it was in my heart like burning fire, pent up in my bones, so that I was weary of containing it and could not" (Jer. 20:9).

What can such experiences offer our inquiry? As believers in free will, who give weight to the introspective consciousness of freedom (even if we don't take it as a proof of free will), it behooves us to recognize that freedom is eclipsed in a variety of ways by unfreedom or the negation of freedom. If having an adequate awareness of what it means to act freely is conducive to the proper development of free will, then an awareness of the experiential reverse of freedom may be an integral component of the kind of consciousness we are after. One wonders whether it is possible to see freedom steadily and to see it whole without acknowledging its shadow side.

The Imagination of Free Will

Conceiving Free Will

Since the nineteenth century, determinist philosophers have claimed that the concept of free will is an impossible one. Free will is clearly inconsistent with determinism. If it is also inconsistent with indeterminism, then it is incoherent. Most people who have not thought about the subject believe that free will is consistent with indeterminism; in fact, they treat the two concepts as synonymous. The adversaries of free will, however, insist that indeterminism is incapable of supporting the kind of freedom that is needed for free will. Why not? Indeterminism refers to uncaused events. Genuinely uncaused events are random. Freely willed events cannot occur randomly, because the entire point of holding an agent responsible for his actions is that those actions are an expression of his character. Random, uncaused events thus cannot be what the libertarian is looking for; such events are not actions but flukes. (That is why Heisenberg's principle doesn't really help. The indeterminacy of subatomic particles has nothing to do with the idea of persons who are the authors of their actions. Under-determination is not enough for freedom.)

The advocate of free will seems to imagine the agent as occupying a perch above the battlefield where conflicting natural impulses vie for superiority, like the figure in the epigraph from Kafka. What is baffling about all this is how the agent manages to remove himself from the fighting line in order to set himself up in an executive capacity. In line with the above, the advocate of free will does not describe the freely willed act as uncaused. Of course the act is caused; the cause is none other than the agent herself. The idea of the agent (the self, the "soul") as a separate substance, affecting the material world but untouched by it, is not in favor these days.[18]

More important for my purposes than the philosophical disdain for dualism is the limited effectiveness of such a theory in resolving the challenge. If my actions are caused (determined) by my soul, or by my character, the question arises: what causes the soul to desire and pursue the goals that it does? At this point, the libertarian is often sucked into a recursive account. If the soul is not a self-creating substance, then how the soul is and what it desires does not depend on the soul but on those agents or events that brought the soul into being in its present constitution, and the soul is not, in the final analysis, free. If the soul has spontaneously created itself, it is indeed free. But this alternative, in the form just stated, is theologically dubious and philosophically unappetizing. If the soul is some kind of composite substance, part of which is self-creating, and part of which is created by "God in conjunction with the earth,"[19] then the nature of that mixture becomes a puzzle. What exactly is the relationship of the self-creating component, of which the human being is the responsible author, to the components contributed by God and nature? Does it make sense to say that man's responsibility is limited to those "indeterminate pieces" that he has contributed to the amalgam? The confusion about the nature of free will defines an element of confusion about the nature of man.

The way out of these confusions is to posit a self that is different in kind from the paradigmatic substances that normally function as causes.[20] This self cannot be conceived as the aloof commander-in-chief controlling the human being below, master of its domain, unaffected by the tangible human being that obeys it. Nor can it be conceived as the sum total of conflicting desires and constraints struggling for possession of the creature. If we wish to think of the free will on the model of an executive or a general, it would not be that of the chief who barks his orders and awaits their fulfillment, but rather that of the Druckerian manager who must enlist the compliance, loyalty and enthusiasm of his underlings, creating an interactive and integrated human being in which the fragments come to wholesomeness.

This is close to what Kierkegaard meant when he said that man is spirit:

> The self is a relation that relates itself back to its own self, or it is the fact that in the relation that the relation relates itself to its own self; the self is not the relation but the fact that the relation relates itself to its own self. Man is a synthesis of the infinite and the finite, of the temporal and the eternal, of freedom and necessity, in short a synthesis. A synthesis is a relation between two factors. So regarded, man is not yet a self.[21]

The language sounds unnecessarily obscure until one realizes that Kierkegaard is attempting to articulate a psychological reality that is both complex and mysterious. It may help to consider that a relation that relates itself to itself corresponds to what we would call "feedback," and that the notion of man as a synthesis that is "not yet" a self conveys the indeterminate quality of human agency and existence. Spirit is not a traditional substance that can be delimited and defined. It is the very essence of spirit that it is not complete and measurable. If the use of Heisenberg's principle to clear a realm of freedom is usually discredited, it may still be borrowed as an analogy: Just as quantum mechanics teaches that it is impossible to fix both the velocity and the exact location of a particle at a particular time, so it is impossible to define the self in the moment of its self-defining. Another analogy: For the human being to fix the fluent inner discourse of the self would be akin to the human eye gazing at itself directly.

Admittedly these formulations leave the nature of the self, and the nature of the human being, an abiding mystery. The unsympathetic sigh: one person's mystery is another's muddle. Daniel Dennett complains: "It is hard to take seriously the idea that something that could matter so much could be so magnificently beyond our ken.[22] "Religious readers, of course, would respond that it is hard to take seriously the idea that something that matters so much can be translucent to finite, sinful human beings. The problem is to make sure that we have located the right mystery, and that we indeed humble ourselves before the elements of reality that are genuinely beyond our capacity. The moral imagination of free will, penetrating to the core of our self-awareness, touching upon the paradox of being, change and becoming in time, fulfilling itself in the shadow that falls between thought and intention, between the passion and the action, is an appropriate locus for the sense of mystery. Recognizing this brings us closer to the truth about ourselves and about our world.

Another way to put the mystery in context: consider for a moment the activity of the creative artist. The work of art is new and unpredictable; it cannot be manufactured according to formula. The subtle touches of genius that distinguish Beethoven at his best from a highly accomplished musical technician or craftsman, or from the work of another composer of the same rank, are hardly random, nor do they seem to come into existence mechanically through the necessary operations of Beethoven's talent and other forces of nature. The successful work of art is the polar opposite of randomness; the greatest works proceed so intimately from the author's soul that they are experienced, in retrospect, as virtually inevitable. And yet they seem to slip through the nets of universal causality. I do not dispute that mechanists will deny this account and opt for some causal chain or happy accident to explain the creative process. I submit, however, that the upholder of free will can discover in the phenomena of artistic creativity a paradigm, or an experiential model, for the gesture of freedom.[23]

Two Ways of Imagining Free Will

To some of you, my proposal that it would be insightful to think of freely willed actions along the lines of imaginative creativity will seem natural. Others will find the analogy odd. Presumably this is because they operate intuitively with a different model of free will. In this model, the person confronts a menu of choices. Not to act freely means that one option is more appealing than the other, and that one does not choose to act contrary to that path of least resistance. To act freely means that one is able to, and in fact frequently does, ignore the pressure and choose one of the less appealing options. Freedom is thus bound up with inner struggle and overcoming. If a person who thought in this way were told to substitute a more concrete word for the word "will," he would speak of effort: successful willing would be the triumph of effort.

The other model, the one that draws on the analogy with creativity, posits a situation in which the agent does not face an exhaustive list of choices. When an individual does not act freely he drifts passively, acted upon by his circumstances. Freedom means sizing up the situation, either deliberately or intuitively, and being able to conceive and follow, a path different from the obvious one. The individual who acts freely molds his situation and takes responsibility for a new resolution. There may be powerful incentives against his initiative, and in such a case freedom will be accompanied by a sense of struggle and effort. But the nexus between freedom and alienated overcoming is contingent. Freedom may be just as dramatically manifested in strenuous and refreshing joyousness. For the proponent of this model the effective exercise of imagination is at the root of the triumphant free will.

In order to establish the value of the creativity model it is unnecessary to show its superiority over the effort model. Nonetheless, the relative usefulness of the two models can be judged by testing the adequacy of each in clarifying central issues in the free will debate. As already noted, the creative model includes situations where action involves effort and the overcoming of obstacles. The effort model sheds little light on situations in which concentration of effort is secondary to the initiation of a new vision. Likewise, the creative model obviously does a better job of capturing the element of unpredictability that is part of our intuition about freedom. Moreover, the idea of a creative self, despite (or because of) its open recognition of mystery, enables us more easily to conceive what it is like for the individual somehow to transcend the causal factors warring for his soul. The struggling self, by contrast, resembles a *deus ex machina*, itself detached from the blood and sweat of the arena, putting its thumb on the scales of motivation to ensure the victory of the preferred causal gladiator.

Surprisingly, the creative model may even succeed in explaining some of the phenomena connected to conflict and struggle better than the effort model. When we attend carefully to the way we act in such situations it often seems that we succeed in overcoming resistance, not by gritting our teeth and demanding of ourselves increased effort, but by re-directing our attention and imagination. Take, for instance, an individual who finds himself, on the morning of 9/11/2003, more than 150 blocks from where he is supposed to be in less than two hours. By pushing his body to the limit, he can arrive reasonably close to the set time and get on with his day's work. What is driving him, after the initial moment of resolve, as he half-walks, half-runs, darting, and weaving like a running back, hurtling through thick pedestrian traffic, counting the blocks that separate him from his goal? Most likely his imagination is as active as his legs: he may be fired by a vision of the kind of person he would like to be, or he may be impelled at the prospect of losing a day and having to reschedule his activities. The difference between this person and someone in the same position, with the same physical capabilities, who cancels the day and glues himself to the TV, is not that they differ in their effort thresholds, but that they differ in the way they conceive of the situation and their responsiveness to it. The proximate explanation of the fact that he made it and his colleague did not is that he made the effort, but the crucial factor is that he chose to imagine the situation as he did.

William James's classic analysis of that most homely of nineteenth-century predicaments—getting out of bed on a freezing winter morning, illustrates the limits of effort as an explanatory principle:

> We think how late we shall be . . . but still the warm couch feels too delicious, the cold outside too cruel, and resolution faints away and postpones itself again and again just as it seemed on the verge of bursting the resistance and passing over into the decisive act. Now how do we *ever* get up under such circumstances? If I may generalize from my own experience, we more often than not get up without any struggle or decision at all. We suddenly find that we *have* got up. A fortunate lapse of consciousness occurs . . . the idea flashes across us, "Hollo! I must lie here no longer"—an idea which at that lucky instant awakens no contradictory or paralyzing suggestions, and consequently produces immediately its appropriate motor affects.[24]

We earlier noted that free will is significant, whether as a psychological reality or as a moral-juridical principle, only when freedom is a substantial component of our lives. Whether one thinks primarily in terms of the effort model or the creativity model has important implications for this question. To see how this is so, we shall look at a problem raised by Peter van Inwagen. Not long after publishing his *Essay on Free Will*, a powerful metaphysical argument for incompatibilism, van Inwagen wondered how much free will he had secured. In the vast majority of situations, van Inwagen argued, it is not correct to say that an agent really could have chosen to do anything other than what she actually chose to do. That is because agents have well-established identities that entail a set of preferences, and the sensible thing for them is to act in accordance with those preferences. The agent may have the physical capacity to act against her preferences but, unless she changes her character, such behavior would not make sense.

Van Inwagen recognizes three classes of actions that are free and rational: 1) Cases where there is no strong reason to prefer one option to another. The classic case is Buridan's Ass, the proverbial donkey who starved because he had no reason to prefer one bale of hay to another, but can also extend to close calls. It is unclear how frequently we face such choices, but it is clear that this is of no importance. Only an advertisement junkie can vest momentousness in the choice between two brands of soft drink. 2) Conflict between choice and inclination. These are often (but not always) moral conflicts, where my duty struggles against my strong first-order desires. Such situations, van Inwagen maintains, are normally rare: for better or for worse, one side or the other predominates, establishing itself as habitual, winning every battle. 3) Cases where the individual chooses, not how to achieve the ends he has already committed himself to, but the fundamental commitments that determine the rest of one's life. Because the competing options are incommensurate, there is no standard from which to infer the sensible course of action. Van Inwagen refers, in this context, to the Kierkegaardian choice between the aesthetic sphere of existence and the ethical. Such choices are indeed momentous, but they are surely rare, and many an individual may never experience the opportunity or the burden of choosing himself in this way.[25]

With the benefit of the creative model, the second and third categories turn out to be broader than they might initially appear. In artistic creativity, the challenge is not to overcome one inclination with another motive, or to be converted into an entirely different kind of person. It is rather to articulate a vision, to resolve a unique aesthetic problem, in the best possible way. How is this pertinent to the characteristic case of free choice in general, and moral choice in particular? After all, most of us do not spend our lives composing string quartets. Picking one alternative from a finite menu, choosing, let us say, to return someone's lost property as opposed to appropriating it, does not seem like a creative activity.

The answer is that most everyday moral challenges are more like artistic creativity than they are like checking off the right box on an ethical multiple choice exam and then backing up one's selection with a little will power. Every choice, to begin with, is an act of self-creation. Returning lost property confirms certain admirable character traits, such as unselfishness, honesty, empathy, and/or fidelity to halakhic imperatives (or less admirable character traits like desire for applause). The degree to which these various motivations and moral emotions come into play in any particular episode also has a formative influence on the agent's character and often on the moral tone of those with whom he or she interacts.

The need for creativity is not limited to the character-building aspect of moral activity. Often the exact nature of the action required is not given beforehand. Consider, for example, the manner in which a person might express agreement or disagreement with another, or argue a point, or relay some criticism, or praise someone, or request a favor, or ask for pardon after recognizing that one has offended another, or negotiate forgiveness for an offense against oneself. I daresay that such activities are part and parcel or our daily lives and make a crucial contribution to the moral quality of our lives. Such encounters (and their parallels in our relationship with God) cannot be scripted in advance, yet they require the utmost concentration and preparation; they can hardly leave us unchanged. If these activities ought not call forth an intensity of creative imagination equal to that which the artist commits to the work of art, there is something wrong with our priorities.

Before putting ourselves in the hands of R. Dessler and R. Kook, one last distinction between the effort model and the creative model. Free will, for the effort model, means that the agent is the author of that particular action. Creativity, by contrast, pertains directly to the agent, not the particular action; an agent acting freely is inevitably the author of free actions. Ascribing freedom to the agent first and foremost will sound apt to disciples of R. Soloveitchik, for whom repentance, in *Halakhic Man*, is identified with the gesture of self-creation. Elsewhere, speaking of the Jewish sense of mission (*shelihut*), he distinguishes between the human being who is an agent (*gavra*) and the human being who is an object (*heftza*).[26] His categories correspond well to the kind of freedom, or failure of freedom, inherent in the creative model.

R. Dessler

R. Dessler is famous for presenting the notion of limited free will to the Musar community. This approach, in a nutshell, maintains that at any particular moment individuals do not have unlimited control over their actions. Think of the battlefront in a war between nations, he suggests. Each combatant controls territory in which its sovereignty is unchallenged. Between the armies lies a no-man's-land, as it was termed during the Great War of R. Dessler's youth. This area is where the battle rages; control passes back and forth as first one army, then the other, goes over the top. The same is true of the human heart, where the kingdom of truth struggles against each woman and each man's imaginary realm of truth. At this moment, the highest rungs of saintliness may be inaccessible to me; likewise there are depths of degradation that are not live options to me. There is, however, a range of behavior and attitude that is open to me. This is the point of choice (*nekuddat ha-behirah*). The exercise of free will moves the battlefront: when the individual acts well, his *nekuddat ha-behirah* shifts accordingly; when he lapses into sin, his range of freedom is likewise affected. Praise and blame, reward and punishment, would seem appropriate, on this theory, only insofar as the actions involved are within the range of choice. To the extent that one's options are constrained by factors prior to his own agency, he is not responsible for his choices.[27]

This last point constitutes R. Dessler's Achilles' heel, from the standpoint of the traditional Jewish mentality. Judaism is usually taken to teach that people are responsible for their actions in general and not merely for the small amount of wiggle room that R. Dessler seems to grant them; most people do not assume that they get no credit for good actions that can be explained causally on the basis of "moral luck," their God-given temperament, education, and so forth. In his best-known discussion of free will R. Dessler reckons that "most people, even if they were educated to *mitzvot* and good deeds, are doing so by rote." He then adds, anticipating the objection: "Although God does not begrudge any creature its reward, and even for this they will be rewarded, nevertheless they hardly have any acquisition of higher levels, that is to say, the choice of the good, for they are rarely involved in the struggle against their inclinations (*milhemet ha-yetzer*)."[28] He cannot deny these individuals their reward, even though the philosophy presented here cannot explain why, and so he seems to wave the question away. Disciples of R. Dessler are perturbed by this difficulty, as by the converse problem of sinful behavior caused by factors beyond the individual's control.[29]

I believe that the nucleus of an answer to this problem is found in another of R. Dessler's essays on free will.[30] Here he discusses situations where God has foreordained that a person do a certain action in order to affect another person, as for example when he causes the death of an individual whose demise has been divinely ordained. It may happen, writes R. Dessler, that God wills the continued life of a person as a means toward affecting the other, as an "instrument" serving the free actions of the other. Of this instrumental individual R. Dessler writes: "His purpose is fulfilled whether the person upon whom he is dependent chooses the good or not. If he chooses the good, he creates a [divine] manifestation through his good choice. If not, in any event there is actualized the manifestation of God's judgment regarding his evil choice."[31] Implicit in R. Dessler's analysis is the thesis that the actualization of God's will is a value, even when the alternative would have been an even greater value, namely the actualization of the divine will via the free human choice of the good.

This phrasing may offer a clue as to why R. Dessler (and the entire Jewish tradition) is committed to the reward of the person who fulfills God's will by performing His commandments, even when the performance is routine and does not reflect genuine free will. One way or another the person has brought about the manifestation of God's will. This insight, in itself, is not really startling. For instance, the patent fact that many unintentional transgressions require atonement implies that the halakhic conception of responsibility is broader than the category of intentional activity freely chosen.[32] In this respect, the halakhic way of thinking, like common sense morality, shares features with the classical Greek view. R. Dessler, like many modern thinkers, is strongly attracted to the Kantian outlook that values the moral act by the intention invested in it; if the intention is derived from prior causal factors, and is therefore not free, the act loses its value. That is why he, and his disciples, flounder about when confronting the gap between traditional Judaism and their intention-based philosophy. The formulation in the later essay restores the place of intention-independent assessment of actions: what gives an action value (though that value is minimal in terms of character-building) is the fact that it manifests, one way or another, the divine judgment. Despite this resolution, the tenor of R. Dessler's conception is conducive to the conviction that moral value is pretty much confined to a dramatic, and fairly narrow, arena of struggle and effort.

Let us round out our discussion of R. Dessler by commenting on three aspects of the phenomenology of free will that emerges from his writings:

1. R. Dessler takes self-awareness seriously as a significant psychological datum. At times he states that the *nekuddat ha-behirah* can be discerned as the point at which the individual encounters resistance, a statement that fits the effort model of freedom.[33] His major pronouncement, however, in keeping with our observations at the beginning of this paper, does not entail regarding every specimen of introspection as probative. He writes:

   > a human being who has overcome his inclination (*yetzer*) even once, feels clearly in his heart that he is distancing himself from fantasy, and resolving to grasp truth vigorously. But he who has never overcome his inclination, cannot discern this, for his experience teaches him that whenever he desired an evil thing, the desire enslaved him with no escape; therefore he cannot understand the idea of free choice, because in examining the sources of his actions he sees only external causes and effects.[34]

   The point is not that we are convinced that a particular action is free, but that we know ourselves to be capable of freedom. As noted already, this insight has enormous educational implications.
2. R. Dessler's vivid comparison of the human being's inner struggle to the action on a battlefield has rightly fascinated his admirers. But the analogy is more insidious than it appears. The sense that war is about the conquest of territory or that occupying more or less land is the key to victory runs up against more decisive factors, most prominently the economic ones. Armies must, in the final analysis, limit their casualties; they

require lines of supply and the manufacture of munitions; civilians need food. The trench warfare that marked World War I was a waste of life on an epic scale. Germany was defeated while still occupying French soil, brought to starvation and economic ruin one step ahead of her enemies. These complications become even more evident in a post-modern war, without frontiers, against faceless foes possessing nothing beyond the apparatus of destruction and terror. Taking R. Dessler's figure of speech at its face value, we are led to reflect that winning one encounter with the evil inclination does not always result in the permanent strengthening of the forces of good. Sometimes we overstretch our psychological resources, so that momentary victory is followed by fatal prostration. Sometimes we fortify habits that eventually harden into obstacles preventing our further progress or, like the Maginot line, hide weaknesses in our defenses. Often we operate in a psychological or moral fog, unable to identify clearly the dangers we must overcome. In a word, R. Dessler's application of the image articulates the one-dimensional effort model of free will, while a thicker description of our predicament calls for the kind of perspective we have identified with the creativity model.

3. In another essay, R. Dessler talks about the apparent contradiction between free will and cause and effect determinism.[35] The conviction of free will grows out of our inner experience of successfully overcoming resistance through effort, as noted above. The conviction of cause and effect grows out of the external perspective, which discovers causal relationships in the world and universalizes the regime of cause and effect as governing all events. (The inner perspective, R. Dessler remarks, has primacy, because the subject who adopts the external perspective must perforce be a subjective self as well.[36]) R. Dessler would be inclined to agree with Dr. Johnson's saying (at the end of a long conversation on the paradox of divine foreknowledge and free will): "All theory is against the freedom of the will; all experience for it."[37] Our own examination of these issues concluded that neither belief in free will nor faith in determinism is logically necessary or incorrigibly intuitive. Conviction of freedom may be illusory and a methodological commitment to causal investigation can survive exceptions. R. Dessler, however, seems fixated on his version of the Kantian antinomy. Hence his solution to the problem is to deny the ultimate reality of both outlooks. The inner perspective of self-awareness and free choice is needed to inculcate the sense of "obligation and responsibility, and reward and punishment." The external perspective, tending to determinism, is necessary to enable us to recognize the sovereignty of God, as we proceed from the concept of cause in its phenomenal sense to the idea of a divine First Cause. This closing

insight, of course, goes beyond the Kantian analysis by moralizing both the libertarian and the deterministic strands in human consciousness.[38]

R. Kook

R. Dessler's breakthrough was his recognition of the limitations on free will in many situations and his occasional references to the phenomenological aspects of willing. Overall he remained anchored to the effort model, and preoccupied with the question of reward and punishment. R. Kook's two letters on the subject are not as lucid as R. Dessler's essays.[39] Writing to his disciple Moshe Seidel, he appears to be struggling obscurely to articulate his ideas, and in the second he is impelled to correct Seidel's misunderstanding of the first. Nonetheless, it is possible to extract from these letters and from other scattered remarks, a set of insights that advance our inquiry.

The first letter, dated 26 Adar 5670 (March 9, 1910), responds to Seidel's query about the principle of *zekhut avot* (ancestral merit). R. Kook proposes that we take pleasure in the correspondence between the freely willed good act and its successful actualization because our sense of justice is satisfied. He immediately adds that many enormous benefits have nothing to do with our choice, both external goods like wealth or beauty and internal goods like character and wisdom. Our desire to ascribe justice to the product of free will is grounded in the recognition that absolute, unconstrained freedom reveals the essence of life, an essence that is ontologically prior to the distinction between good and evil. Free will, as we find it in the world, is not absolute: every gesture of freedom is fettered by many elements of necessity. R. Kook speaks of a hidden divine freedom, whose worldly manifestations are attenuated echoes of its essential nature. He also refers to this ultimate freedom as *behirah tzefuyah* (anticipated, or foreseen choice) that is most strongly represented in the great initiatives of the Patriarchs, reverberating, in a derivative way, in the actions of their descendants. The ultimate divine will also purifies the consequences of the evil tendency of human free choice.

The second letter, of 1 Elul 5671 (August 25, 1911), opens with a criticism of Seidel's response. R. Kook accuses Seidel of missing the point of his statement about our approval of the justice inherent in the connection between free will and responsibility. Seidel takes it as an axiom about moral desert. R. Kook regards this understanding as the effect of obtuse thinking mired in the routine notions of "practical reason" and "political morality." He reiterates that even when an act is free, human existence is not chosen: the human being can reasonably claim that he has been inserted unwillingly into the world of choice. If existence is good this judgment is not derived from the familiar "practical" calculus of good and evil in the social or juridical sense, but rather it grows out of a more fundamental perception of reality.[40] What we seek is absolute being, and our ideal is that of absolute essence (*mahutiyyut muhletet*) as revealed in free will.

The above summary does not do justice to the richness or the difficulty of R. Kook's formulations. It can, nonetheless, provide a sketch of his original contribution to our discussion. Like R. Dessler, R. Kook acknowledges the limitations on human freedom. His angle of vision, however, constitutes a radically different set of ideas on the subject.

1. R. Dessler has difficulty making room for the concept of moral luck. It was only via a close reading of his text that we were able to salvage a Desslerian version of the traditional Jewish doctrine that God rewards good actions, even when they are not done in an atmosphere of moral struggle and resistance. R. Kook values the coincidence of freedom and spiritual achievement, but he displays no embarrassment at the fact that much in our material and spiritual lives is independent of earned merit.
2. For R. Dessler, in line with the previous point, free will is important for its role in defining moral desert. R. Kook goes out of his way to deny the centrality of juridical considerations in assessing the doctrine of free will. Freedom is manifested in the battle between good and evil in this world, but its source is beyond these categories. According to R. Dessler, one might say, an action is ascribed to me because I am responsible; according to R. Kook, I am responsible because the action expresses who I am.
3. For R. Dessler, freedom is defined psychologically by the overcoming of resistance. For R. Kook, freedom is understood metaphysically as the lack of constraint. R. Kook's concept of freedom is rooted in the human being as the image of God, imitating, in a dim and diminished manner, the divine creativity.
4. For R. Dessler, freedom is predicated primarily of actions, specifically those actions that must be chosen in the war zone where truth and falsehood fight it out. R. Kook is concerned primarily with the freedom of the person, seeking to participate in the essential liberty of the Creator.
5. For R. Dessler, once the exact scope of an action has been defined, the agent is either responsible for doing it or not, depending on his *nekuddat ha-behirah*. R. Kook's approach seems more compatible with the common intuition that responsibility admits gradations.

In the passages we have examined, R. Kook does not extol the unpredictability, indeterminateness and freshness of free actions in the manner of James or

Kierkegaard. However, the paradigmatic status of the divine creative activity, the preference for harmonious willing over the effort model of free will, the rejection of the juridical calculus as the major concern motivating free will, all move him in the direction of the creative model.

A reflection on the function and scope of human causality will further show how R. Kook's orientation is helpful to the creative model. Let us take R. Yosef Albo's account as typical of the medieval philosophical mindset.[41] In categorizing actions, Albo makes room for those in which freedom is partial. What he means by this is that the agent fully and sincerely intends to do the act, only external impediments prevent him from succeeding: one thinks he is eating tuna, but is in fact partaking of shrimp; one intends to throw an object two cubits, but instead flings it four cubits. That the intention itself, and the subtle way in which it is translated into action, may not be completely within the power of the agent, is foreign to this analysis. R. Dessler acknowledges that the agent may not be responsible for the character traits behind the intention. Nevertheless, he posits a boundary line (whether or not it can always be readily discerned by us) between those actions that are free and those that are not. R. Kook need not be committed to the idea of such demarcation between the free and the unfree. Only God is absolutely free; human beings, affected by factors of moral luck, the merit of forefathers, and divine guidance, are only relatively free. This dialectic of freedom and necessity is not confined to any particular set of acts. It pervades experience.

Additional texts exhibiting R. Kook's outlook on freedom shed light on its practical applications. He addresses, for example, the modern scientific propensity to regard certain natural inclinations as congenital, and thus to silence any moral condemnation of the practices involved (Freud?), alluding to both homosexual impulses and aberrant heterosexual appetites. Although he insists that "the word of our God will stand forever (Isaiah 40:8)," so that Judaism's prohibitions are not vulnerable to revision based on new scientific knowledge, and also suggests that the corpus of Halakhah does, to some degree, make allowances for some sexual idiosyncrasies, he acknowledges that it may be virtually impossible to eliminate individual suffering: "some resentment (*tina*) [at a bad tendency] that may be encountered in an individual in such a manner that it cannot be uprooted"[42] Here R. Kook recognizes that bad moral luck neither exempts the individual from struggle against his desires nor society from rejecting the practices.

R. Kook's original contribution to the problem of repentance, and how it alters the past actions of the sinner is of a piece with his letters to Seidel. Insofar as God is sovereign over the universe, all occurrences (including sins) conform to His will. To that extent, all events can be viewed as a fulfillment of the divine plan. Insofar as the sin can be classified as an act of rebellion against God, it violates the divine plan. Both descriptions are equally true, but they are diametric opposites with respect to moral evaluation. Confession of sin means that the individual takes responsibility for his actions; he casts blame on nobody but himself. Repentance entails that the person dissociate himself from the description of his actions as rebellion. The penitent now hands over his actions to God; he subsumes his actions within the larger context of God's actions. It is only under the description that ascribes the actions to man alone, and not to God, that they warrant the predicate "evil."[43] Please note that both here and in the letters R. Kook is quite willing to bring the metaphysical and experiential awareness of absolute divine freedom into his discussion of human responsibility.[44]

Recover It or Lose It

Hilary Bok is a determinist (though her term is "mechanist"). Keenly aware of the ethical problems posed by her philosophical position, she argues that libertarians have not solved these problems either. Whether or not our choices are determined by natural events, "we may already be in the situation libertarians feared that mechanism would produce: one in which we cannot justify some of our most fundamental beliefs about ourselves."[45] The major thrust of this essay has been, if not to justify certain ideas about the self, at least to reimagine these ideas in a plausible manner.

Whether or not I have succeeded, the greater challenge is to make intellectually plausible ideas real in the imaginations and actions of living men and women. For free will is not an organ with which we are born, like an arm or a leg. It is a capacity to respond to the world and its challenges in a certain way. If it is not nurtured properly, it will never develop or it will wither away. The remainder of my paper will focus on the areas in which our robust experiential sense of a realistic freedom of the will is especially vulnerable. If we still retain it, we must use it or lose it; if we have already drifted into the camp of de facto determinism, we must go about recovering it. We shall examine three crucial areas: the use of language; the rehabilitation of the individual as active agent; the quest for a realistic and creative balance between our knowledge of the self as determinate and the experience of being a self as a relation relating itself to itself—the self as an unfinished task and perpetual adventure.

Talking Free Will

The passive mentality is nowhere more lethargic than in its readiness to accept discourse off the rack, so to speak, without fully considering the connotations of terms and habits of speech. In a society where both belief in free will and the living experience of freedom are in decline among the cultural elites and those who take their cues from them, it would not be surprising if our jargon and clichés were not saturated with deterministic messages. The kind of stories we tell each other, whether about ourselves or about others, naturally take on the form of a deterministic narrative, as if any other were impossible. A story of freedom is indeed difficult to tell, because all plausible stories contain causal explanations, and it is a matter of art and sensitivity to hint at the limits of causal explanation, to clear a space for freedom, without giving the impression that one has arbitrarily curtailed the causal account. Free will, in discourse, must always skate on thin ice: a little less causal pressure, and the story fails to progress plausibly, because it is not adequately motivated; a trifle more, and the concept of agency breaks down and becomes mechanical.

All of this can be illustrated ad infinitum. Within the confines of this essay, we will have to be satisfied with a glance at two classic works of fiction by two significantly different masters.

Exhibit A

For on the one hand all her conventional training was now urging her to stand firm—not to belittle herself in this way—whereas on the other, all her desires for love, understanding, companionship, urged her to run after him before it was too late, and he was gone. . . . And yet so binding were the conventions which had been urged upon her up to this time that, though suffering horribly, a balance between the two forces was struck, and she paused, feeling that she could neither go forward nor stand still—understand or endure this sudden rift in their wonderful friendship.[46]

Theodore Dreiser was a determinist. In his novels, he deliberately shaped a world congenial to his philosophical vision. It is a Darwinian universe in which the strong devour the weak, and the weak, when they attempt to better their lot, are devoured by their own hapless attempts to escape the nets in which they are enmeshed. Dreiser's characters, desperate and needy, are fit vessels for his favored plots. His fiction solidly and brutally depicts a real social world and captures the thought processes of people trapped in that world, but it is not a world that most of us would care to inhabit. The section quoted above presents Roberta's reaction to her boyfriend's demand for sexual intimacy. Her social and economic circumstances, as an impecunious shop girl who has attracted the attentions of her supervisor, the boss's nephew, make her eventual acquiescence a foregone conclusion, though a pious novelist would have allowed her to say no, and might even have made her refusal psychologically convincing.

What interests me about the scene is not the outcome of the plot but the language used to describe Roberta's deliberations: We are told about two colliding forces—the moral and religious norms urging her not to "belittle herself," and her physical desire for Clyde. These forces are measured quantitatively, so that at the moment inspected they are equal and she is therefore unable to act. It is not simply that she has difficulty making up her mind about a matter where morality's dictates are unambiguous, or that she ends up making the wrong choice. The very language in which the decision is couched is that of determinism. There is no perspective other than that of ignorant motives clashing in the dark. Dreiser has often been criticized as an inelegant and clunky stylist, but his words in this instance reflect perfectly the outlook he upholds. One may remove Dreiser from the curriculum or from the library on the grounds that his stories make immorality inevitable and acceptable. But one cannot remove his way of speaking from the street, from the radio, from everyday discourse. And if that inner monologue undermines the imagination and practice of freedom, the appropriate prophylactic is to become more thoroughly aware of the language we hear and use and more committed to developing our own alternatives.

Exhibit B

> The question of the settlement loomed larger to her now: it depended, she had learned, so completely on herself. Her choice, as her friend had called it, was there before her like an impossible sum on a slate, a sum that in spite of her plea for consideration she simply got off from doing while she walked about with him. She must see Mrs Wix before she could do her sum; therefore the longer before she saw her the more distant would be the ordeal.[47]

The novels of Henry James are not the repository of an explicit philosophy. Many of his novels hinge on the exercise of free choice. Typically the protagonist is faced with a choice, given time to reflect on it, and emerges having made the more difficult and morally satisfactory decision. That James is able to carry this off is due, in part, to his manipulation of plot, and also to the social world he re-creates. His characters do not labor under the economic constraints and social airlessness that make Dreiser's frantic with despair; they can make a lonely choice and survive its consequences. The alternatives are thus reasonable resolutions to their dilemmas. Though James is preoccupied with moral authenticity, most of the time his world is as vacant of traditional religious constraint or rigorous philosophical morality as it is free of overwhelming economic pressures. Were this not the case, the process of deliberation would be loaded with "ideological" baggage, and the essential freedom of the agent would be obscured. All this helps to engender the conviction that if agents are free, their freedom is something like what James writes about.[48]

While the vicissitudes of plot and character that enable James to be an effective narrator of free will are surely pertinent to our inquiry, let me again call attention to the micro-narrative in the passage quoted. I have chosen *What Maisie Knew* in part because of the popular contemporary appeal of the subject matter—the child whose fate is entangled in a tale of multiple divorces. Looking at the situation from the child's perspective complicates James's work. Not only must he try to chronicle faithfully the growth in Maisie's awareness of her plight and the moral ambiguities surrounding the adults with whom she is involved; he must also make her a credible moral agent despite the limitations on her knowledge. Her decision, whether to live with her two step-parents, who are together after the collapse of their marriages to Maisie's natural parents, despite the fact that it is not morally acceptable in the eyes of her less attractive tutor, Mrs. Wix, who insists that they are not truly "free" to marry each other, is based on a relatively abstract conception of what "free [to marry]" and "moral" may mean. Nevertheless, James attempts to present her as a free agent, despite her limitations. He does not picture her either as the toy of mechanical forces, actuated by whichever happens to be stronger at the crucial moment, or as capriciously picking one option at random. Instead she understands that she is facing a problem that has correct and incorrect solutions (like a complicated sum in arithmetic); though she has an intimation of the insight that this problem does not yield to calculation like a sum, she is nevertheless bound to work it out. Not surprisingly, she hopes that Mrs. Wix will help her to get it straight, and readers who experience the novel's denouement may wonder whether Mrs. Wix exercises too much influence in getting the child to follow the path of morality as she sees it. (One wonders what Maisie will think of her decision later on in life.)

These two examples should illustrate the importance of language in fostering or stifling the possibility and nature of moral discourse, and especially the inner experience of freedom accompanying moral deliberation. When, as a youngster, I first became aware of the implicit messages embedded in discourse and the need to be vigilant in one's reading and listening and speech, it seemed to me a simple thing to purify the moral dialect of the tribe through rigorous scrutiny of our language, weeding out Dreiserisms, commending James. In retrospect, the hope of ethical authenticity through the analysis of literary style is as illusory as the dream of psychoanalytically induced transparency that had by then already passed its peak.

Language can be either friend or foe of moral honesty and responsibility. But it cannot do the work alone. (Does Alcoholics Anonymous's preference for "illness" over "disease" succeed in promoting a traditional sense of responsibility? I hope so.) No amount of monitoring can eliminate the inauthentic except to replace it with an invariant, philosophically kosher vocabulary that will inevitably falsify human experience even before it hardens into cliché. The modest conclusions of an eminent historian of alcoholism should be taken as a continuing challenge rather than a surrender:

> The less dogmatic historian of the present will admit that, in everyday conversations, and in some self-help groups on their better days, it is occasionally possible to discuss one's feelings, hopes, and regrets without either rationalizing oneself in the name of the free will or psychologizing oneself in the name of some diagnosis. These inconclusive conversations will never amount to a public sphere, of course; but they may be the only indications we have that the habits and theories about the will and its compulsions that we have inherited were not historically inevitable and are now not unshakeable.[49]

Master or Slave?

We have noted more than once that free will is experientially meaningful only when the individual believes that his or her (free) actions are important. To exercise freedom on rare occasions, to little effect, is tantamount to not having freedom at all. How much freedom we have, in the meaningful sense, depends on the importance we ascribe to the actions in our power. This is not a matter of scientific fact, but of moral-spiritual evaluation. With the scope of freedom under hostile scrutiny, the importance with which Judaism endows human agency must be made explicit.

"Who has contempt for the day of small things?" asks the prophet (Zechariah 4:10). The answer is: we all do. From time to time we question the significance of what we accomplish. At the end of the day, the greatest of days, we sum up our attainments and are left with nothing: "What are we, what is our life . . . for the multitude of their deeds are nullity, and the days of their lives are vanity before You."[50] Note well that whether or not our actions have exclusively natural causes seems irrelevant to the time-honored, soul-wrenching confession of helplessness. And note further that our culminating words of return do not boast of our assured fidelity to God in the future, but instead entreat that it be His Will that we do not sin again, and that we be cleansed of our sin through the application of divine mercy.

Despite this real sense of humility and recognition of our insignificance, we accept responsibility for our endeavors, and we are not to feel contempt for the small days that fill our lives. Great accomplishments, to begin with, are built up incrementally. Both R. Kook and R. Soloveitchik find a place, in their phenomenologies of repentance, for "sudden" change and overnight transformation, but they too know that the life of repentance normally consists of innumerable steps gradually re-defining the person.[51] R. Dessler's battlefront image conceives of spiritually significant action as a kind of trench warfare, where victory is inches and the heart's changes.

This value judgment addresses the concerns of those who are disturbed by their increased awareness, via science and scientism, of the limits on freedom. Such individuals need to be reminded that character traits and other factors that are not within our control can be modified or re-directed to suitable goals, and that while such efforts may not transform the personality as radically as one might wish, they are nonetheless spiritually and practically valuable.

More importantly, by honoring the day of small things we defy the materialistic dogma that rules over our secular culture. The utilitarian mentality knows nothing but the bottom line, and values nothing but the external consequences of acts. Refinement of character, from this viewpoint, is merely a means to an end. The religious counter-culture will not succeed in challenging and subverting secularism unless we internalize the fundamental religious conviction that the intangible, seemingly minor, progress of the religious life toward the perfection of our characters is important to God, and important to us.

As R. Dessler observed, the robust conviction of free will arises from the experience of self-control and the overcoming of resistance. Where this is absent, the experience of freedom will be absent too. There is an educational lesson here. To the degree that we can help it, we should ensure that children have the opportunity to enjoy the competence that engenders the free will "habit." A parent, or a school, that adopts as its text, so to speak, the moral equivalent of "the little engine that couldn't," inculcates the belief that whenever one is tested, one fails. The minimal conception of free will is thus effectively nipped in the bud.

Where a healthy conviction of free will is not properly developed two different philosophically interesting reactions may take place. One phenomenon is the cult of "self-esteem" of which so many enlightened people are nowadays enamored. The kernel of truth in this movement is the need to educate children and other immature individuals by letting them taste the satisfaction of success. The error is that the success is unearned. The authority figure rewards the child regardless of whether he or she has actually accomplished anything. There is little sense of genuinely overcoming resistance, or correspondence between acclamation and accomplishment: "all the men are strong, all the women are good-looking and all the children are above average" (Garrison Keillor describing the fictional Lake Wobegon, Minnesota). The individual so encouraged is ill prepared for the challenges that await once he steps outside the protective cocoon of family and school. Even before that, the essential vacuity of the praise that is distributed routinely as if it were a bribe becomes evident, and the empty verbiage, lacking credibility, feeds convictions of cynicism and despair. Self-esteem is a shabby substitute for self-respect.

Irrational behavior is a flashier substitute for freedom. Its classic incarnation is Dostoevsky's Underground Man. The logic here is that freedom is displayed when the person acts contrary to his or her self-interest, or when enormous resistance is overcome (as is the case in acting against one's interests). The only way to prove one's freedom is therefore to defy convention and the utilitarian calculus of routine happiness. In this situation, the individual's behavior is determined by his desire to be free and to demonstrate his freedom. A free person would not be so in the grip of anxiety about whether he is free or not, and would not be subject to the impulse to exhibit his putative freedom in a spectacular manner. In contemporary psychological lingo, the Underground Man, and others of Dostoevsky's characters, are "acting out."

Offhand, it would seem that the Orthodox community has little to fear from the last phenomenon. After all, our lives are punctuated with great and small acts of self-control. Many of these acts also manifest a rejection of the secular culture against which we define ourselves. Thus, it would be unnecessary to invent bizarre, self-harming modes of behavior as a desperate attempt to demonstrate one's freedom. Unfortunately, the psychological dialectic is not as uncomplicated as we might prefer. On the one hand, the liberal wing of Orthodoxy has tended to downplay the conflict between Torah and secular culture, thus projecting an image of gray conformity varied only by a few minor idiosyncratic departures from the pieties of the upper middle class American dream. On the other hand, even where Orthodox deviation from conventional norms is celebrated, the style of deviation and rejection takes on a rigidity of its own. As we have noted, conformism bodes ill for the performance of those *mitzvot* that require creativity, primarily those dealing with personal relations between human beings and God and between human beings. For free will, conformism poses an additional danger, insofar as the atrophy of the creative impulse, the formulation of the individual as *heftza* rather than *gavra*, denies the individual the self-forming opportunities to experience himself as a free agent.

A factor, often unnoticed, that undermines the sense of responsibility among modern people is the deprecation of intellectual deliberation and steadfast belief. Religious acceptability is made a matter of external behavior, detached from the psychological life that includes intellectual conviction and commitment. Real assent, in addition to its normative role in religious life, is also the manifestation of resolute character. An ideology that views shaky, vague beliefs with equanimity, and that avoids formulating and expressing fundamental religious and ethical commitments, thereby encourages the cultivation of personalities without a center, and thus without responsibility. Accountability is replaced with a kind of evasiveness in which the therapeutic and bureaucratic jargons become almost indistinguishable.

The prominent American Catholic intellectual George Weigel, referring to the therapeutic history of an odious priest, shows how a deficiency of dogmatic and moral clarity affects the integrity of a religious community. Perhaps we can learn something from his warning:

> A man has been raping children for decades and a Catholic therapeutic center has "no particular recommendations concerning his spiritual life," because the man's breviaries show signs of use and he talks with a spiritual director? Please. . . . It was also striking that the 1995 "spiritual assessment" of John Geoghan by St. Luke's Institute did not probe the man's beliefs, even at the elementary level: Did Geoghan believe in God? Did he believe that God can make his will known to us? Did he accept the creeds of the Church and the Church's teaching on sexual ethics? Did he believe in sin? In punishment for sin? . . . What is even more striking, however, is the seeming assumption by the priest-interviewer . . . that these questions of belief have absolutely nothing to do with the "spiritual assessment" of a clerical sexual predator. Here was the triumph of the therapeutic at its most disturbing.
>
> Why did the bishops of the United States repose such trust in psychology and psychiatry—among the most fluctuating of disciplines—in assessing cases of clerical sexual abuse? Why, in more than a few cases, did the bishops take counsel from therapists who were on the record as being in opposition to aspects of the Church's sexual ethic, particularly on questions of homosexuality? No serious person doubts that psychology and psychiatry can be of help in handling problems of clerical sexual abuse. The question is why so many bishops seemed to be willing to hand over such a high degree of analytic and decision-making authority in these cases to therapists.[52]

The pharmacy is a much more frequently discussed threat to the consciousness of freedom, if not to freedom itself. The arguments, if I understand them correctly, are that the frequent efficacy of drugs for a variety of emotional disorders implies that these conditions are beyond our control, thus expanding the domain of determinism. Or, it is claimed, choices made under the influence of drugs cannot be considered to be free decisions because it's the "Prozac talking, not the person," thus undermining the autonomous free self.

These arguments, as stated, are logically flawed. It is doubtful whether the first formulation, the argument for determinism, adds any relevant new knowledge to the debate. It is not as if we had previously believed in a self that chooses actions in a vacuum, untouched by natural causation. And it is not as if we now are swept away by a comprehensive account of our mental and spiritual functioning that leaves no room for freedom. Before Prozac (which I will use as a shorthand for the whole pharmacological menu) we knew that various substances could affect our moods and judgment. With Prozac, as with other internal and external causes, we are left with a complicated picture: no drug, unless it brings about unconsciousness, totally usurps our mental functioning. The new biochemistry may define more precisely the interaction of the drug with the individual, and this may be disturbing to some. It may lead people who had not thought about the issue to consider whether their freedom is far more limited in certain situations than they had previously imagined. But these discoveries do not inherently change the terms of debate.[53]

Likewise with the argument that drug use necessarily entails loss of the self as a moral agent. Would anybody insist that supping on salmon (lately dubbed "nature's Prozac") undermines my moral freedom and resolve, or that when my spirits have been restored due to my having indulged in the consolations of Brahms's "Academic Festival Overture" or Schubert's "Trout Quintet" my thoughts and feelings do not belong to me but to the fish or the composer/performers respectively? It is dismaying when individuals opt reflexively for the more passive path, represented by drugs, shunning more effortful, but ultimately more satisfying spiritual regimens, even preferring the pill to such depression fighters as physical exercise. Yet at times it is a justified choice. Is it not the case that taking my mind off painful and distracting psychological states liberates one for the task of making free, creative decisions in other areas of life? Can it not be said that the free decision to undertake a drug regimen is as free an act as any other?

Andrew Solomon in his best-selling survey of depression has eloquently articulated this position:

> I think that all of us have from nature a thing called will. I reject the notion of chemical predestination, and I reject the moral loophole it creates. There is a unity that includes who we are and how we strive to be good people and how we go to pieces and how we put ourselves back together again. It includes taking medication and getting electric shock and falling in love and worshiping gods and sciences.[54]

This is strong enough to warm the cockles of every libertarian heart. But maybe it's just a bit too glib. Solomon's optimism presupposes a great deal, both about individuals and about contemporary society. It may be reasonable, perhaps even necessary, for people suffering certain kinds of mental distress to control their moods through resort to prescription drugs, just as is the case for physical pain or illness. But we should not pretend that the prevalence of such practice has no effect on individuals and on the society as a whole.

The radical helplessness of physical disability or disease affects sufferers' experience of freedom in a variety of ways: at one extreme, total retreat from action and responsibility; at the other extreme, the strenuous activism of the two Presidents Roosevelt. Perhaps the same range of response is available to those who make an intelligent choice for drugs. Faced by testimony to urgent, intolerable suffering one has little appetite to dispute John Lennon: "Whatever gets you through the night/is all right, all right." I am not in a position to assess the inner lives of others differently situated than I; certainly not to judge the depth of pain and dysfunction that makes medication an optimal solution. The social impact of the boom in pharmacology may be easier to discern, especially in a conformist utilitarian society where people, and HMOs, are trained to identify the most economic solution as the only rational option. The resulting devaluation and delegitimizing of spiritually "labor intensive" approaches to character development may make it more and more difficult for individuals or families to resist the pressure to take the less strenuous, and increasingly more traveled road to mental health.[55] In any event, we cannot avoid the simple logic of R. Dessler's principle: a person whose experience teaches him that he is powerless over himself is liable to act accordingly. If we continue down that path, as we probably shall, we cannot close our eyes to the price, and cannot fail to consider ways of mitigating it, by respecting and fortifying the individual's sense of freedom. To be a patient does not mean being passive; where responsibility has not been suppressed, the *gavra* has not been eclipsed.[56]

Am I the Captain of My Soul?

The relationship between the Orthodox community and the therapeutic society exhibits an all too familiar pattern—initial obliviousness or dismissal melting, after an appropriate tape delay, into uncritical surrender. When it comes to free will, there is an abstract and unqualified affirmation of free will on ceremonial occasions, when a boundless freedom is trotted out, decked out in its festive kittel. The rest of the time, people sluggishly or enthusiastically take their cue from the faceless opinion-shaping "they" whose job it is to know about these things.

A critical stance toward the regnant culture may impel a concurrent reexamination of our understanding of Judaism. If the popular conception of the Jewish doctrine of free will assumes a sovereign executive self, master of its domain, the reformulated ideas presented in this essay revise that conception, not merely in order to provide a truce between religion and social science, but to recapture certain dimensions of the Torah's truth.

In one of his oral discourses of the 1970s, R. Soloveitchik asked his audience to propose the one adjective that best sums up the human condition. His own answer was "vulnerable."[57] On one occasion he asked why free will is not enumerated as one of the thirteen principles, and suggested that it is because people do not always have free will.[58] As we have already noted, we petition God not only for our material welfare, but also for His help in our spiritual strivings. For Judaism the myth of the Victorian "self-made man who worships his maker," is as alien as the "depraved because deprived" jingle of the social workers.[59] The former is the heresy of the successful; the latter, the heresy of those who are seen as failures. The two illusions can consolidate their appeal: a line runs from the grandeur of Emerson's self-reliance to the contemporary gospel of unmerited self-esteem: the self, confident in its autonomous innocence, because whatever is not innocent can be blamed on something outside the self. Much of the optimism we have detected in R. Kook derives from the realization that the restraints on untrammeled human willfulness are not a curse but often a blessing. We may emulate divine creativity, and we ought to accept full responsibility for who we are and what we have done, but we do not truly fashion ourselves unaided and unguided. If we cannot make peace with this insight, we will have trouble retaining, or recovering, the experience of freedom, and we will be blind to our condition and our situation before God.

A prominent feature of our intuitive conception of freedom is that it involves novelty and creativity. The mysteriousness of the creative gesture means that we not only do not exercise total control over our actions and our motivations, but also that we are substantially opaque to ourselves. We do not fully understand ourselves nor are the actions of which we are the authors and for which we are responsible perspicuous to us. Depending on temperament and circumstances this awareness can be exhilarating or terrifying. If we intend to cultivate free will as essential to our moral experience, we must learn to enter into and tolerate, if not exult in, the mysteriousness and uncertainty at the center of our lives.

To do so we must resist ideological impulses that reject contingency in the name of theological rationalism. The science-intoxicated philosophers of the "block universe" satirized by James, who feared that "the smallest modicum of independence, the faintest tremor of ambiguity about the future" spelled chaos, the utilitarians with their dream of infallible axiological calculations, have their counterpart among our religious folk. For the rationalist there is only one possible way for the world to be and the one optimal way for the world to be permits not the slightest deviation.[60] As we have seen above, this rigid, militantly unimaginative approach to living and thinking is incompatible with the imagination of free will. Hence an outlook that fully embraces the experience of freedom can help to liberate us from an emotional and experiential narrowness in other regions of religious life.

Judging Others

It may be objected that our preoccupation, in this essay, with the creative and hence ambiguous aspects of the experience of freedom, ignores the role of free will as the foundation of juridical responsibility. R. Kook, as we have seen, dismisses these considerations as relatively unimportant. Even R. Dessler, with his conception of freedom that is tied to the borderline conflict, offers little help in explaining why and how the apparatus of punishment should operate. The analysis of limited realistic freedom that we have sketched may be useful in illuminating our personal lives, the way we think about the choices we make, our attitudes toward others, our experiences of competence and vulnerability. Like R. Kook, I deem the cultivation of the self in its relation to God and man as a more profound undertaking than the rationalization of a political-juridical order. Yet assessing blame in a court of law, and the conclusions drawn from such verdicts, cannot be eliminated from the analysis of freedom and responsibility. Thus the question remains: Can our discussion coexist with the role of free will in propping up legal accountability?

If legal responsibility cannot survive without the absolute assurance that the individual could have acted otherwise, and if factors of moral luck endlessly complicate the calculation of the individual's freedom respecting the action under consideration, then our analysis does nothing to help. The criminal law often, if not always, requires an all-or-nothing assignment of responsibility, and our analysis doesn't provide one. The only obvious, though unsatisfactory evasion would be a soft determinism according to which moral sanctions and punishments motivate agents who are without freedom in the metaphysical sense.

The creative model we have worked with may provide another, less obvious way of imagining the problem. For the creative model, freedom inheres less in the action than in the agent. The question is not whether we can delimit with any precision the degree of power exercised by the person's Desslerian will over the action. It is whether this person is capable of creative involvement in the challenges and opportunities that face him or her: Is this individual a *gavra*, so to speak? If the person is a competent human agent, this entails some measure of responsibility and a concomitant presumption of accountability. The human evaluator (halakhic or secular) can only pass judgment based on his own eyes: A person who qualifies as an agent is then treated as free, so long as there is no conclusive reason to deny the presumption.[61] The indeterminacy and fluidity of freedom, our inability to arrive at fixed, immutable judgments regarding the mysteries of free agency, are thus compatible with the workings of the legal system. But only so long as we remember: man sees with his eyes; only God sees to the heart.[62]

At the risk of resorting to homiletics, let me elaborate on this point by reference to a puzzling halakha. R. Kahana taught that if the members of the Sanhedrin failed to raise any argument in favor of the accused, he is exempt from capital punishment. R. Zvi Hirsh Chajes adduces his knowledge of nineteenth-century Western European jurisprudence in an attempt to explain the paradox:

> In every affair there are far-fetched considerations for acquittal from the death penalty. One may say that perhaps his mind was not clear then, and without intelligence his act does not deserve the punishment of death, and other far-fetched excuses. We observe that in France and Great Britain each criminal and rebel and murderer are allowed a defense attorney to argue on his behalf and intervene for him even if he perpetrated the greatest wickedness. Elsewhere it is the judges who seek exonerating factors. Since this court immediately determined guilt and did not discover any distant reason for acquittal, this demonstrates that the court is not well versed, so that their judgment is not law, or that they decided the law hastily, without proper investigation and criticism and discrimination.[63]

In the light of our analysis, we may suggest that a competent court, according to R. Chajes' reading, must acknowledge the mystery and uncertainty that shadow all attempts to judge the actions of other human beings, although the final verdict holds the defendant fully responsible. Although the court must determine guilt and responsibility on the assumption that a responsible person is held accountable for his, or her, acts, so long as they are not demonstrably unfree, the judges are not exempt from bringing to bear the same considerations that would influence their response in ordinary life. If they fail to do so, their judgment is invalid.

Creativity and freedom grow out of the indeterminate and flourish in the soil that is hidden from all mortal supervision. The moment what should be indeterminate becomes hardened and rigidly fixed, the creative gesture has lost its raw material. Whence the sadness in the earnest letter to the editor asking how to utilize R. Dessler's thought to determine with precision one's own *nekuddat ha-behirah.*[64] That is why a tacit belief in determinism is the necessary prelude to the life of determinism. That is also why privacy is essential to freedom. A soul that has been stretched out on the rack and forced to lay bare its secrets, a soul that has been defined and classified out of existence, no longer informs a person capable of the mysterious gesture of creation. It has succumbed to an evil panoptic eye.

Avoiding the Evil Eye

R. Yohanan did not worry about becoming the object of the evil eye (*ayin ha-ra*); he was, he said, descended from Joseph, who was immune to the evil eye. The Talmud (*Berakhot* 20a) offers two prooftexts to establish this. One (Gen. 49:22) implies that Joseph was superior to the evil eye; the other infers his immunity from Jacob's blessing to Joseph's sons: "May they multiply like fish in the midst of the earth (Gen. 48:16)." R. Kook explains the dual source as reflecting the two ways whereby the individual can withstand the intrusions of malicious looks and influences: One is that the soul knows its great value. The action of another has no effect because the person is "concentrated and fortified in his perfection." This corresponds to the notion of Joseph being superior to the evil eye. The second is when the soul's perfection is inward and hidden from others. This corresponds to the comparison of Joseph's children to fish that are hidden from view and therefore safe from hostile glances.[65]

Today's fish are not always so lucky. Consider the prolific cod, endangered by the even more prolific fisheries. One disadvantage of the cod is its tameness. When hooked, it tends to give in, rather than to strive mightily for its freedom. Unlike Joseph, it is not able to maintain its superiority over those who would exploit it. The other is the development of sonar, which enables fishermen to penetrate the aquatic wall of privacy that had, from time immemorial, shielded the fish from prying eyes.[66]

R. Kook's discussion of the evil eye and the fish can serve as a metaphor for our own existential struggle against the spirit of determinism, which is, I have argued, a more urgent task than defusing the metaphysical challenge posed by purely philosophical arguments. If we feel overcome by mechanistic ideologies and lack a robust religious sense of self, it is because we have not emulated the Joseph who was invulnerable to the evil eye. We have not sufficiently sustained the existential superiority that is ours as self-respecting participants in *Torat Emet*. Nor have we safeguarded the realm of privacy and individuality from calculating glances and sterile conformity. Having been plunged into these uncharted waters, we confront the urgent need to rethink and re-experience the categories of freedom and responsibility: the problem is before us, "like an impossible sum on a slate." The spiritual destiny of humanity depends on whether our generation can muster the right kind of creativity and imagination to recover its soul. "Death of cod" ecologists are desperate to do for the threatened fish what the fish cannot do for itself. Those of us who dread what C. S. Lewis called "the abolition of man" must accept responsibility and do what we can to reverse our own predicament, even as we pray that God do for us what we cannot do for ourselves.[67]

Notes

1. James Boswell, *The Life of Samuel Johnson L.L.D.* (New York, nd), 352 (October 10, 1769).

2. "Notes from the Year 1920," cited by Hannah Arendt, *Between Past and Future* (New York, 1961), 7.

3. *Pahad Yitzhak, Iggerot u-Ketavim* (Brooklyn, 1981), 71.

4. Mark Steiner: "Philosophizing in Yiddish: Rabbi Reuven Agushewitz on Freedom of the Will," *The Torah U-Madda Journal* 11 (2003):24. It is too bad that Prof. Steiner's juicy translation of Agushewitz's idiomatic Yiddish appeared after I completed this essay.

5. *The Wanderer and His Shadow* §9. Elsewhere, of course, Nietzsche regards free will as a superstitious consequence of the moralistic desire to impose responsibility, to praise and to blame. For a recent discussion, see R. Abbey, *Nietzsche's Middle Period*, 24–33.

6. See *Shemonah Perakim*, ch. 8.

7. See the title essay in P. F. Strawson, *Freedom and Resentment*, reprinted in *Free Will*, ed. Derek Pereboom (Indianapolis, 1997), 119–42. On the question of how broadly to classify the reactive emotions pertinent to the Strawsonian judgment, and the allied question as to whether such resentment is a virtue or a vice, see R. J. Wallace, *Responsibility and the Moral Sentiments* (Cambridge, MA: Harvard University Press, 1994) and Gary Watson, "Responsibility and the Limits of Evil: Variations on a Strawsonian Theme," in F. Schoeman, ed., *Responsibility, Character and the Emotions: New Essays in Moral Psychology* (Cambridge, 1987). One can also imagine a Levinasian version of this argument. Here the moral agent would become aware of his freedom, not through the experience of confident, successful exercise of will, but as a phenomenological feature of responsibility for the other.

8. See the title essay in Bernard Williams, *Moral Luck* and Thomas Nagel, "Moral Luck," in his *Mortal Questions*, for two influential discussions.

9. G. Ainslie, *Breakdown of Will* (Cambridge University Press, 2001), 12.

10. This would appear to be Kant's position in the Third Antinomy of the *Critique of Pure Reason*.

11. See R. Sorabji, *Necessity, Cause and Blame: Perspectives on Aristotle's Theory* (Ithaca, NY: Cornell University Press, 1980), chap. 1.

12. W. James, "The Dilemma of Determinism," in *Essays on Faith and Morals* (New York, 1962), 154ff. Cf. Agushewitz (21–22):

> By contrast, the indeterminist does not regard himself as such an authority [*halt zikh nit far aza mumkhe*] on the interpretation of the rule of causality; he holds that our cognition is insufficient to give the rule of causality such a far-reaching exegesis . . . we are presented with a cognition concerning the universality of the rule of causality, while we know that our reason cannot be in any position to draw a conclusion like this for every corner of the universe.

13. John Searle's recent *Rationality in Action* revives quantum uncertainty as part of the incompatibilist position despite these reasons for discomfort.

14. For the idea of the therapeutic as the radical foundation of contemporary secular culture, see Phillip Reiff, *Triumph of the Therapeutic*, especially chap. 1.

15. The term in quotes is Asher Friedman's. My conversations with him were particularly valuable in working out the ideas in this section.

16. For a systematic attack on the idea of addiction as disease, see Stanton Peele, *Diseasing of America: Addiction Treatment Out of Control* (Lexington, MA, 1989) a work well regarded in some philosophical circles.

17. Consider, as an example, David Donald's interpretation of character in his acclaimed biography *Lincoln*. Donald maintains that Lincoln never acted on his own initiative, but only in reaction to events. He could not have produced the Emancipation Proclamation any earlier than he did, for instance, because there was insufficient support for it; nor could he have delayed any longer because by then the pressure was too great. Though Lincoln may have justified his actions in this way, for political or psychological reasons, the only evidence that Lincoln's behavior was determined in this way is the fact that he acted in that way. The story can be thickened with all the knowledge at our disposal, and the various factors can be analyzed to the last decimal point, so to speak. But the conclusion follows from the premises by circular reasoning.

18. Mind/body dualism is not extinct. See, for example, *Evolution of the Soul*, by the Christian philosopher Richard Swinburne. See also the book-length debate between Sydney Shoemaker and Swinburne, *Personal Identity* (Oxford, 1984). I am happy to leave the distillation and elucidation of this vexed topic to David Shatz's chapter.

19. See Ramban to Genesis 1:26.

20. Compare to the discussion in Robert Kane, *The Significance of Free Will* (Oxford, 1996), especially chap. 6.

21. Kierkegaard, *Sickness Unto Death*, beginning of First Part, *Fear and Trembling and Sickness Unto Death*, trans. Walter Lowrie (Princeton, NJ: Princeton University Press, 1941), 146. I have combined Lowrie's translation with that of the Hongs for readability. The early German translator skipped the passage.

22. D. Dennett, *Elbow Room: The Varieties of Free Will Worth Wanting* (Oxford, 1984), 136.

23. Agushewitz, 24, presents a similar obstacle to the imagination of free will by "reason." He writes:

> [I]n order that a feeling should be heard from we must have at the time at least a clear recollection of it—but how can we have a clear recollection of something that we do not understand, something against reason? . . . What kind of a freedom should we have in mind when we speculate concerning determinism and indeterminism? Only a fiction, a freedom, which is fabricated momentarily [*beshasmayse*]. This kind of freedom is in fact nothing but an illusion; worse, it is a speculation of something of which we know only that it is against our reason.

24. W. James, *The Principles of Psychology*, vol. 2 of *The Works of William James* (Cambridge, MA: Harvard University Press, 1981), 1032–33.

25. Peter Van Inwagen, "When Is the Will Free?," in *Agents, Causes and Events*, ed. T. O'Connor (Oxford, 1995), 219–38.

26. See R. Joseph Soloveitchik, *Halakhic Man*, Part II and *Yemei Zikkaron*, chap. 1.

27. "Treatise on Choice: Part I," in *Mikhtav Me-Eliyyahu*, vol. 1 (Bnei Berak, 1964), 111–116.

28. "Treatise on Choice," 116.

29. See discussion by Benzion Sorotzkin, "Bechirah: How Free Is Free Will," *Jewish Observer* (April 1996), 17–21, with subsequent reactions, September 1996, 40–44.

30. "Treatise on Choice: Parts III–IV," in *Mikhtav Me-Eliyyahu*, vol. 4 (Jerusalem, 1983), 93–108. The time lag between the publication of the first three volumes of *Mikhtav Me-Eliyyahu* and the appearance of the subsequent volumes may be responsible for the paucity of references to this piece.

31. "Treatise on Choice: Parts III–IV," 95.

32. See, *inter alia*, *Nazir* 23a.

33. vol. 1, 307.

34. vol. 1, 112.

35. "Treatise on Choice, Part II," vol. 1, 278–83.

36. In another text (vol. 3, 263ff.) R. Dessler implies that the inner perspective, as opposed to the external, does provide direct access to reality.

37. 807 (April 15, 1778).

38. See also T. Ross, "Man and His Moral Choice in the Thought of R. Dessler," *Daat* 12 (1984): 111–26 and particularly 119 n. 39 on elements of Kantian metaphysics. Ross attributes some of R. Dessler's deviations from Kant to Kabbalistic themes in his thought.

39. *Iggerot ha-Reiyah*, vol. 1 §283 (Jerusalem, 1985), 319–22 and vol. 2 §379, 40–44. For English translation, see Rav A. Y. Kook, *Selected Letters*, trans. Tzvi Feldman (Ma'aleh Adumim, 1986), 139–61.

40. For R. Kook's commitment to the goodness of life as an ontological perception, see S. Carmy (ed.), *Jewish Perspectives on the Experience of Suffering* (Jason Aronson, 1999) 112ff.

41. *Sefer ha-Ikkarim* IV, 5.

42. *Orot ha-Kodesh*, vol. 3 (Jerusalem, 1964), 3:35, 297. R. Kook mentions Talmudic statements (*Nedarim* 51a and *Nedarim* 20b) referring to both kinds of desire. In the phrase quoted, it is unclear whether it is the resentment that cannot be uprooted or the cause of the resentment, i.e. the sexual desire. For a practical ruling in a case of (suspected) homosexual behavior, see Kook, *Daat Kohen*, §3.

43. *Orot ha-Teshuvah* 16:1, note added beginning with the 5th ed. On the question of repentance transforming the past in other twentieth-century Jewish thinkers, see Yitzchak Blau, "Creative Repentance: On Rabbi Joseph B. Soloveitchik's Concept of Teshuvah," in *Exploring the Thought of Rabbi Joseph B. Soloveitchik*, ed. M. Angel (Hoboken, NJ: 1997), in particular p. 274 n. 30. The description of R. Kook as a seeming Davidsonianism *avant la lettre* draws on S. Carmy, "On Repentance," in E. Gellman, *Essays on the Thought and Philosophy of Rabbi Kook* (Rutherford, NJ, 1991), 114–20.

44. Compare R. Kook's remarks about the kernel of truth in Schopenhauer's doctrine of the blind will that impels the universe. See *Orot ha-Kodesh*, vol. II, 4:2:15–16, 482ff. Though R. Kook's integration of his discussion on free will with his strong sense, and optimism about, divine providence indicate affinity with Hasidic teaching, in particular with the doctrines of Izbic and R. Zadok of Lublin, his philosophy is committed

to free will in a way that the Hasidic views apparently are not. See J. Gellman, "The Denial of Freedom in Hasidic Thought," and M. Nehorai, "Determinism and Ethics in the Teachings of Rav Hasdai Crescas and Rav Abraham Isaac Hacohen Kook," in C. Manekin and M. Kellner, *Freedom and Moral Responsibility: General and Jewish Perspectives* (College Park: University of Maryland Press, 1997), 111–32 and 205–16 respectively.

45. Hilary Bok, *Freedom and Responsibility* (Princeton, NJ: Princeton University Press, 1998), 51.

46. Theodore Dreiser, *An American Tragedy*, II:20 (New York, 1964), 292.

47. Henry James, *What Maisie Knew* (Oxford, 1996), chap. 31, 259.

48. Robert Pippin, in his *Henry James and Modern Moral Life* (Cambridge, 2000), has also noted the absence of traditional transcendent motivations in James. For this reason he regards James's fiction as prophetic of a post-modern age in which these factors have ceased to be definitive.

49. M. Valverde, *Diseases of the Will: Alcohol and the Dilemmas of Freedom* (Cambridge, 1998), 205.

50. Meditation at the end of *Neilah* service concluding Yom Kippur.

51. R. Kook, *Orot ha-Teshuvah*, chap. 2; R. Soloveitchik, *Al ha-Teshuvah*, ed. P. Peli, commenting on Rambam, *Hil. Teshuvah*, ch. 7.

52. George Weigel, *The Courage to Be Catholic: Crisis, Reform and the Future of the Church* (New York, 2002) 103ff.

53. See on this, and on the parallel discussion of genetics, P. S. Greenspan, "Free Will and the Genome Project," *Philosophy and Public Affairs* 21, 31–43.

54. Andrew Solomon, *The Noonday Demon: An Atlas of Depression* (New York, 2001), 432.

55. Similar concerns attach to prospective advances in genetics. It is conceivable that the designer child may become the ultimate status symbol of the upper middle classes. Such children, if the scenario holds, will perpetuate themselves as a controlling elite, leaving behind those sectors of the population who, because of economic constraints, a conviction of the importance of old-fashioned character development, or taboos against abortion and infanticide, are unable to compete with the wealthy enlightened classes. The fact that genetic engineering or selection is voluntary does not entail that it serves a genuine expansion of freedom.

56. If this appears overly abstract, set aside the situation of adults capable of informed consent, who come into treatment with a reasonably well-established sense of personal autonomy and Desslerian freedom. Consider the patient who is a child, and who has no other paradigm of action than to be diagnosed, doctored, and dosed, questioned before bedtime, whispered about after, intruded on and unalone.

57. It is astonishing that most twentieth-century moral theory posits fully competent human beings as participants in ethical discourse, and treats those who are not as exceptions, oblivious to the fact that the young and the old and the disabled make up a large share of humanity, and that most of us will spend a major part of our lives in these categories. For the recent philosophical discovery of this fact, see Alastair MacIntyre, *Dependent Rational Beings* (London, 1999).

58. At the time I took this as a reference to Rambam, *Hil. Teshuvah*, ch. 6. On other occasions, the Rav presented R. Yosef Albo's answer: free will is not a dogma of Judaism, because it is an axiom of all moral systems, not only those founded on revelation (*Sefer ha-Ikkarim* I, 9).

59. The last quoted phrase is from Stephen Sondheim's lyrics to "West Side Story," Act II, Scene II, in *Romeo and Juliet/West Side Story* (New York, 1965), 207.

60. Often Rambam's view on choice without preference is met with utter disbelief in these circles. Rambam, as is well known, held that it is futile to explain the specific details of certain *mitzvot*, because there is no reason to prefer one option to another (*Guide* III:26; see also A. Hyman, "A Note on Maimonides' Classification of Law," *PAAJR* 46–47, 339ff). This is taken by rationalists to be scandalous because God must always have a reason. God, on this view (which, by the way, raises problems regarding divine omnipotence) cannot create a world that is no better or worse than some alternate possible world. If there were two such worlds, God would be unable to choose, on rational grounds, which one of them to create. This is Spinozism in overdrive!

61. In evaluating extenuating factors, of course, the court may well end up incorporating aspects of our analysis.

62. I Samuel 16:7 and *Pesahim* 54b. See also *Noda bi-Yehudah, Mahadura Kamma, Orah Hayyim* §35, about punishment at the hands of a human court where there is evidence of repentance and expiation, and the editor's references, p. 362 n. 24.

63. R. Z. H. Chajes, Notes to *Sanhedrin* 17a. See also the dismissive comments of R. Barukh Epstein, *Torah Temimah* to Exodus 23:2, n. 17.

64. *Jewish Observer*, September 1996, p. 44.

65. R. Kook, *Ein Ayah* to *Berakhot* 20a, 3:37. It is worth noting that while R. Kook ascribes piscine immunity to *ayin ha-ra* to the symbolically valuable quality of being concealed from the eye, Maharal regards fish as spiritually inferior, beneath the level at which rational discourse (and the evil eye) can operate (see *Gur Aryeh* to Gen. 7:2 and *Hiddushei Aggadah* to *Kiddushin* 13a). Maharal's comments place him closer than R. Kook to the spirit of medieval rationalism.

66. See Mark Kurlansky, *Cod: A Biography of the Fish that Changed the World* (Toronto, 1997).

67. Without my discussions with David Shatz over the years and during the writing itself, my essay would not have been as good, though it would have been longer. Judith Bleich made useful comments on the draft. I have gained much from conversations with Yitzchak Blau and especially Asher Friedman. Benyamin Ron was my partner for a close reading of the key passages in R. Dessler and R. Kook. I had looked forward to the comments of my revered mentor R. Walter Wurzburger *z"l* and was discomfited by his unexpected absence from the Forum; later I learned that he had been hospitalized the night before.

II

PERSONAL ACCOUNTABILITY AND THE RELIGIOUS COMMUNITY

5

Choice-Diminished Behavior and Religious-Communal Policy

Basil Herring

Man Is Free to Choose

THIS PAPER ACCEPTS THE VIEW THAT as far as the Torah and the rabbinic tradition are concerned, whether in the realm of halakhah or *hashkafah*, man is ultimately, and uniquely, a free agent. That is to say, he is fully endowed by his Creator with the will and ability to exercise genuine choices when confronted with a variety of alternatives and options. There have been various formulations of the parameters and consequences of this freedom, from Talmudic times on.[1] None, however, is more precise or authoritative than the classic summation found in Rambam, *Hilkhot Teshuvah*, chapters five through seven, which begins: "Choice is granted to every person. If he wishes to turn himself toward the good path and be a righteous person, he has the choice. And if he wishes to turn himself toward the bad path and become a wicked person, he has the choice" (5:1).[2]

Such predominant affirmation of man's freedom to choose was never shaken in any serious fashion. There were only occasional attempts to demur by affirming that human behavior is predetermined, whether by reason of natural causation or divine foreknowledge.[3] Paradoxically, it might be argued, the very ability of such lonely voices to argue their position against the received doctrine of their forebears and contemporaries, in itself attests to the ability of man to choose to dissent from, or act in spite of, any allegedly determinative causal chains or circumstances. In any event, the consensus of belief in free will remained steadfast, and its consequence was a broadly held affirmation that man, being free, is therefore to be held responsible, both by God and by man, for his actions, his beliefs, and the consequences of his choices. The notable exceptions only prove the rabbinic rule. Thus, the scriptural descriptions of the removal of free choice and the ability to repent and the punishment meted out to Pharoah, Sichon, the Canaanites, Hofni and Phinehas (the sons of Eli), and the unrepentant Israelites in time of Elijah merely serve to affirm the essential truth that for everyone else choice and repentance remain intact and in place.[4] [Parenthetically, one might note a particularly elegant formulation of this human capacity for freedom and choice, albeit not from a rabbinic source, to be found in a celebrated mid-twentieth-century letter by Isaiah Berlin, on the subject of human dignity in an age of materialistic determinism (as encountered particularly in fascist and communist doctrines). There he states "that every human is assumed to possess the capacity to choose what to do, and what to be, however narrow the limits within which his choice may lie, however hemmed in by circumstances beyond his control . . . capable of pursuing ends for their own sakes by deliberate acts of choice, which alone make nobility noble and sacrifices sacrifices."[5]]

Since man is viewed as an ontologically free agent, ultimately unshackled from the chains of causal determinism or divine foreknowledge, we should not be surprised to find that the tradition affirms the ethical corollary of such liberty; in a pronounced and repeated demand, expressed in both halakhic and hashkafic teachings, the tradition stresses that man (and the Jew in particular) prevent, control, discipline, and subdue any illicit or improper thoughts, feelings, and behaviors—whether emanating from within himself or from his environment—and face the consequences of a failure to do so.

The Geography of the Soul: The *Yetzer ha-Ra* and the *Yetzer ha-Tov*

In the early rabbinic tradition, such control and self-discipline were not seen as being beyond man's ability to exercise. Unlike their Christian contemporaries, who saw post-Edenic man as a morally flawed agent whose corrupted nature required an external "savior" to redeem him from the effects of the "original sin," the talmudic sages affirmed that it is indeed within man's unaided power to choose good over evil, whether that evil emanates from within or without.[6] Evil forces were seen neither as irresistible nor dominant, but merely as representing one side of the coin of the human personality. That side was embodied in the notion of the *yetzer ha-ra* (as in Bereishit 8:21 "*yetzer lev ha-adam ra mi-ne'urav*"). It is man's great challenge, and glory, to overcome the *yetzer ha-ra*, by harnessing the liberating energies of his higher self, as embodied in the *yetzer ha-tov*. After all, says God, in the repercussive words of *Hazal*—"*barati yetzer ha-ra, barati lo Torah tavlin*" (Kiddushin 30b, Bava

Batra 16a). It is Torah, in all its multi-faceted complexity, that is the antidote and serum that fortifies man in his unceasing confrontation with evil in all its forms, whether they emanate from God, human impulse, or nature.

This is not to say that *Hazal* underestimated the difficulties involved at every step of the way, even for the greatest of men. As they put it, "*kol ha-gadol me-haveiro yitzro gadol hemennu*"—"the greater the man and the higher his spiritual elevation, the stronger the enticements of his *yetzer ha-ra*." The only exception to this rule, as Maimonides famously put it in the sixth chapter of *Shemonah Perakim*, is the *hasid* who is entirely free of the world's seductions. But for the rest of us, including the *tzaddik*, the challenge of dealing with the *yetzer ha-ra* is never absent.[7] And, indeed, *Hazal* duly recognized that the *yetzer ha-ra* does chalk up temporary victories leaving unredeemed, ethically compromised man to take himself, in one celebrated passage, when all else failed, to a distant city to indulge his *yetzer ha-ra* in the least damaging manner possible (Moed Katan 17a). But even then, it is to be noted, such behavior, while indulged, was not granted legitimacy: it was simply the lesser of two evils.

But such realism did not interfere with the persistence of the Sages' demand that man must continue the ceaseless confrontation between his higher and lower selves, with the precious and inestimable fortification provided by the *tavlin*, or antidote, of Torah. With all the difficulties and challenges thus encountered, every man and woman is seen as fully capable of achieving moral greatness and redemption, by virtue of his or her freely choosing to live in the embrace of, and submission to, a life suffused with Torah and *mitzvot*, blended together in splendid halakhic and ethical excellence.

And in case one retains any doubts as to the seriousness and scope of the demand of *Hazal* that man control his *yetzer ha-ra*, at least in regard to certain *aveirot*, one need recall only the talmudic discussion of an individual who fell passionately in love with a woman, and was overcome by sexual desire to the point that his life was hanging in the balance, unless his desires were requited.[8] Even with his life thus *in extremis*, still he was required to control rather than indulge his *yetzer ha-ra*, inflamed as it was by the prospect of consuming "forbidden fruit," or, as the Talmud there puts it, *mayim genuvim* (literally "stolen waters"), through this liaison with an unmarried woman. In fact, he was not even permitted to simply converse with her from behind a partition, consequences be damned.

This perspective did not change in the time of the Geonim. As understood by the Netziv, in his introduction to Shemot, the *Baal Halakhot Gedolot* reflected this very outlook when he referred to Sefer Shemot by the unconventional title *Sefer Sheni*, literally "The Second Book (of the Torah)." The Behag intended, according to the Netziv, to teach the intertwined and complementary nature of the first and second books of the Torah. That is, the natural phenomena associated with Creation and *maaseh bereishit* remained incomplete and tentative until Sinai, when the obligations of God's revealed law, synonymous with *Shemot* and *mattan Torah*, were given to Israel and mankind. By giving that Torah to man, God charged him with the responsibility of redeeming the causal, instinctual, and conditioned phenomena of the physical world, especially those that are found within man himself, by means of the Torah and its commandments. Human nature may be a given, but it is a given that it is man's responsibility to control and transform.

A contemporary formulation of this principle is encountered in the recently published writings of Rav Joseph B. Soloveitchik. "Humans may be subject to the operation of causal law, but they are also confronted by divine norms which require them to exercise their freedom by liberating themselves from complete domination by instinctual drives . . . not by repression (of instincts) but by their sanctification and redemption."[9] Indeed, in this view of the Rav, there are actually two causal frameworks: the first is in the physical and instinctual universe inhabited by natural laws, while the second inheres in the divine norms and imperatives that are encountered by man in God's Torah. Man, and particularly the Jew, achieves his redemption at the interstices of these two domains and causal chains, when he repeatedly finds the strength of character to empower his higher self as expressed in his *yetzer ha-tov*, rather than indulge the enticements of his lower and instinctual self, epitomized by the *yetzer ha-ra.*[10]

This same principle that empowers man to overcome his lower nature, leads the Rav to posit in powerful terms the conviction that man can overpower, and indeed harness, the very forces that might in the past have led to his moral failures, and direct them to a higher, redemptive, purpose—habit and past practice notwithstanding. As developed in detail in his delineation of the processes of *teshuvah*, the internal dynamics (libidinal energies) that led to the illicit behaviors themselves can become engines for growth, personal redemption, and transcendence.[11] As *Hazal* put it, "*Ilmalei yetzer ha-ra lo banah adam bayit, ve-lo nasa ishah*" (Bereishit Rabbah 9:8). It is the sublimation and channeling of the libidinal instinct at every stage that results in the building and maintenance of family, community, and society at large.

Accordingly, and this is the crux of the matter, there can be no justification (whether in the eyes of God or man) in the claim that a particular act forbidden by the Torah is innate, instinctual, or reflective of an acquired behavior trait, for even if true, the behavior itself is not thereby rendered "proper," "good," "legitimate," or "acceptable." Put somewhat differently, an individual may be understood as, and even *excused* for, acting in certain improper ways beyond his will, but that does not mean that the individual was in fact *justi-*

fied in so behaving. It remains the responsibility of the individual or the society to sanctify and redeem, via individual self-discipline, enlightened communal censure, social rejection, and institutional education, such forbidden, albeit "natural" and even understandable, urges, no matter how widespread or condoned they may be within the larger societal sphere.

The Consequences of Failure to Exercise Self-Discipline

As a result of this strong proclivity within Judaism to hold man responsible for his behavior, or for his failure to overcome his *yetzer ha-ra*, there was always a decided inclination to place sanctions of various kinds on such improper behaviors. These took various forms.

When the Jewish people were sovereign, and lived under the rule of the Sanhedrin and its associated courts and penal institutions, the whole panoply of biblical and rabbinic punishments and sanctions were meted out, including the death penalty, *makkot*, incarceration, and financial penalties. Once the powers of judicial authority and their legally enforceable punishments were removed, informal sanctions were put in place, such as social ostracizing via *herem/nidduy*. The specific form these took would of course depend on period, location, cultural setting, and particular circumstance.

Recognition of Choice-Diminished Behavior as a Mitigating Factor

All of the above notwithstanding, the rabbinic tradition also recognized certain limits and constraints on human behavior that, while not ultimately removing freedom of choice and its consequences, greatly complicate and diminish human freedom. As a result there was a concomitant readiness to accept a measured diminution of culpability for their being acted upon, at any given time or situation. We can refer to such actions or behavior as CDB, "choice-diminished behavior."

Among these constraints are instinct and habit. Man is a natural creature, sharing much with the animal world. Thus, he possesses, and is driven by, natural instincts as well as nurtured or acquired behavioral patterns. Since we are creatures of instinct, habit, ignorance, and socially accepted wrongful practices, the halakhah certainly recognized that men and women are not always as a practical matter able to choose a theoretically possible path.[12] Such people engage in CDB that is judged in the halakhah by a different standard.[13]

The halakhah similarly recognizes that some situations involve immediate conditions or circumstances that directly affect how that illicit behavior will be judged or punished, and how that individual or group will be treated by others or by the community, as in behavior that results from *ones*, or duress. This might even describe life-long patterns or ostensibly freely accepted choices.[14]

It is possible, in fact, to reach the point where an individual is not held liable at all for what might on the surface appear to be freely chosen illicit actions or behavior. A more familiar instance of this approach occurs in the context of suicide, whereby suicides as a class are treated as committed unintentionally, and therefore are to be exonerated.[15]

Cases in which, after sincere and sustained effort, certain illicit behaviors cannot be controlled, and are at the same time not injurious to others (e.g., consensual non-public sexual offenses), then become candidates for a judicial turning of the proverbial blind eye, and opting for a policy that chooses the lesser of two evils.[16]

Practical Policy Consequences and Alternatives

All of the above conceptualizations can be seen as leading to a number of principles and practical approaches in formulating communal responses to illicit or immoral CDB. First, we must accept unequivocally that the fact that an illicit behavior is seen as CDB does not effect the judgment of whether it is right or wrong, moral or immoral. We cannot accept the so-called naturalistic fallacy, which argues that because "that's the way it is," one can conclude that "that's the way it ought to be." Neither does the number of people who engage in that behavior change that evaluation or make any difference in its halakhic acceptance or rejection; if in a given locale eight out of ten men abuse their wives, or molest minors, or act with CDB, that is not a reason to say "so many do it, it must be natural/good, and therefore we should permit it." To the contrary, the more widespread the incidence of the problematic behavior, the more reason one could adduce to be *less* receptive, and more rejecting, of that behavior in a moral context.[17] The only question, from the communal policy perspective, is a tactical one: what is the most beneficial method or response in dealing with those inappropriate behaviors or broader social realities?[18]

Even in those situations in which CDB behaviors are excused, although not justified, few will deny that under certain circumstances CDB is not to be tolerated at all. Thus, for instance, where the behavior is not only prohibited but harmful to others, either physically or emotionally, there ought to be no tolerance or leniency whatsoever. Even though one might argue that such behaviors have their roots in profound psychological disability or pathology and are therefore not as deserving of punishment as those completely freely chosen, these behaviors must be anathematized nonetheless.[19] When properly identified or diagnosed, such behavior (and this would include sexual harassment, kleptomania, and physical or emotional abuse), should be forcibly, and swiftly, dealt with, via actions that would include termination of employment, notification of governmental authorities, and a refusal to facilitate future employment via positive references, etc., especially when minors are involved.[20]

The truly difficult issues arise when the CDB is not directly harmful to others. In such cases, absent the full halakhic judicial framework of sanctions and punishments, and in a contemporary "open" societal setting, a carefully constructed case can be made favoring restraint and caution in responding to such behaviors, especially when trying to identify the motivations and causes from which they emanate. If this is true of punishments in general, it is especially true of CDB.[21] Moreover, the concern and goal would not be to establish or carry out *punishments*, but rather to formulate and apply appropriate *sanctions* that will prevent such improper behaviors in the future, whether by the individual concerned, or by others whose choices might be negatively influenced, whether directly or indirectly, were such behavior to go unchecked.

Before constructing such cases, it is important and useful to make a clear demarcation between adult synagogue settings on the one hand, and schools/youth group settings on the other. Take the cases of the non-violent drug addict or an alcoholic, both of whom are likely to die a premature death. If such individuals wish to participate fully in synagogue life, including attendance at synagogue *kiddushim*, *semahot*, and social events, what should synagogue policy toward them be?

It would appear that the institution can and should properly permit and encourage their participation in synagogue life, albeit with appropriate conditions and safeguards to ensure proper maintenance of public health, personal dignity, and communal decorum. As long as this person's outward behavior conforms to the range of institutionally accepted public behavior standards, what he or she does with his private life should remain private and confidential. Thus, an alcoholic or substance abuser could and should be welcomed, on condition that while in the confines of the synagogue or its services and programs, there can be no recognizable drunkenness or chemically altered state of mind. Consequently, as long as these agreements are honored, there need be no institutional sanctions against him or her on account of such behavior, halakhically improper though it may be.

There is a correlative privacy issue to be dealt with here. Assuming adherence to such agreements, it would seem that on grounds of privacy there would be no need to, and every reason not to, notify the general membership of the synagogue of the problem at hand. At the same time, the question does arise if the family of the individual concerned is unaware of the problem, or if the individual becomes romantically involved with another in the synagogue:

is there an obligation to share this information with, or somehow alert, the other party? Without getting into the specifics, the guiding principle would seem to be dictated by an honest evaluation of the likely impact and material effect of the particular CDB on these other parties, whether measured by halakhic, financial, mental health, or similar criteria and considerations. It would appear that there is no generalized "right to know," while respect for the individual's privacy in such matters would be an important consideration.[22]

Perhaps the most obvious implication of CDB is with regard to homosexual behavior or feelings. On the one hand, this paper accepts the argument, made by a growing number of therapists, that homosexual thoughts and behaviors can be successfully treated and permanently changed (or, at the very least, individuals can significantly reduce their homosexual feelings or behavior) in as many as two-thirds of those homosexuals who seek help by entering appropriate therapy settings, typically accomplished over a two- to three-year period.[23] Such treatment and transition is clearly the optimal response, personally as well as communally, to homosexuality. On the other hand, however, there is a segment of the homosexual population that fails to transition successfully to a heterosexual orientation (for reasons that may include ignorance of, lack of access to, or less than complete motivation to finish, the available therapeutic options). For them, and for those who are in an interim or transitional state, responsible and sensitive policy guidelines must be articulated.

To that end, the following guidelines present themselves. They assume that the individuals under discussion are self-identifying Orthodox Jews, observant of halakhah in the other "usual" respects associated with a life of Torah and *mitzvot,* and also assume that there are no infectious health issues (i.e., AIDS) involved, or if there are, that they can be safely addressed.

When such individuals seek to be "accepted" within institutional Orthodoxy, a careful distinction can be made. On the one hand, if acceptance means that a homosexual individual who wishes to participate in synagogue life should be welcomed and not be discriminated against or excluded, then such acceptance is quite correct, appropriate, and feasible. In this sense, acceptance can be extended as long as he or she understands that this does not in any way validate, legitimate, or grant communal blessing to private homosexual sexual behavior, feelings, or preferences. Acceptance of such persons as full-fledged members of the community, however, does not extend to acceptance of what the Torah and rabbinic tradition unalterably consider deviant behavior, which should be *sans* public expression in the synagogue. While comparisons are always invidious, there is a fundamental parallel here to other religiously deviant behaviors: just as we can insist that a synagogue member who is a Sabbath violator should not park in front of the *shul* on *Shabbat,* or that one who has profound doubts about *tehiyyat ha-meitim* not articulate them in public so as to sway others to his point of view, so, too, a homosexual can properly be welcomed into communal life, at the same time that he is expected to keep his halakhically illicit feelings or behaviors to himself.[24] It is the public behavior, or overt efforts to have it recognized as "an alternative lifestyle," which should be avoided and appropriately censured. This distinction between acceptance of the person, as opposed to acceptance of problematic behaviors, with the former being welcomed while the latter is placed under sanction, is parallel to the principle enunciated by Beruriah, wife of R. Meir, in the familiar words *yittammu hata'im ve-lo hote'im*, "let sin be eradicated, not sinners."

Yet, even at this point, in dealing with "sin," a case could be made that it is not always necessary to institute sanctions against, or make judgments of, problematic behavior. After all, when faced with inordinate societal moral turpitude (whether of a violent or sexual nature), the Sanhedrin as a tactical matter refrained from sitting in judgment over such matters, and thereby suspended at least temporarily what would otherwise have been strict judicial sanctions for such taboo behavior.[25] Insistence on the strict application of the halakhah in such matters would need to be tempered with consideration of prevailing psychological and social realities. And potentially self-defeating or counter-productive short-term policies and actions, might need to be balanced, in classic halakhic fashion, by longer-term considerations, in the interests of promoting better halakhic, moral, as well as spiritual, communities.

In this regard, it is important to avoid practicing selective sanctions. For example, how does an institution justify being more tolerant or inclusive to those who are known to commit certain serious illicit behaviors, such as denial of elements of the thirteen *ikkarim*, desecration of *Shabbat*, engaging in pre- or extramarital cohabitation, or dishonest business practices that could potentially bring shame upon the community, while excluding or placing sanctions on others, such as alcoholics, drug abusers, or acknowledged homosexuals? Moreover, from a public policy perspective, whatever arguments may be marshaled in their support, the rationale behind exclusionary policies must not only be cogent, but must also be perceived by the community or those effected as being fair and reasonable. And this is true even though synagogues are legally considered private associations that can exclude individuals for defined reasons. Civil courts in the United States have upheld the right of a synagogue to exclude certain individuals, such as disruptive or non-*shomer Shabbat* Jews. The issue in this regard, therefore, is not a legal one, but one of a halakhic/moral/public policy nature. If the membership is "uncomfortable" having a privately homosexual individual in their midst, they may legally ask him or her to leave. The question, however, is whether this is the "right" and appropriate thing to do.

The school setting involves additional crucial considerations. While synagogues do, of course, promote educational efforts and projects, a yeshiva is overwhelmingly, and by its very nature, a place in which character formation of impressionable young men and women is an integral part of its mission. Moreover, social and environmental influences are central to fulfillment of that mission. This demands special care and concern in formulating educational institutional responses to CDB. Thus, even if such illicit behavior is not directly harmful to others, individuals who practice it might have to be excluded or placed under serious sanctions.[26]

In light of these many complex and sensitive considerations, involving conflicting issues and interests in educational institutions, it is of utmost importance that such institutions formulate policies and guidelines *ab initio*, in a comprehensive and structured fashion, rather than by merely responding on an ad hoc, ex post facto basis. Furthermore, given the sensitive nature of the issues involved, it would be most helpful if this were accomplished and endorsed by as many institutions as possible, acting in concert with each other, thus diminishing any undeserved stigma, while also facilitating broad implementation and compliance. Such policies and standards should be widely distributed and publicly acknowledged, by each institution, its staff, the parent body, and of course the students themselves. And, following the point made earlier, it would be imperative to make clear that the behavior in question is wrong and halakhically forbidden, even if, for various practical or educational reasons, punishments or sanctions may not always be instituted.

A further consideration: Even though a given behavior might be the result of CDB, a school might want or feel the need to make "an example" of an individual for the good of the student body, or the institution or the community. By overlooking egregious behavior, it can be argued, one risks a slippery slope by which it might become "permissible" for others to imitate or tolerate this behavior, or behaviors that are even worse. While such arguments are surely cogent, it is also true that there should be recognition of the best interests of the individual concerned, as well as a realistic evaluation of what may or may not ensue if certain behaviors are forgiven, or temporarily overlooked.[27] As applied to CDB behavior in a school, part of any such calculus will involve evaluation of whether the case reflects repeated behavior by this individual, and whether or not it is a widespread or merely isolated phenomenon within the school body.[28] Indeed, the very investigation of allegations or suspicions raised should be undertaken by unbiased and independent investigators, given the vortex of competing concerns, considerations, and inexperience that might affect an "in-house" investigator.

Added to the above, it is to be recognized that if local civil or judicial authorities have requirements in dealing with such behaviors that differ with indicated halakhic policies, the range of choices available to an institution or its leadership becomes correspondingly limited or more complex. Where the behavior in question is both illegal and dangerous (e.g., selling hard drugs), the school would be duty-bound to report it to the authorities.[29]

In formulating such policies, institutions must consider the criteria by which behavior will be evaluated and treated. The roles and guidance of counseling professionals, as separate from rabbinic/halakhic authority, need to be factored in and determined, so as to integrate differing perspectives in diagnosis and response. While the halakhic process and perspective allow for scientific and clinical determination of psycho-social realities and the treatment or policy response that might ensue from them, it is not always clear how far a school's response might extend beyond classic halakhic punishment or treatment and counseling modalities. For instance, if the most effective available drug rehabilitation program involves likely Shabbat or *kashrut* violation (as in the popular "Wilderness" programs), one could argue that this would qualify for inclusion in the permissive stance of Rabbi Moshe Feinstein toward using a non-Jewish sanatorium for the benefit of the emotionally disturbed.[30]

Another dynamic would involve a claim of having done *teshuvah*, or repentance. At which point might an individual be considered rehabilitated by virtue of genuine remorse and resolve to pursue a new path? Making a determination in such a situation that would readmit an individual who has suffered with CDB would perforce need to involve a judicious mix of enlightened rabbinic authority and professional evaluation. A similar determination would also need to occur at the other end of the spectrum: that is, in diagnosing a case of incorrigible CDB, when do we apply a "one strike and you're out" policy? Does, for instance, one act of sexual harassment suffice to trigger the chain of consequence and preventive policy implementation? Multiple perspectives would need to be consulted in making such a determination.

Educational institutions must also make serious efforts to ensure that parents are part of the solution, especially when it is believed that they may be part of the problem. Wherever possible, they should be made aware of their responsibilities, as the primary influence and agent in forming and influencing their children's character and behavior, in partnership with the school, both in prevention and intervention of CDB behaviors.

In summation, effective communal responses to what we have called "choice-diminished behavior" must necessarily be complex and nuanced. They need to reflect both principled judgment and practical considerations. These responses must be sensitive to individual needs, even while mindful of short- and long-term communal consequences; and they must be faithful to classic halakhic/hashkafic teachings and precedents while remaining keenly aware of the insights of contemporary social sciences in understanding human behavior and popular cultural mores. Only then will they serve to raise the standards of personal and communal behavior so as to fulfill the Sinaitic mandate to be a *goy kadosh*, a sacred nation and a just society, distinguished by the Jew's faithfulness and loyalty to God, His Torah, and its challenge to preserve and enhance the highest possible embodiment of ethical and spiritual excellence.

Notes

I would like to thank Dr. David Shatz for a number of very helpful and constructive suggestions.

1. See, for instance, the sources referenced in Efraim Urbach, *Hazal, Emunot ve-Deiot* (Jerusalem: Magnes Press, 1975), 229ff.

2. This interpretation of Rambam follows the well-traveled traditionalist road of Maimonidean research, as opposed to some of the more recent approaches, such as the so-called Compatibilistic School of Altman, Pines, Stern, and Sokol.

3. This point was most famously argued in the doctrines of Hasdai Crescas—see Guttmann, *Philosophies of Judaism*, trans. David W. Silverman (Garden City, NY: Anchor Books, 1966), 238.

4. See Rambam, *Hilkhot Teshuvah* 6:3.

5. Isaiah Berlin, letter to George Kennan, reprinted in *The New Republic*, 28 January 2002, 23ff.

6. Urbach, *Hazal*, 415, 423.

7. For a moving description of that inner struggle, see R. Bahya ibn Pakuda's *Hovot ha-Levavot*, vol. 2, *Shaar Yihud Hama'aseh*, 51–53.

8. See Sanhedrin 75a, and Rambam, *Yesodei ha-Torah* 5:9.

9. Rabbi Joseph Soloveitchik, "The Redemption of Sexual Life," in *Family Redeemed*, ed. David Shatz and Joel B. Wolowelsky (New York: Toras HoRav Foundation, 2000), 73ff.

10. On this, see Walter Wurzburger, "Review of *Family Redeemed*," *Tradition* 35, no. 2 (Summer 2001): 75. See also the analysis by Marvin Fox, in *Interpreting Maimonides* (Chicago: University of Chicago Press, 1990), 141–51, especially p. 149, in which he elaborates on Maimonides' view (as opposed to the medieval Christian view, associated with Aquinas) that the divine law is absolutely authoritative, protects people from debilitating bodily passions, orders proper relations between men in society, and frees human reason in everyone to rise to the highest level of self-realization.

11. See the Rav, *Al ha-Teshuvah*, ed. Pinchas Peli (Jerusalem: World Zionist Organization, 1974), especially the entire chapter titled "*Beur ha-Ra O Haalato*" ("Evil: Its Destruction or Elevation?"), 149–87.

12. Such as in the cases of the *tinok she-nishbah* ("the abducted baby") and acts undertaken *be-shogeg*, or through inadvertence.

13. A more recent formulation that significantly expands the scope of CDB is found in Eliyahu Dessler, *Mikhtav me-Eliyyahu*, ed. Aryeh Carmel (Jerusalem, 1994), vol. 1, 111ff.

14. See Maimonides, *Iggeret Teiman* and *Iggeret ha-Shemad*, for his evaluation of those Jews and communities who had "converted" to Islam under duress when the conventional requirements of halakhah would have required submission unto death. More recently, instances of similar halakhic attitudes can be encountered in the Holocaust response literature.

15. See Benjamin Gesundheit, "A Halakhic and Moral Analysis of *Masekhet Semahot*, chap. 2, Laws 1–6," *Tradition* 35, no. 3, (Fall 2001): 45, in which he concludes that

> although his act is completely unjustified from the objective perspective of principle, *Hazal* in *Masekhet Semahot* do not condemn him for it, nor even raise the possibility of judging him as falling into the category of "one who kills himself intentionally." On the contrary, they instituted a halakhic-social practice aimed at preventing any recurrence of this phenomenon in the future.

16. See Urbach, *Hazal*, 314, on choosing between sin with *Hillul Hashem*, or sin without *Hillul Hashem* (as with uncontrollable impulses that are permitted by anonymous indulgence in a distant city). See above.

17. See my *Jewish Ethics and Halakhah for Our Time: Sources and Commentary* (New York: Ktav Publishing House and Yeshiva University Press, 1984), vol. 1, on capital punishment, in the context of Rambam's extra-judicial punishment "*mippenei tikkun ha-olam*" that is applied precisely when abuses are more widespread. There is also the notion of emergency communal enactments, under the rubric of *eit la-asot*, "it is a time for precipitous action."

18. See R. Aharon Lichtenstein, "Formulating Responses in an Egalitarian Age" (paper presented at the annual meeting of the Orthodox Forum, New York, 2001), 17–18.

19. Even while subjected to therapy wherever feasible. See Abraham Twerski, *The Shame Borne in Silence: Spouse Abuse in the Jewish Community* (Pittsburgh: Mirkov Publications, 1996), 134ff. Even here, however, as Twerski points out, we must be careful in defining exactly what is included in emotional abuse. Not every manipulative argument or cruel insult automatically qualifies.

20. One can imagine the following exchange: "Judge, I have been diagnosed as a pedophile, in fact I myself was abused as a child, therefore I had no choice but to do what I did, and don't deserve to be sent to jail." "That may be true, but as a judge I am obligated to protect society from people like you, therefore I too have no choice but to sentence you to a lengthy prison term, to prevent you and others like you from such behavior in the future." A complicating factor arises in situations where the adult victims are themselves "willing" victims. Examples might include battered wives who are either in denial, or unwilling to leave the marriage because of fear, financial insecurity, love, and so on. Can they be involuntarily "saved"? It would appear not. This would be similar to the case of intervention by a third party, the *matzil*, to save a *nirdaf* (one who is being pursued for his life) from a *rodef* (his homicidal pursuer),

where the *nirdaf* refuses the help of the *matzil.* In the case of minors, no such concern would occur, as the Bet Din would be the *apotropos* (protective regent) acting in the best interest of minors, even against their, or their parents' will.

21. See R. Aharon Lichtenstein, "Formulating Responses in an Egalitarian Age," 12–15, with reference to *Bava Kamma* 38b, etc.

22. On this delicate and complex matter, see the recent New York State judicial disclosure rulings regarding halakhically illicit behaviors and rabbinic responsibilities to disclose them to family members.

23. See the article by Arthur Goldberg and Elaine Berk, at www.jonahweb.org, and also the information available at www.narth.com. This is a case in which popular or conventional wisdom has followed a path that is politically correct, rather than scientifically sound or statistically borne out. Indeed, many in the traditional community itself are not aware of findings such as these that go a long way to support the biblical and rabbinic prohibitions in this regard.

24. Indeed, some might argue that this is true of any sexual or romantic behavior and preferences, even of the permitted heterosexual kind.

25. See *Sotah* 9:9, *Avodah Zarah* 8b, etc., on the principle of *keivan de-nefishi rotzehi/mena'afi.*

26. Daniel Sperber, *Minhagei Yisrael* (Hoboken, NJ: Ktav, 1999), vol. 6, regarding *ad de-la yada* on Purim, esp. 214ff.

27. In this regard, see the important considerations raised by Rav Aharon Lichtenstein, "Formulating Responses," 19–20, in evaluating when or whether it is appropriate to invoke the slippery slope argument.

28. In this regard, see also Maimonides' Guide III:34, where he explains that the Torah and its laws do not deal only with individuals, but also with the larger good, even if as a result a given individual may suffer.

29. This, in turn, raises questions related to *mesirah* in our society. On the parameters of *mesirah* in a society such as the contemporary United States, see Michael J. Broyde, *Journal of Halacha and Contemporary Society* 43(2002): 5–48.

30. See *Iggerot Moshe, Yoreh Deah* II: 59, and *Orah Hayyim* II:88. Others would oppose this view.

6

If an Abuser Cannot Control His Impulses, What Is the Responsibility of Other Adults in the Community? A Response to Basil Herring

Rivkah Teitz Blau

Community Responsibilities

THE PROPER RESPONSE OF THE COMMUNITY to the individual whose sexually abusive behavior is, in Rabbi Basil Herring's terms, "choice diminished,"[1] is straightforward: Once a person has abused a youngster, he cannot serve in any position that will present him with another opportunity. We cannot take a chance that he will make the same choice, however free or constrained, a second time.

Any single instance of abusive behavior can set into motion a tragic process. Consider what the youngster experiences when an adult he respects abuses him. Often, he thinks, "I must have done something to deserve this," a reaction that feeds the abuser's need to make the victim feel responsible for the abuse. When a youth advisor suddenly kicks a boy, the child, now in physical and emotional distress, thinks, "When will the pain stop? Why was I chosen for this?" When a teenager who spends *Shabbat* at a teacher's home awakens to find him in her room, she thinks, "Did I do something to attract this unwanted attention?" If he touches her, there may be an involuntary physical response on her part, which will further convince her that she must be guilty. If the teacher now tells her never to share what has happened with anyone else, she will feel that she is a partner in the crime. The secrecy that serves the abuser so well is now firmly in place. Finally, if the abuser is a religious mentor, the foundations of the child's world are shattered by the abuse, a consequence exacerbated by any inadequacy in the religious community's response.

The potential further damage to children in the event that even a one-time abuser is allowed to continue working with them is incalculable.

The responsibility to prevent the abuser from interacting with children does not rest on the community leadership alone. *No one* who becomes aware of the abuse may remain neutral. Experts have persuasively argued that neutrality is actually a moral choice, since "all the perpetrator asks is that the bystander do nothing. He appeals to the universal human desire to see, hear, and speak no evil. The victim, on the contrary, asks the bystander to share the burden of pain. The victim demands action, engagement, and remembering." Because the perpetrator "seeks out situations where his tyrannical behavior will be tolerated, condoned, or admired" and because speaking about the crime invites "public humiliation, ridicule, and disbelief," the victim feels powerless. She cannot talk about the abuse, no less deal with her feelings of shame, her loss of place in the community, and her loss of trust in human goodness.[2] People who have not studied the literature cannot fathom the seriousness of the trauma.

Abuse, furthermore, is not done in front of witnesses, and the failure of adults to act leaves children with no protection. If the child is persuaded that he is "damaged goods" and can never get a hearing, he may even consider suicide the only option. It is for good reason that Rabbi David Cohen has applied the term *rodef* (pursuer) to a molester,[3] as the abuser threatens the child's physical and emotional well-being, and even the child's life.

An adult who disregards children's reports, then, becomes an accomplice in the crime. The longer he disregards the evidence, the more difficult it becomes to admit that his sin of omission is enabling sins of commission. While his indifference tells young victims that they are helpless, it signals to the abuser that he can do as he pleases. Emboldened by what he perceives as acquiescence, the abuser's behavior will worsen.

The argument that the perpetrator's accomplishments balance his crime is unacceptable. If an executive director raises millions of dollars but steals thousands of dollars for himself, would any school or synagogue allow him to keep his job? The price of one child becoming observant cannot be the destruction of another.[4]

Practical Proposals

The religious community needs to address this problem on a number of fronts.

First, a review board consisting of trained professionals must investigate every accusation. Dr. Chaim Kranzler, who has seen this system work for the benefit of both staff and children in a secular setting, has in fact made this suggestion. The benefits are many. Children and parents will know they can report to people outside the school or other environment where an incident has occurred. Furthermore, adults who work with children will be confident that allegations without merit will be probed and dismissed, while accusations that can be substantiated will result in the removal of the perpetrator from contact with children. Employers will also have a reliable source to check before hiring new personnel. If we do not provide a mechanism within the community, the only method to stop abuse will involve enlisting government agencies.

We must also teach our children to recognize abusive behavior, to resist it, and to report it. If we teach children that no adult should hurt them verbally or physically, in "fun" or otherwise, and that when an adult acts inappropriately they should run out of the room and report what happened, we will eliminate surprise and secrecy from the abuser's arsenal. Parents, teachers, principals, summer camp staff, advisers, and heads of youth movements must learn how to respond to any hint of inappropriate behavior and know where to turn when there is evidence of abuse.

Finally, teaching the laws of *yihud*, which mandate that a man and woman who are not married to each other should not be alone in an enclosed space, will help prevent offenses. It should be made clear that under no circumstances may anyone claim that these laws do not apply.

Addressing these problems appropriately will enable us to fulfill the ideals of the Talmud:

> "Distance yourself from ugliness and anything that resembles it" [*Hullin* 44B].
>
> The Torah scholar "should be exacting with himself and do more than the law requires . . . to the point where everyone praises him, loves him and wants to do as he does . . . [thereby fulfilling Isaiah 49:3] 'You are my servant Israel in whom I will be glorified'" [Maimonides. *Hilkhot Yesodei Ha-Torah* 5:11].

Notes

1. Sexual abuse includes any instance in which an individual forces, manipulates, or tricks another into sexual contact of any kind. In the case of a young victim, the limitations of the youngster's ability to understand or to give informed consent are generally decisive in defining sexual interaction as sexual abuse.

2. See Judith L. Herman, *Trauma and Recovery* (New York: Basic Books, 1992), 7 ff., 28, 75.

3. Audiotape of OHEL Conference, September 26, 2000, "Let's Talk about What Never Happened. . . . But It Did: Helping Our Community Deal with Abuse," Side B, at thirty-two minutes. The speakers, R. Avraham Pam, R. Shmuel Kamenetsky, and

R. Yisrael Reisman, had asked R. Cohen to address whether or not one may announce the presence of a molester within the community. Based on decades of dealing with halakhic questions for OHEL and knowing that the abuser poses a mortal threat, R. Cohen replied that one is obligated to warn people.

4. Although not every victim becomes a perpetrator, most perpetrators have suffered abuse themselves. While it is necessary, therefore, to maintain appropriate compassion for perpetrators, everything must be done to prevent their acting out.

III

PSYCHOLOGICAL PERSPECTIVES ON THE PROCESS OF REPENTANCE

7

Psychotherapy and *Teshuvah*: Parallel and Overlapping Systems for Change

Michelle Friedman and Rachel Yehuda

PEOPLE COME TO MENTAL HEALTH TREATMENT because they are in pain. The presentations of their pain vary—they come because they suffer from symptoms that restrict or threaten their lives, because they struggle with inner conflicts that undermine and torment their integrity, or because if they don't come, they will lose their job, their spouse, or their children. People seeking mental health treatments desire change in either outlook or behavior that will result in the removal of their pain. Most people who seek mental health treatment understand—or quickly come to understand—that the alleviation of emotional pain requires active participation in a process that involves understanding the reasons for their suffering and making concrete changes accordingly.

Teshuvah, repentance, can be considered a Judaic analogue of the process through which one's emotional pain is linked with the requirement and inevitability of change. In *teshuvah*, the change takes the form of eradicating delinquent behavior according to the precepts and ideals of Jewish Law. Indeed, observant Jews who believe in a religiously ordered cosmos accept the responsibility to fulfill 613 commandments, *mitzvot*. Transgression of any of these commandments constitutes wrongdoing that the offender is obligated to correct. The observant Jew who commits a transgression presumably feels pain and perhaps guilt, reflecting a disturbance in his/her proper orientation in God's creation.

The religious prescription for *teshuvah* constitutes, according to many authorities, a *mitzvah* in its own right. *Teshuvah* includes several components—confession to God, remorse, restitution, and the capacity to act differently when placed in the same circumstances, including, *inter alia*, the same temptation.

Teshuvah is a profoundly powerful moral goal in and of itself, in part, no doubt, in recognition of how difficult change can be, even if it is motivated by pain or guilt. The process of *teshuvah* is largely outlined in various halakhic texts where different sages address the nature of the transgression that has been committed. Sins against one's fellow man, for example, require that specific behavioral restitution be made. *Teshuvah* following theft, therefore, can only be accomplished after the stolen property has been returned.

Like psychiatric treatment, the process of *teshuvah* is optimally performed when a person expresses a desire and commitment for change. The individual engaging in *teshuvah* has a well-defined idea of the deficiencies that must be corrected and a relatively clear picture of the optimal end result. In contrast, in psychiatric treatment, the unfolding articulation of what needs to be changed or better understood is often the transformative component of the treatment. Indeed, psychiatric treatment begins with a clinical interview, a highly specialized dialogue between therapist and patient. The therapist sits with the patient and invites him to reveal his misery: "Please, tell me what brings you here."

Most therapists will agree on the diagnosis, or the proximal cause of the suffering. The therapist's specific orientation, the theory of human behavior to which she subscribes, however, will determine the clinical formulation, along with suggestions of treatment options. Mental health treatments span the gamut from "talking cures" that delve into the distant past and explore the present situation, to those more strictly focused on cognitive/behavioral goals. Individual, family, or group therapy may be advised. Mood altering medication might also be prescribed.

In this paper we look at psychiatric treatment and *teshuvah* as systems of change. We posit that these systems are distinct, yet overlapping, and we focus on three areas where these points of convergence are manifest: locus of authority, use of the transference, and establishment of goals. We then illustrate these points using case examples. These vignettes will facilitate discussion regarding whether, and to what extent, familiarity with psychological principles might facilitate and enhance the *teshuvah* experience and, conversely, whether, and to what extent, an understanding of *teshuvah* can enrich the psychotherapeutic process.

Factors Affecting Behavioral Change in Psychiatric Treatment and *Teshuvah*

Locus of Authority

Following the initial evaluation, the therapist helps the patient set the specific goals of the treatment designed to alleviate the patient's pain. Sometimes these goals are straightforward—for example, the reduction of symptoms such as insomnia, anxiety, or depression, or the implementation of behavioral changes such as anger management or impulse control. In such cases, patients may or may not wish to understand the origin of the symptoms or behaviors that they now wish to change.

In some cases, however, patients seek to understand their current problems in the context of repeated patterns or as outcomes of earlier adversity before deciding whether it is feasible, or even appropriate, to change their behaviors or circumstances. For some patients, the creation of new circumstances might be too difficult. Instead, they seek a therapeutic process that will help them cope with circumstances as they are. For example, individuals in abusive marriages may not feel able to leave the marriage or to significantly alter their spouses' behavior. They may benefit, however, from insights regarding their own reactions and behaviors in such relationships, or from concrete suggestions of how to augment other support systems to ameliorate the damaging effects of relationships they have chosen to maintain.

Thus, the psychotherapeutic process is one in which the patient is the final arbiter of what is right or wrong for them. Although the therapist may often point out that certain behaviors or attitudes are inconsistent with the patient's articulated goals, the therapist does not direct, but more guides the process of "recovery" or change. He or she is a kind of psychological lamplighter, helping the patient illuminate his or her inner conflicts and desires without imposing or advocating any particular moral or behavioral standards.

Observant Jews presumably have a code of behavior that, if adhered to, implicitly leads to a correct and good life. When a Jew experiences emotional pain, he is taught to reflect on his actions to determine whether he has departed from a set of behaviors that would mitigate the pain, including distress arising from external adversity. Thus, the religious Jew's desire to engage in *teshuvah* may be prompted by feelings similar to those suffered by a secular person seeking behavioral change through psychotherapy.

The process of *teshuvah* in Judaism provides a powerful structure through which a person's emotional pain can be contained. The laws of *teshuvah* not only operationalize the process of change, but also offer a system of social support and destigmatization by delineating a season, the High Holy Days, in which all Jews are expected to be engaged in this activity. *Teshuvah* is also a process that allows people to participate in a dialogue with God, to take responsibility in a world where there is divine oversight.

An extension of this idea also holds great power for many religious Jews. For them, wrongdoing is directly connected with suffering—the punishment is pain. From this perspective, failure to do *teshuvah* carries with it the threat of personal and/or communal tragedy. In this emotional and religious climate, *teshuvah* may be motivated more by the fear of divine reprisal than by the primary desire to realign oneself with God.

Herein lies a major difference between *teshuvah* and psychotherapy systems. Depending on one's understanding of *teshuvah*, the desire for change arises out of fear of, or love of, God. Religious life commends and cultivates such motivation, based, as it is, on awe of God as the ultimate authority. Psychotherapy, however, looks down on incentives primarily founded on avoiding or pleasing a powerful "other." Efforts to get a patient to change "for the therapist" would be regarded with dismay. Such technique would be criticized for serving the therapist's needs more than the patient's. Over time, fostering such dependence could be unhealthy as well as expensive. The patient would become emotionally shackled, relying on treatment indefinitely to maintain beneficial results. The locus of authority rests within the individual; ideally, the patient will grow out of the therapeutic scaffolding and still maintain permanent change.

Transference

Key to both psychotherapy and *teshuvah* is the enormous emotional power of the therapeutic relationship. In psychotherapy, much has been made of the relationship between patient and therapist, known as transference. Transference refers to a process in which the patient invests the therapist with a constellation of feelings whose origins are based on earlier relationships. The therapist may therefore be regarded by the patient as a benevolent or tyrannical parent, an approving or condemning authority, or an admiring or rejecting lover. By maintaining an initially neutral stance, the therapist comes to understand the type of salient interpersonal conflicts regularly experienced by the patient.

Whether the formal contact is through one-time crisis intervention or many-year psychotherapy, the consultation room is a kind of laboratory, a microcosm of the patient's typical relationships, conflicts, and behaviors. More psychoanalytically oriented therapies explicitly set up the treatment situation to encourage transference development. The therapist chooses to maintain relative anonymity and a non-directive stance so as to maximally nurture a transference development that can then be analyzed.

The therapist interprets transference in many ways, depending on his or her particular theoretical orientation and personal style. However, the therapist must be respectful of and responsible to the power of transference, since it is the bedrock of the therapeutic relationship, as well as the patient's template for learning about the self. Therapy is a dyadic experience; the patient's emotional responses to the therapist and the therapist's awareness of feelings catalyzed by him- or herself provide the emotional fuel that ignites change.

Teshuvah also depends on the power of transference. For the penitent Jew, a wide range of relationships may be involved. Another individual may have been wronged—a family member, a friend, a colleague, a stranger. The notion of restitution to a fellow human being, while at times embarrassing or difficult, would not strike any thoughtful person as strange. *Teshuvah,* however, recognizes that infraction of any commandment also damages a Jew's connection with God. That relationship must be restored as well. How, one might reasonably ask, can the power of the transference be harnessed in the absence of a human partner? We suggest that the *teshuvah* process with God also calls upon a series of transferences.

How does an individual human being create and maintain a relationship with God? The psychologist appreciates a person's inner representation of God as a composite of feelings and fantasies that originate in earliest infancy. Our liturgy reflects this multiplicity of images and emotions. We beseech God with words that we use in relationships with those we love and fear. We speak to God as we would to our mother or father, a judge, or a monarch.

The religious wrongdoer's sense of responsibility may shift between self-blame and self-justification: "Surely God, who created and understands human nature can look into my soul and understand what compelled me to behave as I do, particularly if there are untoward circumstances that affected my judgment." Presumably, the religious person feels a vibrant, immediate relationship with God. That person might feel angry or critical of God for allowing him/herself to slip into sin, just as a child might feel anger towards the parent who did not prevent the child from engaging in dangerous play that resulted in physical pain. In this context, at least part of true *teshuvah* involves forgiving God for the unhappiness that causes such people to hurt others or themselves.

On a community level, the penitent Jew may seek guidance from a trusted teacher or rabbi. He or she endows that religious authority with expectations of approval and disapproval that the psychologist would trace to earlier parental attitudes. For many individuals, the rabbi can also serve as a psychological intermediary for the sinner to approach God. The supplicant Jew might feel shame, anger, or any of a myriad of difficult feelings in front of a rabbi. Transference dynamics are activated. The rabbi is in a powerful position to call the congregant's attention to his/her responsibility for infractions and to encourage inquiry as to why that wrong behavior occurred. How much should the rabbi pressure his penitent congregant in a specific direction? We might consider such tactics in a psychotherapist to be manipulative. The extent to which the same standards of neutrality apply to the rabbi is important to consider. The rabbi, even when motivated by the most noble desire to help a Jew struggling to reconcile with God, must be aware of the powers inherent in persuasion.

Goals

The initial goal of the mental health establishment is to relieve acute suffering. Over time, in the context of the therapeutic relationship, other goals may develop. These include cultivation of insight, empathy, and other traits that might loosely be termed "character growth." The therapeutic result, however, is not measured by conformity to any one cultural convention. The predominant value of therapy is to lessen psychic conflict within the rubric of a basic ethical structure. Of course, psychotherapy is practiced by individuals who each bear the stamp of his/her own inborn temperament, upbringing and culture. Much of the work of training is to identify and analyze one's own prejudices and blind spots and to be aware of how they may enter into the therapeutic work.

A question that arises is whether we consider "character growth" that hopefully develops under therapeutic conditions to be instances of *teshuvah.* Does the relinquishing of harmful behaviors and the restoration of well-being pave the way for religious return if there is no genuine remorse? Take a case in which an abusive husband is remanded for treatment. The court-appointed psychiatrist suspects a mood disorder. The husband agrees to take a psychotropic medication and undergo anger management therapy. He no longer beats his wife, but he still believes that she is lazy, wasteful, and flirtatious and deserved every insult or smack he gave her in the past. Is this *teshuvah*?

To the extent that people have neurobiological vulnerabilities that predispose them to cope in maladaptive ways, it is interesting to think about whether the effort to comply with treatments which are time consuming, costly, and possibly accompanied by unpleasant side effects, in itself constitutes *teshuvah.* It is hoped that the clinician participates in a process that determines the underlying source of pain so that it can be controlled or stopped permanently. Many people contribute to their own pain by using and re-using ineffective coping tools. They can take an active role in reducing their own suffering by working to expand their repertoires of responses, strategies, and solutions. The therapeutic relationship allows for this type of exploration. Some clinicians feel that emotional pain can be addressed neurobiologically, but few believe that this is all that needs to be done. For most people, how to get through life with minimal pain is not immediately obvious and requires a lifelong process of exploration and trial and error.

Case Studies

Case 1: The Desperate Wife

Mrs. Cohen, a forty-year-old married Orthodox woman and the mother of an eight-year-old son, called Dr. One to arrange an appointment to discuss her escalating anxiety, which had reached panic proportions. They agreed to meet several days later. Shortly after introducing herself at the first session, Mrs. Cohen announced that she was scheduled for an abortion that same afternoon. Dr. One was surprised that Mrs. Cohen had not mentioned this matter over the phone. The therapist inquired further.

The therapist learned that Mrs. Cohen is the second wife of an older businessman whom she met while he was separated from his first wife, the mother of their four adult children. After a lengthy divorce proceeding, the Cohens married, with the understanding that they would have only one child together. The marriage, which had produced a son that both parents adored, was generally happy. In recent months, however, Mrs. Cohen felt her husband becoming more detached and less sexually interested. She worried that his attentions might be waning, perhaps even wandering to another woman. One night, when he seemed amorous, they had unprotected relations. Mrs. Cohen was now seven weeks pregnant.

Dr. One felt that Mrs. Cohen's hasty decision to have an abortion was fueled by overwhelming, unprocessed feelings. She urged her patient to postpone the procedure until they could discuss it at least a few more times. Mrs. Cohen adamantly refused. She returned the next week having terminated the pregnancy.

In the weeks that followed, Mrs. Cohen became more and more distraught over what she had done. She and Dr. One set up a twice-weekly psychotherapy schedule. Therapy revealed that Mrs. Cohen had told her husband neither about the pregnancy nor about the abortion. She felt rigidly bound to her promise. Though Mrs. Cohen longed for a second child, she insisted that her husband would resent the burden at this point in their life and be angry at her.

Mrs. Cohen, an intelligent and psychologically inquisitive woman, engaged in the psychotherapy deeply. She revealed that as a child, she was the designated, responsible care-taker of a large brood whose parents were self-absorbed and benevolently negligent. Mrs. Cohen alternated between relishing her authority over her siblings and resenting their ceaseless demands. She longed for her father's and mother's full attentions and crafted a fantasy life in which she was the sole, beloved child.

Over the next several months, Mrs. Cohen's anxiety was replaced by a deep sadness. She was consumed with an overwhelming regret over the abortion. Together with Dr. One, she formulated a hypothesis as to why she had become so depressed. Her wish to maintain what she perceived to be precarious status as favored wife was in conflict with her desire to have more children. She had supplanted one wife; perhaps she would be pushed aside for yet another woman. Further, her anger at her own siblings had been unconsciously ignited. Just as she craved being the sole object of her parents' attentions, she unconsciously struggled with her wish to share her husband's affection with another child. Further, even if she were to dare and ask her husband to support her desire to bear the second child, she would be upsetting her own son's privileged position and dooming him to move over for a younger child, just as she had had to do many times.

These insights enabled Mrs. Cohen to tell her husband what had happened. Mr. Cohen expressed his sincere sorrow for her suffering and his unwavering love for her. He did not, however, wish to have more children. While Mrs. Cohen did not challenge her husband's decision, she recognized that her fantasy of doing *teshuvah* by getting pregnant yet again and bearing a child was not a realistic option. The sight of pregnant women and babies on the street brought her to tears.

Mrs. Cohen remained in therapy for several years. Over that time, she achieved significant insights that allowed her to function more fully and feel more deeply in her roles as wife, mother, professional and religious woman. She felt that she had not been fully present in her earlier relationship with God due to the distractions of her inner conflicts. She recognized how the unplanned pregnancy and the abortion represented a dramatic demonstration of long standing self-hatred and insecurity. Part of the therapeutic work was learning to live with some measure of anger and remorse. Mrs. Cohen always became sad around the time that would have been her due date. She wished that she had understood herself better earlier. She would have had more choices; she might not have become pregnant unexpectedly, or, if she did conceive "by accident," she might not have excluded her husband from the dilemma; she might have even had the baby.

For Mrs. Cohen, psychotherapy paved the way for a more full *teshuvah* experience. The insights gleaned through a classical transference analysis allowed her to take responsibility for her behavior, mourn her losses, and cultivate deeper relationships with those significant to her, including God.

Case 2: The Guilty Bookworm

Mrs. Stern, a thirty-eight-year-old Hasidic homemaker and mother of seven, came for psychiatric evaluation complaining of low grade depression of many years duration. She described her symptoms as tearfulness, sadness, and overreacting to situations in which she felt criticized. She was happily married, financially well off, and loved her children, but she felt distant from her own life and unable to fully participate in the bounty that was before her. Her one enjoyment was retiring to her room and reading fiction. There was, however, a catch. Mrs. Stern was a voracious reader whose literary curiosity strayed outside the bounds of the Yiddish language libraries maintained by her community. She wanted to read modern fiction but felt terribly guilty. Hasidic culture, according to her understanding, discouraged and even banned secular literature.

Although Mrs. Stern was a reflective and introspective woman who had spent some time in psychotherapy previously, she found the insights gleaned from that work interesting, but not especially useful in relieving her dysphoria. This time, she wanted to try medication. Dr. Two agreed to a trial of anti-depressant.

Mrs. Stern returned to Dr. Two's office after six weeks. She reported a dramatic response to sertraline, one of the newer generation of anti-depressants in the serotonin re-uptake inhibitor family. She felt more relaxed, alive, and emotionally involved than ever before. Her husband and children remarked how much happier she seemed. Dr. Two, the psychiatrist, also noticed a pleasant brightness in Mrs. Stern's manner. She wondered if Mrs. Stern's reading dilemma was affected and asked her patient. Mrs. Stern was eager to discuss the topic.

Mrs. Stern had been a reader since childhood, when she attended a religious but non-Hasidic school that allowed some secular literature. She married a Hasidic man and began having children. When her eldest child began school, Mrs. Stern became more aware of her Hasidic sect's disapproval of secular reading material. Mr. Stern did not appear to be bothered by his wife's interest, and Mrs. Stern did not have to hide her reading from him. Still, she felt uneasy, but continued to borrow books from the local library.

When Mrs. Stern's fourth child was born, the baby had some minor medical complications. Her toddler was sick as well. Sitting in the hospital, Mrs. Stern came across her New York Public Library card and impulsively ripped it up, vowing that if God allowed her children to recover, she would not take library books out again.

The children got better. Mrs. Stern's desire to read was as strong as ever. With a rueful smile, she told Dr. Two about her compromise. She would allow herself to buy books from Barnes & Noble, thereby not technically violating her promise. She joked that it was ridiculous to spend money when she could borrow the same books for free, but "That is how it is." Further, while her mood was sunnier on the medication, she still felt a keen sense of confinement in her life. Curling up with an English book in her bed, with the door closed, was her greatest pleasure, and, no matter how she procured that book, she felt guilty about it.

Dr. Two was certain that Mrs. Stern's chronic depression had lifted as a result of the medication. She also noted a greater availability on the patient's part to examine and reflect upon complex issues. Dr. Two understood Mrs. Stern's destruction of her library card and vow in the hospital as a kind of spontaneous *teshuvah* act, a bargain forged in the heat of her children's illness but layered over her long-standing guilt about reading. Would it be advisable, wondered Dr. Two, to probe this further? What, at this point, should the goals of the therapy be? She suggested psychotherapy for her patient. Mrs. Stern, however, declined the invitation. While she acknowledged the potential insights that she might gain, she didn't want to rock the boat. She agreed to come in for check-ups at six-month intervals.

With a thoughtfully written prescription, Dr. Two had succeeded in alleviating Mrs. Stern's present complaint of low-grade depression. What about the long-term prognosis? Dr. Two suspected that Mrs. Stern chafed under the yoke of her rigidly defined Hasidic life. Would psychotherapeutic exploration allow her patient to make a more adaptive compromise between her private desires and the beliefs of her community? Should Dr. Two attempt to help Mrs. Stern feel less guilty about her reading? Or would energy best be spent helping the patient to adhere to the norms of her community by abandoning reading? Perhaps the most radical position Dr. Two might take would be to conclude that it was her obligation to "out" her patient, to get Mrs. Stern out of her reading "closet." She then might encourage the patient to make her reading more public and face the consequences, even if this would require altering, or even abandoning, her lifestyle.

Mrs. Stern knew that she could ask a *she'elah* about the reading. Taking that step, however, was risky. Dr. Two conjectured that the seemingly best outcome would be if the Rav told her patient to read discreetly and not feel bad about it. However, Dr. Two was not sure if the transference power of the Rav, the authoritative arbiter of religious law, would neutralize Mrs. Stern's guilt. Also, suspected the therapist, Mrs. Stern had an unconscious attachment to her clandestine identity as a reader. She was a Hasidic woman who was secretly different, worldlier, and more sophisticated than the rest of her community, yet was unwilling to expose herself or her family to potential criticism or scandal.

On the other hand, if the Rav told her to stop reading altogether, she would be obligated to cease the activity that gave her the greatest peace of mind and pleasure. Better to stay in the loophole that she had created and suffer tolerable guilt in silence. Although there had been a time when the patient's emotional pain had caused her to consider doing *teshuvah*, by abandoning a behavior that was considered inappropriate in the eyes of God, once the emotional pain had lifted, she was no longer motivated to change, nor could she determine the relevance of her behavior to her ongoing mood states.

Case 3: The Complacent Philanderer

Mr. Levine, a forty-two-year-old married father of five, consulted Dr. Three on the urging of his old *havruta* and good friend, Dov, who was increasingly worried about Mr. Levine's extra-marital exploits. Dr. Three had treated Dov's adolescent daughter with good results.

In the first of four visits to Dr. Three, Mr. Levine explained his situation. Since age twenty-two, he was married to a woman he described as pious and devoted, but not sexually exciting. Some years ago, he began frequenting massage parlors and became involved with one particular masseuse. His mistress, a Columbian immigrant on an expired tourist visa, was an ambitious young woman. In exchange for sexual favors, Mr. Levine helped her with night school homework and her immigration papers.

Dr. Three was struck by the patient's nonplussed manner in telling his story. Mr. Levine seemed quite pleased with the situation—his proper wife was bringing up their children according to correct, *frum* standards, while he was satisfying his lustier drives discreetly. From his point of view, no one was getting hurt. When Dr. Three inquired as to whether Mr. Levine considered the risk of catching a sexually transmitted disease from his mistress, the patient responded, "I'm careful, and she is only seeing me anyway." When the therapist commented on the disconnect between Mr. Levine's public and private behavior, the patient shrugged his shoulders: "That's what Dov said. He's afraid I'll get in trouble, but so far, it's OK."

Mr. Levine seemed to enjoy his sessions with Dr. Three. His approach to paying bills was idiosyncratic. Although Dr. Three explained that her office policy was to bill monthly, Mr. Levine insisted on paying for each session at the end, with a check from a joint marital checking account. Dr. Three asked whether Mrs. Levine would be curious as to the expenditure. The patient laughed, "She never pays attention to these things. I take care of all the money."

Dr. Three was struck at how Mr. Levine's treatment of the finances fit right in to the transference. In Mr. Levine's eyes, the psychologist was a slightly upgraded prostitute, another interchangeable female who would take care of his immediate needs, get paid, and disappear until the next time he called her. Dr. Three felt that any mention of her insight would push Mr. Levine away from treatment. Instead, she chose to comment on the disconnect between his public and private lives and to gather more history, hoping to find a plane of alliance with her patient that would motivate him to delve more deeply into his behavior.

Mr. Levine came for one more session and then called to cancel his next appointment. He was reluctant to reschedule, stating that he would call Dr. Three if he felt the need. She did not hear from him again.

Dr. Three realized that the essential missing ingredient in this case was anxiety. Mr. Levine's relationships with significant others in his life were generally shallow. People were important to him insofar as they helped him relieve anxiety. Mr. Levine did not feel shame about his immoral behavior. His main transference was to Dov, who served a surrogate parental authority. Mr. Levine's desire to please Dov motivated him to come for a consultation, but was not enough to propel him to go beyond a superficial encounter.

The case was a failure from both *teshuvah* and psychotherapy points of view. Mr. Levine merely went through the preliminary motions of treatment. Once the therapist began to delve into the glaring split between his professed religious commitments and his personal conduct, the patient fled. Dr. Three felt uncomfortable and dissatisfied. She wondered if her neutral, probing manner had frightened Mr. Levine. Perhaps he expected her to reprimand him for his adulterous behavior, threaten him into straightening out. Perhaps if she had treated him like a bad child, reprimanded him and given him specific prescription for correct behavior, he would have felt more connected to her and stayed in the treatment. Psychotherapeutic inquiry had certainly not provided much benefit to Mr. Levine or to his family.

Discussion

We hope that the above three cases enliven the comparison of *teshuvah* and psychotherapy. These two processes are concerned with the cultivation of maximal human potential according to the demands of these respective systems. *Teshuvah* and psychotherapy recognize that people suffer from feelings and behaviors that they are capable of changing. We begin our discussion by reviewing the frames of reference that underlie the prescribed restorative and transformative processes of *teshuvah* and psychotherapy.

Observant Judaism posits a clear frame of reference as to what the proper path in life is. Torah is the blueprint. Failure to adhere to this *derekh* results in dislocation and pain. The *mitzvot* constitute the template for proper behavior, and *teshuvah* offers an authoritative way back from wrongdoing. Conflicts occur when individual desire goes outside the system.

Psychiatry is less unified in its frame of reference for achieving its goal. There are many models of the mind and a smorgasbord of methodologies to choose from. Hence, clinicians' diagnoses and treatment depend on their orientation. There is no universally agreed upon organizing theory, and, therefore, it is far more difficult to know which behaviors are "right" or "wrong." Rather, these constructs are derived on the basis of whether they cause or relieve pain, or increase pleasure. This lack of absolutes can be experienced as liberating and/or threatening. Regardless, this point probably constitutes the most salient difference between the process of psychotherapy and *teshuvah.*

Another point of divergence is *teshuvah*'s notion of transference. Since the process of *teshuvah* operates fundamentally between man and God, the absence of a human collaborator can make emotional restitution a difficult, lonely task. In contrast, psychotherapeutic work deliberately cultivates the relationship between therapist and patient so that it may serve as the primary template for change.

The goals of insight-oriented psychotherapy vs. *teshuvah* may also differ. In *teshuvah*, the goal is not to end one's pain, but to alter one's behavior. Pain relief may occur as a secondary benefit. In classical psychotherapy, the diminution or cessation of suffering is the goal. Insight, behavioral change, and medication may be employed separately or in combination. Theoretically, insight should lead to pain relief. Resultant behavioral changes are useful insofar as they maintain a better functioning system and prophylax against future suffering. *Teshuvah* posits that appropriate behavior will pave the way for higher-level understanding and transformation of feeling states. Cognitive-behavior therapy, which works to correct habitual negative self-perceptions and other cognitive distortions, agrees with this point of view. Classical psychodynamic psychotherapy, however, posits the reverse—properly harnessed insight will motivate correct behavior.

It is important to emphasize that there are many forms of psychotherapy, and that we have presented these processes in a simplistic form for the purposes of discussion. Our intention is not to provide a precise analysis of the processes involved in either psychotherapy or *teshuvah*, but rather to provoke some discussion about how these two processes may work in tandem, or at variance, with one another. This articulation is particularly important in light of the fact that there are numerous rabbis who are involved in pastoral counseling, as well as many religiously observant clinicians who often feel caught in a dilemma of how to honor the patient's goals, even if those goals clash with the therapist's moral code.

A clear understanding of the principles and practices of *teshuvah* and psychotherapy is valuable to rabbis providing pastoral counseling as well as to psychotherapists treating *frum* clients because it will facilitate more honest and ethical negotiations with persons seeking help. Indeed, it is particularly important for clergy and religious educators to understand and disclose the nature and limits of the work they can offer a person seeking to reduce their pain in a therapeutic process with them.

The rabbi can potentially provide enormous support for a troubled congregant. He needs to balance his dual roles as compassionate listener and religious authority. While he needs to help the supplicant Jew find comfort, he must also make clear that not all impulses, solutions, or behaviors are acceptable within the framework of Jewish law. As such, there will be limits to the type of exploration that a rabbi can be expected to engage in. Rabbis should not rely on the congregant in need to implicitly understand this, since those who knock on the rabbi's door come with multiple expectations. Some suppose that the rabbi will offer counseling akin to what they might receive from a mental health professional. Others may harbor the belief that the rabbi can offer a prescription for divine absolution.

Indeed, even in cases where a clinical result would not incur any obvious harm in secular society, a rabbi who is counseling his patient must do so in accordance with the community's understanding of halakhah. Thus, at first response, Mrs. Stern should not read Jane Eyre, a husband may not physically comfort his grieving wife who is in *niddah*, a *kohen* cannot marry a divorced woman. Disclosure of his commitment to religious authority can help the rabbi clarify the transference dynamics and fully inhabit his role as a counselor. He can extend compassion by helping his congregant distinguish between a correctable "wrong behavior" and the corrosive self-hatred of feeling like a "bad person." Perhaps Dov's efforts would have gotten further if he had directed Mr. Levine to a rabbi who could invoke some of these principles. However, to the extent that there are limitations to the type of exploration a rabbi is willing to engage in, he should duly inform the congregant that a mental health professional might be able to explore issues from a more open vantage point. The merits and drawbacks of this approach might then also be discussed.

The therapist has wider latitude; his or her patient will need to chart his own course. The astute clinician respects the power of religious life. She or he recognizes the compelling pull that ancient and sacred traditions have on their adherents, even on those members who have lapsed. She or he can offer nonjudgmental exploration, but cannot advocate one solution. At the same time, the psychotherapist understands that painful feelings may arise when the tenets of tradition collide with individual yearning. If Mrs. Stern truly wants to resolve her dilemma, she will either have to eradicate her secular aesthetic appetite or go public as a reader and reckon with official response to her literary habit.

At the core of this discussion is the idea that, for a religious Jew, the process of psychotherapy might fill a different need than that of *teshuvah*, but that there may be a real need for an individual to engage in both processes at different times in their lives. Therefore, there is a need for a greater collaboration between religious leaders and mental health professionals to ensure a deeper understanding of these processes and the ways in which they work to reduce human suffering.

8

To Whom, to Where, and to When Does One "Return" in *Teshuvah*?

Moshe Halevi Spero

Behold: [you are betrothed to me] on the condition that I am a righteous person [*al menat she-ani tzaddik*]!" [The ruling is:] Even if he is completely wicked, she is betrothed to him, for perchance he had an inner stirring of repentance [*shema hirheir teshuvah be-daato*].

—Talmud, *Kiddushin* 49b

Suddenly I felt a misty consciousness of something forgotten—a thrill of returning thought; and somehow the mystery of language was revealed to me. . . . The living word awakened my soul, gave it light, hope, joy, set it free! I left the well-house eager to learn.

—*Helen Keller* (1903, p. 315)

For this explanation of the origin of infantile anxiety I have to thank a three-year old boy whom I once heard calling out of a dark room: "Auntie, speak to me! I'm frightened because it's so dark." His aunt answered him: "What good would that do? You can't see me." "That doesn't matter," replied the child, "if anyone speaks, it gets light."

—*Sigmund Freud* (1905, p. 224n; 1917, p. 407)

ALL AGREE THAT *TESHUVAH* (repentance) is one of the fundamental tenets of Jewish theology, philosophy, and halakhah (see Shechter, 1909, pp. 293–343; Soloveitchik, 1974; Urbach, 1971, pp. 408–15). The fundamental concepts and mechanics of *teshuvah* divagate across a multitude of aggadic aphorisms and tales, and, being a central element of Jewish praxis, are catalogued in full complexity within the various talmudic commentaries and subsequent codes of practice. In addition, *teshuvah* has merited a large and unique secondary literature that flourished in the middle ages and beyond in the effort to enhance the conception of *teshuvah* as not merely an occasional act required under specified conditions and times but rather as perhaps *the* essential dynamic or pulse of everyday Jewish life (Don, 1975, 1985; Katz, 1974; Pachter, 1985). Solomon Schechter's classic summary probably best encapsulates that which, as we shall see, is both fascinating as well as deeply problematical about *teshuvah*. "Repentance begins in thought, and its effect is instantaneous" (1909, p. 335). How can something so ephemeral as a mental whim, deriving from an entity as putative as that which we are accustomed to call the self, wield such transformational power?

Having said only this much, we might already sense ourselves falling into two of the great, incompletely circumscribed black holes that the *teshuvah* literature tends to pretend do not exist. The first can be partly delineated by a question: Is the self—let us for now define this as the subjective experience, partly conscious and primarily unconscious, of being a reflective ego—a sufficiently comprehensible, predictable, or graspable entity as to enable us to legitimately speak of *self*-knowledge or *self*-transformation in the first place? Might there be amalgams within the self that are sufficiently non-self in origin as to seriously complicate matters, and might the knowledge that is self-knowledge also include what for lack of a better word one might term *unknowables*?

The second black hole can also be delineated by a question: Is past or "forgotten" psychological experience—ranging from semantic-propositional and therefore readily accessible experience (i.e., linguistically coded but repressed) to experience that is preverbal, not yet mentalized, but inferable through psychosomatic pathways (cf. Mitrani, 1995)—actually capable of being located and retrieved with the relative ease and reliability that *teshuvah* writers presuppose? In the technical parlance, we say "return," generally, and not simply "to turn" (*li-fnot*) or "turn" (*peniyyah*). The latter, in fact, implies something potentially more random, certainly not possessed of a specific direction in time or space. The prefix "re" of return is therefore important. If the specific term *teshuvah* was purposefully chosen to connote the towering edifice of repentance, and is no mere linguistic flourish, then one ought to ask: what is it to *re-turn*? Is it to repeat, to remember, to reminisce, to re-enact, to re-present? Finally, presuming some sort of returning is indeed possible, to *what state* does the self return, to *where* (i.e., to what mental locale or quality of mental space), and to *whom*? To *itself*, or to something quite different?

A Postmodern Inquiry into Halakhic Notions

In a sense, then, this will be a postmodern inquiry, but not simply in the literary sense. In order to draw forward the major problematical aspects of *teshuvah*—not only those pertaining to it as a *moral* phenomenon but also to its being an essentially *psychological* process—it is necessary to interrogate the underlying assumptions of the concept of *teshuvah* with the kinds of questions the postmodern observer has learned to pose while listening to the intricacies of the clinical hour. On the whole, postmodern questions are the type that challenge the everyday definitions of constancy, knowability, and reality upon which the assumptive premises of *teshuvah* might rest, though in some instances even these premises have never been made crystal clear, which is precisely why they have escaped scrutiny for so many centuries.

The literature on the postmodern perspective regarding the inherent subjectivity and probable unknowability of any form of Cartesian self and of the indigenously language-bound paradoxes inherent in concepts such as "self-definition" and "self-identification" is a massive one. In the psychotherapy literature, the discussion has even led to the extreme call for the abandonment of the notion of the right of the analyst to speak of knowing the experience of the patient, or the validity of interpretations regarding the patient beyond the momentary shared experience in the analytic consulting room (Stolorow & Atwood, 1992; Viderman, 1979). Largely as a result of the influence of Kohut (1984) and the subsequent schools of self, intersubjectivist, and relational psychology, postmodern psychoanalysts began to speak of the self as existing only in context, depicting the role of the other at the base of self-experience in such a way as to argue that the enterprise of psychotherapy ought to be seen as the study of human *engagement* rather than as the study of the investigation of specious "facts" of perception, the recovery of specific repressed memories, or the analysis of conflicts believed to result from mishaps in the otherwise predictable unfolding of lawful patterns of psychosocial or psychosexual development. "Intersubjective moments" and "coconstructions of reality," as opposed to knowledge of reality, became central tenets of many major new schools. As Stolorow, Orange, and Atwood recently declared (2001, p. 143):

> Epistemologically, we find it impossible to know an isolated moment. Clinically, we find ourselves, our patients, and our psychoanalytic work always embedded in constitutive process. Process means temporality and history. To work contextually is to work developmentally. To work developmentally is to maintain a continuing sensibility to past, present, and future experience.

Of course, I am mindful that the postmodern critique of the Cartesian tradition has undergone its own critique and self-correction, both in the so-called hard sciences and certainly as regards the standards of objectivity in the soft sciences such as psychology, literature, and theology. The current interrogation stems from a milder form of methodological skepticism, grounded, first, in faith;[1] second, in sobering data from experimental psychology and clinical psychoanalytic research; and, third, in the awareness that even the automutative, subjective phenomena of human behavior are to a large degree bounded by unique forms of objectivity and "relative certainties" (Balter, 1999, p. 122; Kriegman, 1998). In psychoanalysis specifically—which is the "science" of self and knowledge that perforce influences my approach to the problem of *teshuvah*—a less radical or "corrected"postmodern ontology has reclaimed much that had been thought forever discarded by the overly enthusiastic relativism of the previous decade and a half.[2] This moderated conception nevertheless approaches very carefully such issues as the definition of self, the dependency of self upon other, the determinants of cure, and the proper domain or locale of such fundamental concepts as transference and countertransference, which were becoming locked in polarized conceptualizations (i.e., do these take place in the mind of the analyst, or the patient, or some alternative "third" space that at once belongs to both analyst and patient yet consists of neither alone [Ogden, 1994, 1997]?).

If fruitful for psychoanalysis, would a crossover from even this chastened postmodern template be appropriate for a reexamination of the religious concept such as *teshuvah*, steeped as the latter is in a language and philosophical system devoted to divine certainties, absolute conditions, established and codified ordinances, and often (or at least traditionally) unquestioned propositions? Psychoanalysis, on the one hand, is an intersubjective speaking and listening venture taking place between two human beings; *teshuvah*, on the other hand—notwithstanding the limited arena of social reparation between persons—takes place on the relational plane between man and God, whose "intersubjective" dimensions, however one might wish to conceive of these, are ostensibly not the same as those that obtain between fellow human beings. This methodological quandary will be addressed indirectly in many different ways in the forthcoming paragraphs.[3]

Defining the Parameters of *Teshuvah*

Having said the above, it is not easy to decide what best defines the most relevant dimensions of the phenomenon known as *teshuvah*. Similarly, what best defines the most relevant aspects of the concept of self, from which *teshuvah* presumably emanates and whose existence *teshuvah* might be expected to presuppose?[4] And what document or which commentator shall best represent the Judaic-Halakhic perspective regarding both entities?

The multitude of rabbinic aphorisms and axiomata regarding *teshuvah* are equally well-known, each with its unique mode of encapsulating in laconic phrasing some deep psychological truth about the pain of sin and the longing for return, many of which, more than any other segment of the aggadic literature, gain expression in the codified compilations of the complex halakhic mechanics of repentance. Even the array of general compilations regarding the basic biblical, aggadic, and halakhic references to *teshuvah* is daunting in number and depth,[5] and defies reiteration. The same is the case regarding the secondary and tertiary interpretive analyses.[6]

As a general distillation, however, it would be fair to say that as the *teshuvah* literature has advanced, particularly with the early nineteenth-century Musar (ethical pietism) authors, it has tended toward the widest definitions of repentance, embracing not only the internal evolutions of the human psyche but also themes concerning fundamental cosmological unification,[7] *with the effective role of the self remaining central and its accessibility for change unquestioned.* None among contemporary thinkers have expressed the link between self and *teshuvah* more clearly than Rabbi Joseph Dov Soloveitchik (1978, p. 62):

> Man is surely aware of many needs, but the needs he is aware of are not always his own. At the very root of this failure to recognize one's truly worthwhile needs lies man's ability to misunderstand and misidentify himself; i.e., to lose himself. Because of this misidentification, man adopts the wrong table of needs which he feels he must gratify. Man responds quickly to the pressure of certain needs, not knowing *whose* needs he is out to gratify. At this juncture sin is born. What is the cause of sin if not the diabolical habit of man to be mistaken about his own self.[8]

Hence—or so must surely be the corollary—repentance means self-discovery, despite the aspersions that had just been cast upon the reference points of self-knowledge! Even most recently, in an essay invitingly entitled "The 'I' and *Teshuvah*," Shalom Rosenberg (2002) expresses awareness that the postmodern self is a far more slippery entity than the Cartesian self, as well as the awareness that the unconscious self adds extensive complexity to the moral dilemma, yet for some reason returns his focus to the conscious self and to rather familiar positivistic reflections about *teshuvah.* Yet if the self is mistaken about its own identity and content, how can it extricate itself from a self-imposed prison whose design it has seemingly forgotten? If the self is the ailing or corrupted agent, from which sub-component of this agent can stirrings of *teshuvah* emanate, and how? Which agency within the self can be trusted to implement the corrective process?

Surveying all that has been said thus far, I must concur with Gruenwald's somewhat ironical summary (1985, p. 297):

> From a comparative and relative point of view, one gets the impression that the act of maintaining the idea of *teshuvah* in its halakhic and ritualistic framework is as child's play against the spiritual demands made of man who seeks to realize *teshuvah* for himself in its spiritual-philosophic and mystical molds.

Gruenwald certainly is not being naive or cavalier about the overwhelming intricacies of halakhic issues that surround the codification and proper fulfillment of the ritual obligations of *teshuvah.* He has simply given appropriate expression to the fact that the far greater conundrum of *teshuvah* is to define what actually transpires within the individual who repents, and under what conditions are salutary transformations most likely to happen.

Four Dimensions of the Definition of *Teshuvah*

To be even more specific, I would isolate four dimensions or properties that are generally deemed part and parcel of the praxis and ethos of *teshuvah* and have yet to be satisfactorily understood. I believe the following disquisition will shed useful light upon each.

The first property is the most basic because it is no less than the definition of the Hebrew term *teshuvah.* What type of "turning" or "re-turning" truly personifies *teshuvah,* and to whom or toward what must one return in order for the movement to be efficacious? I raise these questions with full cognizance of the stock and trade answers that exist in the literature, and I do not mean to imply that they are wrong, but rather that a heightened awareness of the fundamental impossibilities I am about to delineate render all previous responses inadequate or trivial. For I insist on spurring the reader to wonder whether re-turning is possible at all.

Second, it is well known (as taught by many references in the Talmud and midrashic sources, including the *locus classicus,*[9] to be discussed below) that halakhic and aggadic traditions have always attributed a certain uniqueness to *teshuvah mei-ahavah,* repentance motivated by love, and considered it more effective and qualitatively mature than *teshuvah mi-yirah,* repentance motivated by fear. This distinction seems intuitively reasonable, and yet it has never been abundantly clear exactly what *teshuvah mi-yirah* lacks that enables it only to reduce the gravity of intentional transgression to the less egregious level of accidental transgression and what *teshuvah mei-ahavah* contains that enables it to completely transform intentional transgression into meritorious deeds.

Third, upon what psychological transformation or internal mental accomplishment did the rabbis see fit to lend full legitimacy to the apparently whimsical and ephemeral phenomenon of *teshuvah be-shaah ahat* or so-called instantaneous *teshuvah*? Is this form of repentance, presuming one grants its legitimacy, essentially a quirk, designed to dramatically emphasize the extent of divine *hesed*, or does it point to something very fundamental regarding the relation between the element of time and the initial dawn of self-awareness?

The fourth point concerns an anomalous stipulation recorded by Maimonides in his *Mishneh Torah* code concerning repentance. Enumerating the criteria for satisfactory *teshuvah*, Maimonides states (*Hilkhot Teshuvah*, 2:2):

> And what is *teshuvah?* It entails that the sinner abandons his sin, and removes it from his thoughts, and [that] he resolves in his heart that he shall never commit [the sin] again . . . and also that he be consoled [*yitnahem*] over his sin . . . *and [that] the One who knows the secrets will be able to testify regarding him that he will never return to that sin again.*

First, note that the term *yitnahem* carries the connotation of simple "regret" as well as a deeper process of consolation, which would by definition include a mourning process.[10] Given this, Maimonides would be the first to include a complex emotional process as an element in repentance. Second, and more striking, is the fact that in the case of no other religious precept does halakhah require direct divine testimony as to the status of the individual's mental set or disposition. Such a super-objective requirement would seem irrelevant on the grounds of the prevailing Judaic philosophical view of human freedom; indeed, it would seem self-defeating to need to invoke God's foreknowledge as a basis for evaluating the status of human motivation.[11] What might be the purpose of including this criterion as an index of *teshuvah*?

Teshuvah Is Impossible

Given the density of the literature, what remains to be said? Upon careful consideration, what has not yet been said is that which would seem unsayable, but must be articulated: upon careful consideration, *teshuvah* is impossible.

Now one might want to suppose that when I say that *teshuvah* is impossible, I surely mean only to underscore, somewhat dramatically, that *teshuvah* is exceedingly *difficult*. However, I do not mean to say merely that *teshuvah* is difficult—that much is obvious and has been said by all. I mean quite strictly that *teshuvah* is impossible, or, has a special relationship with impossibility.

Some may wish to soften the impact of my assertion by other means. After all, have not all the great minds of the *teshuvah* literature, each in their own fashion, mentioned the obvious fact that *teshuvah* defies the concept of the irreversibility of time! In its most ironical form of expression, *teshuvah* has even been described as somehow "unnatural," *neged ha-teva* or even *neged ha-sekhel* (see, for example, Rabbi Bloch in *Shiurei Daat* [1976], p. 23ff).[12] As Rabbi Meir Simhah of Dvinsk saw the matter, for instance, the bounty of atonement (*kapparah*) that is vouchsafed for even lower grades of repentance (i.e., repentance that does rise to the level of complete *teshuvat ha-mishkal*) is nothing short of a miracle, "*she-lo al pi derekh ha-teva.*"[13] Previous authors have therefore emphasized the fact that *teshuvah* is first and foremost a *gezeirat ha-katuv*, a divine decree, inherently mysterious even were one to garner some modicum of understanding of it. That is, there is something so *hadash*, so novel and even idiosyncratic about *teshuvah* that one could not even sustain the thought of *teshuvah* were it not for the divine impetus set into the very order of Creation,[14] rendering *teshuvah* as a kind of *beri'ah hadashah.*[15] Surely, one might hope it is this purely metaphoric meaning that I have in mind when I speak of the impossibility of *teshuvah*?

Unfortunately, the intention of the aforementioned rabbinic attitude seems essentially rhetorical since the foregone conclusion of these authors is always that divine *hesed* makes *teshuvah* possible.[16] The few previous attempts at explaining the paradox are mystical at best, tautological at worst. For example, Rabbi Yosef Horowitz, the Sage of Navaradok, entertained no philosophical or psychological questions regarding the possibility of repentance, arguing that halakhah would not have granted legitimacy to so-called instantaneous *teshuvah*—that is, the famous ruling of the Talmud, *Kiddushin*, 49b ["*shema hirher teshuvah be-daato*"],[17] which we shall expand upon later—unless *teshuvah* was in fact inherently possible and effective (1950, p. 151):

> Therefore one can consider the law of instantaneous *teshuvah*, and if the law is so, then presumably reality is so, for if [such transformation] were impossible, the law of repentance could never be imagined, for how could one imagine a law without reality? The law is repentance and repentance is the law!

Under the influence of this kind of circular reasoning, the intuitive grasp that there is something about *teshuvah* that is inescapably paradoxical has never been fully problematized.[18]

Therefore, by stating that *teshuvah* is impossible I mean to say that *teshuvah circumscribes the impossible.* That is, I am essentially trying to place in bold relief the existence of a broad, complex religious-psychological dimension whose chief ontological properties are remnants of that which no longer is, properties based on negativization, properties that are *of the impossible*, or that *submit to the impossible*—however vexing it may be to hold this thought in mind. For it *is* the case, I intend to argue, that the restoration of self, and the "return" implicit in repentance, viewed so commonly as the commendable and comprehensible goals of *teshuvah*, are in certain fundamental, existential senses, quite impossible. Moreover, the religious personality loses something very significant, of relevance not limited to *teshuvah*, if this paradoxical dimension is ignored or magically inverted into concepts or practices so concrete or facile that the important traces of the "impossible" can no longer be discerned. For my theoretical purposes—and ultimately for the sake of the enhancement of the relevance of *teshuvah—I propose that the impossibility of teshuvah is its essence*, and not its traditional "possibility."

Analogy to the Impossibility of Psychoanalysis

Teshuvah is not the sole impossible enterprise. Sigmund Freud himself came to the conclusion, penned thoughtfully in "Analysis terminable and interminable," that psychoanalysis was an "impossible profession" (1937, p. 248). For, indeed, many phases of psychoanalytic work—those characterized by impasse, stagnation, floundering about in black holes, chuntering and aimless monologue, frigid emptiness, autistic encapsulation, chronic and perverse undermining of every modicum of progress—give impressive testimony to a redoubtable impossibility.

And, yet, does not psychoanalysis offer ample testimony to the possibility of personality change, return, and reparation? As with *teshuvah*, isn't the probability of change the fulcrum of the entire endeavor? Well, if one means simply, "Does psychoanalysis achieve cure or relief of some reasonable kind?" our answer would definitely affirm the realm of *possibility* (in the opinion of all but the most philistine), the lamentable hours of stalemate notwithstanding (though the seasoned clinician knows that these, too, may turn out to be phases of important creative turns in work). And if one means only that it is exceedingly *difficult* to get to the "bedrock" of certain neuroses, as Freud himself stated, then the force of the term "impossible" would become diluted with each advance in analytic technique that takes the contemporary psychoanalyst beyond Freud's idiosyncratic level of therapeutic skill or scope. If *impossible* can be challenged simply by the ethos "*Try harder!*" then there is little indeed to recommend the term.

However, by "impossible" Freud intended to allude to the existence of certain paradoxical dimensions that lie at the root of what psychoanalysis discovered and the kind of practice that it is. That is to say, Freud *primarily* intended this characterization of psychoanalysis to enhance, even if only incrementally, the mind's capacity to lend some conceptual expression to these paradoxical elements.[19] Chief among these are the polarity between the life and death instincts, the repetition compulsion (Freud, 1920), and certain principles governing the mechanisms of early repression, according to which some of the deepest, preverbal memory traces of the bedrock of emotional experience are not recoverable, properly speaking, because they were never registered in semantic propositional form in the first place.[20] Often these elements will only become visible within the pressurized framework of the "insistent transference."[21]

Perhaps the impossibility most relevant to *teshuvah*, hinted at by Freud in the 1937 essay and carried further by Jacques Lacan, is based on the awareness that the images of the mind can successfully represent only a small fraction of the vital energies and wishes of the psyche, and that the mind fulfills its function of unity by always leaving out, always creating gaps in the experience of desire, which in themselves perpetuate the sense of a fundamental alienation in human experience (cf. Boothby, 1991, p. 58–59). What Freud was calling the "death drive" in these latter essays essentially expressed the impossible tension between the real and the imaginary, the pressing toward expression of somatic energies alienated by imaginary identification that ultimately threatens the integrity of that identity. The compulsion to repeat, in this view, could occasionally be seen as a constructive effort to return to something that has been excluded by the subject, or which never entered into self experience, and even to its own unconscious pre-identities from which it had become separated by virtue of the limitations imposed upon it by the defiles of language, symbolization, Oedipal incest barriers, and the bounds of temporality.

Suffice it to say that Freud's farsighted initiative has wisely been adopted as an ever-present counterweight to all illusions about therapeutic certainty, long before the cavils of postmodernity could be identified by name (Becker, 1973; Kohon, 1999). The point I want to stress at this juncture is that Freud did not conceive of psychoanalysis as *absolutely, practically* impossible. Rather, upon careful methodological self-criticism, he identified certain inherent *ontological* and *existential* "impossibilities." Chief among these impossibilities is the fact that, since analysis operationalizes the transference and is operationalized by the transference, it evokes the deepest roots of human relationship and the earliest emotional tie (*Gefuehlsbindung*) with the other. Yet this earliest, primordial tie is itself ineluctably pre-ego, a deeply latent, "direct and immediate identification"[22] that is not capable of being representationalized, and hence cannot be dissolved. It would, indeed, merely repeat, if not for the analyst's penchant for showing the patient the way out of the thicket of his being perpetually signified by an other with whom he has become identified, willingly, neurotically, sadomasochistically . . . sinfully. These impossibilities, one way or another, are an essential companion both to the accomplishments of the analytic couple during a psychoanalysis (in addition to any other clinical- or diagnostic-type phenomena that might preoccupy their psychotherapeutic attention) as well as their doldrums. One way or another, these same kinds of impossibilities account both for the chronic misidentifications of self that ultimately desymbolize repentance (turning it into ritualistic undoing), seduce the self to reproduction instead of remembering, and reduce *teshuvah* to debilitating *shevi,* and also account for any alternative mode of *teshuvah* worthy of the name.

The Contribution of the Concept of Clinical Reality

The change in my understanding of *teshuvah* derived from years spent grappling with phenomena observed in clinical reality. Yet, since this is not a clinical paper, I can make only a brief statement about the contribution of that special kind of reality to our current examination.

Clinical reality, for me, denotes a specific kind of literature rich with observations about sudden and gradual forms of religious transformation.[23] More important, and more directly, it denotes a unique kind of work context wherein interpersonal or intersubjective interactions are converted into "analytic speaking" and perceived by "analytic listening" (see Bion, 1963; Green, 1971, 1973), a mode of speaking and listening that is in some senses more free than conventional speaking and listening and in some senses much more rigorous, bound by different schematics. It is only such listening and speaking that reveal the very processes by which a latent identity emerges from under that which seems new, and where something incipient and subtle can be traced within that which is ostensibly radical and intermeshes or clashes with current identities, and how a clodlike, concrete perception of a thing qua thing gradually transmutates into a symbolic apperception. These kinds of processes involve object and subject representations, time elements and various other ego functions—all of which comprise what has rightly been described as the "religious landscape" of the mind (McDargh, 1986; Randour, 1993; Rizzuto, 1979). These are also the elements that largely determine whether a hostile or benevolent nutriment awaits the seeds of religious growth, and whether conditions can be maintained for transformational emotions such as awe, reverence, remorse, penitence, atonement, and solace. But one will not discover any discussion of these elements in the extant *teshuvah* literature.

In my opinion, only clinical reality enables one to participate in and refract at sufficient intensity those aspects of human experience that account, simultaneously, for the elusive and alluring possibility of *teshuvah* in "one moment" (the famous *teshuvah be-shaah ahat*) *as well as* the observable fact that a large proportion of *teshuvah* projects, sooner or later, fragment, grind into sandstone, or take on infantile, concrete character in the desperate effort to stave off conflict and regression even if it is at the expense of creativity, happiness, and relationship.

Furthermore, I am convinced that the peculiarities and irregularities, unconscious conflicts, and perverse and passive-aggressive impasses seen in clinical work give the better portrait of what goes on during what is ordinarily referred to as sin, and that the complex levels of interaction experienced during psychoanalytic psychotherapy—ranging from autistic encapsulation to schizoid isolation and loneliness to cloying, oppressive symbiosis, and all manner of distancing and attachment in between—give a more palpable and profound portrait of what goes on during what is ordinarily referred to as repentance and return to God. I so state not in order to valorize these portraits of psychic percussion, but in order to isolate their common denominator: massive and pervasive self-alienation that only exacerbates the dilemma of alienation from God. The vehicle of such alienation, if one can put it this way, far prior to any specific act of rebellion or egregious "sin," are what I consider the "domains of impossibility" inherent to the mechanisms of language and self experience. These domains can be thus characterized from the first instants of life, and they exert an inherently, and perhaps necessarily, alienating effect every time a human attempts to represent some experience or sentiment in thought or express it through speech.

It is distinctly troubling that these dimensions were never the focus of theological or halakhic writing, and have not been fully articulated even in the otherwise metaphorically rich and sensitive treatises on *teshuvah* by writers such as Rabbis Kook, Soloveitchik, and Steinsaltz.

Against my temerarious evaluation, one could reasonably argue that these intrapsychic or intersubjective dimensions are not the bailiwick of religious writers. But one would be wrong to so state, for the following reason. *Teshuvah* is, at root, nothing more than a cognomen for a process far more complex and even contradictory than previously imagined, one that is probably *impossible* in the senses that I shall define without a fuller understanding of the psychological dimension.

Even a cursory reading of the talmudic and subsequent literature on *teshuvah* makes clear that the focus of the entire percept is a meaningful transformation of the "heart" or mind of man and certainly not only the modification of external behavior, except insofar as the latter expresses the former or conduces toward it—*Rahamana libba ba'ei* (Talmud, *Sanhedrin*, 106b [Rashi]; I Sam. 16:17).[24] For example, throughout the Talmud and codes one finds a clear delineation of specific objective criteria for ascertaining whether personalities habituated in vice have undergone sufficient internal personality change in order that one might be able to rely upon evidential testimony such individuals may have to offer (Talmud, *Sanhedrin*, 25b). In each of the instances described, the form of *teshuvah* prescribed only makes sense if one *presumes* a rabbinic awareness of a link between the effect of repentance and—as for example in the case of the addiction to gambling—formerly perverse perceptions of time and space, self-awareness, and so forth. And yet the micro-links between the *specific mental processes* through which repentance *actually alters* formerly disturbed or pathological behavioral or emotional patterns—that is, what is the exact nature of the transformation that restores moral and perceptual veridicality to a former gambler's word—remain unarticulated.

Consolation and *Teshuvah*

The phenomenon of "return" must clearly transcend elementary or arithmetic reversal to a prior state of mind, and cannot be defined as a mere undoing of inappropriate, proscribed, "bad" behavior or personality traits[25] and their substitution by appropriate, prescribed, "good" behavior or traits. It is not in response to such "return" that God would return to man in a full relationship. Although it might seem at first blush that the practical art of repentance so painstakingly itemized by Maimonides in his definitive "Laws of Repentance" *could* be satisfactorily accomplished simply by rote learning, memory, and imitation—since by these expedients alone it is possible to imagine an automaton resisting temptations, committing no sin, and omitting no meritorious act—in fact, the "returning" that may be truly considered *teshuvah* must evidence some qualitative transformation of self, and be achieved *through* the vehicle of self-awareness and self-knowledge.

From a voluminous literature, the single proof text that best establishes the need for transformation of self is that which adds an additional operator to the familiar operation of return: "For after I returned *I was consoled*" [*ki aharei shuvi nihamti*] (Jer. 31:19)].[26] This and related texts insists that the frequent appearance of the concept of a uniquely consolatory effect of *nehamah* in the context of *teshuvah* (e.g., Job 62:6–10; Jonah 3:9–10) is the key to a multifaceted world of internal psychological changes that are the determinants of and the consequences of the repentance process. But whereas previous essays may reach this point and rest, I cannot. A complete understanding of *teshuvah* will require the most complete comprehension of *nehamah* available, especially the link between consolation—which would seem implicitly to suggest some kind of prior or concurrent process of mourning—and language. One hundred years ago, the suggestion of such a link would have seemed completely peculiar, and yet anyone with some background in contemporary linguistics and psychoanalysis can already sense the direction I will be taking. In fact, from the standpoint, say, of the Kleinian concept of mourning, reparation, and symbolization, the notion of *nehamah* may advance our comprehension of *teshuvah* beyond all previous understandings.

Understanding the work of consolation and mourning in *teshuvah* begins with the divine word emanating from the Lord that created the "fruit of the lips" (*niv sefatayim*, lit., dialects, speech) (Isa. 57:18–19): *derakhav ra'iti ve-erpa'eihu, ve-anheihu va-ashallem nihumim lo ve-la-aveilav.*

> I have seen his ways, and will heal him;
> I will lead him also, and requite with comforts,
> him and his mourners.

It continues with Maimonides' selection of the aforementioned passage in Jeremiah 31 as a basis for his discussion of repentance, as well as with the details we discover in his atypically prolix descriptiveness of the phenomenology of *teshuvah.* And as we saw above, Maimonides uses the term *yitnahem* to describe the pulling away from sin, a term that implies both simple "regret" as well as the more complex process of some kind of consolation, with its implication of a mourning process. Examined objectively, as Rabbi Joseph Soloveitchik has carefully outlined (1974), a clinical portrait of mourning unfolds (*Hilkhot Teshuvah,* 2:4, 7:8): "It is characteristic of *teshuvah* that the penitent should cry out constantly before God in crying and supplication. . . . It is the way of the penitents to be extremely humble."

Yet, at precisely this juncture, one begins to sense a dilemma, especially if we agree to not being misled by the powerful attraction of metaphors, and if we accept my insistence that we demand clinical, or at least semi-clinical delineation. The above is hardly an *extensive* clinical portrait of mourning, nor are the key mechanics of mourning and grief-work articulated. Indeed, terms such as "mourning" and *nehamah* in the hands of these authors seem to be essentially metaphoric motifs rather than definitive operators and determinants of repentance. Sometimes Maimonides' characterization is interpreted as recommended praxis (i.e., one *ought* to despair and agonize in order to properly repent), but more often than not his depiction is considered simply a portrait of the possible or even common consequences of *teshuvah.*

Now, Maimonides' characterization certainly does accord with the familiar portrait of the tormented penitent during the dark night of the soul, melancholy from awareness of the defects of the personality and alienation of the self, and it personifies innumerable individual biographical testimonies encountered clinically.[27] Indeed, this characterization became the prevailing image cast by many of the authors of the so-called Musar literature of the mid-nineteenth century and even during the earlier gaonic period (Katz, 1964).[28] Yet this is not the only portrait of repentance, either as recommended praxis or consequence, and as our psychological understanding of intra- and intersubjective processes becomes more comprehensive, such a portrait becomes increasingly less complete. Discounting for the moment the obviously psychopathological contrast seen in manic or hysterical elation, and the rhapsodic fugues of sexually excited penitents, a goodly proportion of those who turn in *teshuvah* present a different picture—distinctly thoughtful, sober, pensive, even somber, but neither depressive, obsessive, nor otherwise tormented by guilt and shame. Among this group, relief and even a modicum of joy prevails, a certain stillness or inner peace, not particularly contaminated by odious supremacism; rightness coupled with skepticism, as opposed to self-righteousness; an embracing and shareable piety, as opposed to cloying piousness.

In view of the second portrait, and the intuition that such "joyful" forms of repentance must be the intended goal of the concept of *teshuvah mei-ahavah*, what then becomes of Maimonides' earlier description? What becomes of the emphasis on repentance and mourning that serves as the central motif of Rabbi Soloveitchik's contemporary reinstatement of existential guilt and anxiety in *teshuvah*? Is true *teshuvah* conceivable where gut-wrenching emotions of pain, loss, and sadness are nowhere in sight? Or, is Maimonides providing what amounts to an accurate description of a phase or stage of all *teshuvah*, one without which even *teshuvah mei-ahavah* may be considered incomplete or even defective?

One could say that Maimonides' precise wording—*mi-darkei ha-teshuvah*, "it is typical for" or "characteristic of" the repentance process—intentionally leaves ample room for the halakhic and psychological possibility that this description was not proposed as necessarily characterizing all *baalei teshuvah*. However, it may be more useful to demonstrate that repentance partakes of an extended process during which a distinct phase of despair, mourning, and reparation are sine qua non for further mature growth, even when emotional torment and sadness are not plainly visible, even when the penitent dances in the joyful embrace of God. That is to say, perhaps a *yirah* phase of *teshuvah* must precede and ensconce the *ahavah* phase of *teshuvah* in a manner similar to the way in which, developmentally, the "paranoid-schizoid position" precedes and continuously interlaps with the more mature "depressive position," as outlined by the object-relations school of psychoanalysis influenced by Melanie Klein (see esp. Klein, 1937, 1940, 1946; Caper, 2000).

The Contribution of the Kleinian Psychoanalytic School

Working from a Kleinian psychoanalytic framework, the traditional concepts of an "inferior" and "superior" form of *teshuvah* become much less informative than thinking in terms of two interdigitating phases of mental or emotional experience—the former emphasizing a state of mind characterized by fragmentation and near-paranoid "relations" among disorganized and sharply split partial representations of experience; the latter emphasizing a state of mind characterized by a deeper, caring interaction among more integrated, whole representations, including the necessary "price": a new propensity for sadness and despair in the face of the new cognizance, increasingly incapable of being ignored, that good, evil, and the ambivalent mixtures of both are all components of the same self.

In summary form, according to contemporary Kleinian theory, the following are the main transformations in the movement toward the depressive position and beyond (i.e., these changes are initially seen in the developing infant but continue to be seen throughout life[29]) (see Caper, 2000, pp. 113–18):

(a) A decrease in fragmentation and splitting defenses within the mind, enabling a more realistic sense of self, due to the novel experience that hostile or malevolent forces may be arising not only from "bad" objects pit against the self, but also from within the self;
(b) A tendency toward less idealization and sharply-divided part-object representations, and an increase of well-internalized "whole" object representations (which, in turn, tend to militate against the ability or need to rely upon the splitting defense, etc.).
(c) An enhanced emotional attitude of caring and love toward these newly shaped and evaluated internal objects, though necessarily accompanied by the capacity for mourning and grief (i.e., split-off representations tend not to be mourned, strictly speaking, but merely shunted off, denied, or introjected into the self in the form of permanently melancholy aspects of the ego).
(d) The anxieties that characterize the new phases of development are now experienced on behalf of a loved object and not just a vulnerable ego, and, since they are predicated upon the new capacities for grief and mourning, consist of true mental pain and anguish (known as "depressive anxiety") and not abject terror.
(e) A more mature, less totalitarian superego can now be instated (in concert with more advanced oedipal developments), the central feature of which is that, since the self perceives this internal agent as more human, limited, and to some degree dependent upon the self and not simply its "master," it must begin to deal with emotions such as gratitude and concern, injury and repair, remorse and not mere vengeance.
(f) A logic of integration now holds increasing power, especially since the change in the nature of the anxieties of the self and its defenses enable

> much more interest in consolidation and symbolization, even at the cost of the loss of sheer number of concrete "things." Along with this element there emerges a new sense of reality, much more sensitive to ambivalence, to a broad span of history and contextualization, and to a bountiful sense of future and hope.

According to this approach, then, there might be no single state of *teshuvah mi-yirah* or *teshuvah mei-ahavah,* but rather always *both* states, operating at different wavelengths or upon dual tracks, the one nurturing the other. This type of interpretation makes clear that in order for an instance of *teshuvah mi-yirah* to be considered to have any value whatsoever it would need to be a type of *teshuvah* that emerges from an as of yet unconscious, or as of yet insufficiently representationalized mentality disposed toward *ahavah,* or else it will not even take the form of that *teshuvah mi-yirah* that Jewish thinkers found laudatory. Equally, the outcome of a *teshuvah* project could conceivably yield joy, but not without undergoing an internal, unconscious transformation that includes the types of turmoil described earlier by Maimonides, in the best language available to him in his era, as a form of combined anxiety and mourning, *but that we would have to now acknowledge as incomplete without some further explanation in terms of the contemporary interpretation just offered.*

Thus, if one can accept a theoretical reconceptualization of *teshuvah* as a special form of advance in the process of symbolization,[30] then it may be possible to also argue for the singular role of mourning, not so much as commonly understood as the painfully expressive reaction to death and loss, but rather as a more general psychological process by which emotional experience is denegated, deconcretized, and gradually transformed into symbolic mental representations (Segal, 1957). Mourning understood in this light enables one to comprehend, for example, that the very development of language—which at first blush has nothing to do with death or grief, but in fact is dependent upon the constant "murder" of "thing" presentations, as Hegel put it, and their subsequent transmutation into representations—must involve a mourning process of some kind. For as one's capacity for abstraction and symbolization expands with every increase in depth and grasp of language, come losses that must be accounted for and accommodated through the process of mourning.

Disappointing Postclassical Definitions of *Teshuvah*

One can readily acknowledge that some contemporary rabbinic masters evince even greater psychological appreciation of the potential depths of the *teshuvah* process than were expressed in the more classical, medieval, and even pre–World War I literature, owing among other things to the infiltration of select psychological concepts, from a variety of sources, into their perception, as well as to their literary gifts of expression. The great works on *teshuvah* of such authors as Rabbis Dessler, Kook, and Soloveitchik are replete with reference to concepts such as "will," "automatic behavior," "dark shadows of the spirit," "mystery of thought," "lower and higher impulses of the soul," "wrinkles in time" [*kematei ha-zeman*], and so on. Who could deny the psychologically mindedness of the classical phrase of Rav Kook's, "The horizon of perception is broadened by means of repentance" (*Orot* [1924], 9:2), or the sophistication of the following assessment by Rabbi Soloveitchik (1944, p. 69):

> These [ethical] judgments and norms lying in the immediate proximity of the psychophysical threshold tend to externalize themselves. They find their concrete expression in articles of faith, in petitions, in physical acts of worship, and in other practices and observances, all of which lie in the external world. Ostensibly, religion, though flowing in the deepest subliminal ego-strata, is in eternal quest for spatialization and corporeal manifestation.

These and other psychological elements are tantalizingly picked up by Rabbis Kook and Soloveitchik, yet when following their subsequent paragraphs one finds that they suffer from the same fate outlined above. In the thought of Rabbi Kook, for example, the fear of death and other "dark impulses" within the soul are distinctly marked as prestages, at best, and at worst, as imperfections in the grander accessions in repentance, potentially destructive flaws indicating lesser faith and distance from God: "As there is a deepening of repentance, so does fear of death decrease, till it ceases completely. Its place is taken by the condition of 'and she laugheth of the last day' [Prov. 31:25]" (*Orot* [1924], 11:3, see also 16:5). The intuition is correct, but the specific role and meaning of death as a psychological component is never spelled out, and therefore the fullness of its bearing upon *teshuvah* cannot be discerned.

In Rabbi Joseph Dov Soloveitchik's great work *On Repentance* (1974), on the other hand, one at first obtains strong evidence that the psychological dimension will receive more prolonged and detailed treatment. For example, we begin to read (p. 318):

> Sometimes the Almighty aids the penitent and causes a sudden revolution to take place in his way of thinking, transforming his whole worldview, affecting the whole essence of his being. For many years his personality had been schizophrenic and fragmented; he had been pulled along by dark impulses and instincts; all his life he had been an outsider everywhere he went. . . . And then, through a sudden ray of illumination, he discovered the focal point of his existence, and he was transformed into another person. A new light shone in the

depths of his soul, new aspirations filled his being, he was released from the bonds that held him back, the fragments of his personality were collected and reunited to form an integrated person.

This is vintage Soloveitchik, delivering a moving portrait of the consolation of the almost palpable hand on the shoulder—sometimes divine, though sometimes the memory of just a blush of something humanly maternal or paternal (pp. 259, 303). And yet even R. Soloveitchik, in this and in other works published before and subsequent to this classical citation, leaves incomplete many of the important links between the two great domains.

For another relevant example, while R. Salanter's approach to the inner dynamics of human personality suggest a somewhat systematic appreciation of the "beyond" that is so powerfully active deep below the surface of human behavior, he, as all other authors of the genre, outlined this comprehension in terms that are never declassified from their essentially theological-religious definitions, resulting in the *paradoxical removal of these relevant dimensions from the realm of the humanly psychological!* It is thus not a minor point Aron Rabinowitz makes when he concludes (1999, p. 65):

> Salanter did not develop a theory of primary process thinking. He does not, therefore, mention in his writing distortions of thought processes associated with abnormal behavior. This does not mean that he didn't realize that thought processes can be distorted. It means that the concept of the primary process as a phenomenon beginning at birth and continuing to wield great influence in adulthood is not present in his thinking.

But this is unfortunate because I cannot see "crediting" these authors with any higher understanding of the intricate underlying psychological process when the outcomes they espouse have yielded so little practically or heuristically. That is to say, without a *psychological* theory of primary process thought—or any other aspect of contemporary psychoanalytic theory—one is less able to advance the definition of *teshuvah* or resolve its consummate riddles (e.g., just why *is* repentance motivated by fear so significantly different than repentance motivated by love?). For an interpretive text or conceptual approach to repentance to be considered professional (i.e., usefully) psychological, one anticipates that the respective texts would sustain *this level* of exposition for as long as it takes to present a complete thesis that is recognizably psychological for the express purpose of clarifying the relevant dimension of religious experience (cognition, conation, emotion) *in psychological language* (as is more generally, and successfully the case in the domain of halakhic biomedical ethical writings!). Frustratingly, the scattered phrases and statements that so attract the attention of the reader who senses that the possibility of a deepened understanding of *teshuvah* might be nigh at hand—precisely because they are so richly steeped in their formulators' deep introspection and intuitions regarding the workings of the mind—general turn out to be unspecific, and certainly not organized within a larger psychological (or psychotherapeutic) framework.

The Halakhic Metapsychology as a Framework for Reconsidering the Meaning of *Teshuvah*

It is due precisely to the preceding considerations that I proposed a "halakhic metapsychology" (1980, 1982, 1986) as a framework for perceiving and amplifying suitable analogical structures that combine psychiatric, psychoanalytic, and halakhic dimensions, allowing cross-fertilization among these concepts at a variety of theoretical and practical levels. The chief merit of the halakhic metapsychology—and upon which criterion it ought to be evaluated—lies in its ability to formulate paradigms with distinctly halakhic structures and implications that allow for a more rich "fit" with contemporary psychological, psychiatric, and psychotherapeutic paradigms. There are several dimensions that are critical to the types of halakhic models that might be seen as governing the *teshuvah*-psychotherapy paradigm, and these are not to be construed as merely theoretical or heuristic; in an important sense they make hypothetical demands and impose limits upon psychotherapy. If, for example (see further Spero, 1980, 1986), one refers to the *vidduy* paradigm or to the complex biblical *azov taazov/hakeim takim immo* paradigm (see *M.T.*, *Hilkhot Rotzeiah u-Shemirat ha-Nefesh*, 13 re: Ex. 23:5 and Deut. 22:4) in relation to psychotherapy, one expects specific aspects that are pertinent to the detailed halakhic parameters of *vidduy* and *azov taazov* to be relevant to psychotherapy as defined by psychotherapists, and also the reverse: specific aspects of the detailed parameters of psychotherapy ought to prove relevant to the practice or implementation of *vidduy* and *azov taazov* as defined by halakhah.[31]

I would interpret the halakhic interception of the psychological dimensions of moral enterprises such as *vidduy* and *teshuvah*—with the special emphasis each insists upon securing the prescribed framework for representationalizing the finer elements of the introspective processes, intentionality, and perception of spatiotemporal dimensions—as an analogue to the psychoanalytic conceptualization of the "analytic frame" and the manner in which this frame alone enables the analyst to discern elements of the mental processes taking place between oneself and the patient (Bleger, 1967; Langs, 1978). As Wilfred R. Bion articulated it (1963, 1965, 1967, 1992), the truly psychoanalytic elements of the process of psychic transformation might go unnoticed—some would say that in some sense these elements do not even *exist* (Schafer, 1983; Viderman, 1979)—outside of the analytic frame, since it is the frame alone, psychological or halakhic, that enables the critical properties and dynamics that characterize the psychoanalytic process to be contained in the joint mind of analyst and patient and to be registered upon the visible/audible threshold.

One might form the mistaken impression from the above that the notion of the analytic frame is *simply* conceptual. In fact, the analytic framework is a living structure. It is a *container* that, under the proper conditions, imitates the manner in which the early maternal matrix provided the infant with the first psychic space within which to envelop the complex dimensions of psychic experience, including the mimetic, semiotic, temporal, and representational, in addition to the net content of the patient's monologue (Caper, 1999; P. Goldberg, 1990). Accordingly, without the frame, without the anticipated lawfulness imposed *a priori* by its symbolic structures, the potential for establishing meaningful, salutary relationships among the elements—and the ability, therefore, to accomplish therapeutically relevant transformations—could not become active within the mind of the analyst and the patient, and would not foster the creation of the vital "third" intentional space that ensconces both therapeutic participants, and is ultimately internalized to some degree by each of them (Ogden, 1986). I believe that the same applies to the manner in which the psychological dimension of *teshuvah* is comprehended intellectually and ultimately takes hold effectively.

Tracing the Term *Teshuvah*

The very term *teshuvah* in and of itself holds its own mysteries. Can we compel the term *teshuvah* to inform on the impossibility of that which it solicits?

One immediately notes, as Urbach points out (1982, p. 408, esp. n. 57), that the term *teshuvah* as such, philologically speaking, is a rabbinic coin, though obviously the terminological carriage for the concept of moral return, *la-shuv*, and its root *shuv*, are well anchored in a multitude of biblical citations, each of which depicts either man returning to God or God to man (e.g., Deut. 30:1–13; Ezek. 18:21, 30; 33:11; Isa. 55:6; Jer. 3:24; 18:8, 31:18; Hos. 14:2; Lam. 5:21), and, less frequently, a self-reflective return within the orbit of the self (e.g., Lev. 25:27; Deut. 23:15; Jon. 3:9; II Chron. 18:26). Aside from numerous transitive and intransitive forms such as *shav, tashuv, hashiveinu*, and so forth, which circulate throughout the Writ, there exists no Hebrew term that characterizes the essence of "return" or internal revolution in quite the same way that the term *le-hitvaddot* (to confess) captures the necessarily introflexive quality of *vidduy* (lit., outpouring), or the way the term *le-hitpallel* (to pray) conveys the inherently intra-penetrative quality of *pillel* (lit., judgment).[32] It is difficult, therefore, to accurately fix the exact grammatical form of the term *teshuvah*, which is quite contrary to what might be expected for so introspective a project as repentance.

A breakthrough of sorts comes in the form of Rabbi Soloveitchik's own novel definition of *teshuvah*. Notwithstanding Urbach's valid assertion (to which he does not refer), Soloveitchik indicates that the morphological structure "*teshuvah*" does actually appear in a single location in the Prophets (1974, pp. 98–99), and he is the sole author to interpret this idiosyncratic instance in a manner relevant to our topic.

The biblical text in question describes that after the prophet Samuel completed a yearly circuit of service to the various outlying communities of Israel (I Sam. 7:15–17; cf. II Chron. 36:10):

> And Samuel judged Israel all the days of his life.
> And he went from year to year,
> and he circuited (*ve-savav*) Bet-El and Gilgal and Mitzpeh;
> And he judged Israel in all those places.
> And his return (*u-teshuvato*) was to Ramah for there was his home,
> and there he judged Israel

In Rabbi Soloveitchik's exposition, the *teshuvah* referred to here signifies circular motion, concentric circles, yet always moving inward closer toward the home in Ramah, where Samuel was born and his mother still lived. This, in Soloveitchik's interpretation, is endemic of *teshuvah* as repentance. Now it must be noted straightaway that the Talmud (*Berakhot* 10b) interprets the passage quite differently than Soloveitchik. As recorded by the Talmud, it struck Rabbi Yohanan at least that geographical information could not be the gravamen of the biblical text since it is common knowledge that Samuel lived in Ramah. In Rabbi Yohanan's view, the emphasis is that Samuel did not, in fact, return to his natural home every year, but rather took care to always bring his personal effects and supplies with him to whatever lodgings he was offered so as to not burden his hosts, and in that sense rendered any and every lodging *as if* his home.

Later we shall return to the deepest application of this difference of opinion since its resolution requires further elucidation of the concept of return and repetition. We shall approach this elucidation by first considering what I consider to be the root limitations of *teshuvah*. The audacious pronouncement "*Teshuvah* is impossible!" is intended to provoke direct collision between our received notions and at least four fundamental limitations indigenous to a human mind that are so troubling as to make one wonder how it is that one could ever have contemplated the possibility of *teshuvah* in the first place.

These limitations are consequences of certain inalienable and paradoxical properties of (a) the self, (b) memory and mental representation, (c) time, and (d) language.

The Problematic Dimension of Self

The first problem concerns the dimension of *self.* As paradoxical as it may seem—though humankind has always sensed as much intuitively, long before Hegel (1810–1814), Sartre (1943), and Freud rendered it unavoidable—that which is termed "self" is an illusionary structure predominantly comprised of imaginary, internalized interpretations of myriad interactions with the *other* from which a "self" is, somehow, initially differentiated and eventually (hopefully) consolidated.[33] "Self," even after we attribute to it its boundaries and privacies, is always "of the other," inasmuch as to be a self is to be ever contextualizing oneself in the orbit of an other, or others. And yet, every impression of other involves some form of nihilation of self, such that what one blithely refers to as a "state of self" is in fact a highly complex state of conflict, involving constant dehabilitation and rehabilitation of intersubjective and intrasubjective (intrapsychic) impressions.

Rabbi Joseph Dov Soloveitchik himself echoes this concern about the unknowable self, though he does not directly problematize this aspect of *teshuvah* (1944, p. 73), when he writes:

> Through the individual subjective philosophizing the clandestine ego emerges from its involuted and ramified recesses. We delve persistently into the enigmatic, subjective mists. Yet, however far the regressive movement continues, we are never quite able to fathom subjectivity.

Being this is the case, how can we continue to evoke the aid of this agent in the process of *teshuvah* without a great deal of additional information about the multiple divided interests within the self? In the far different language of the Lacanian school, Ragland summarizes best much that has been written on this topic (1996, p. 192):

> A person is not a "self" [which Lacan views as] a totalizing concept of the person that denies the reality of how human subjectivity is structured in the first place, and how it functions thereafter. A person is, rather, a semblance or seemingness that tries to stop up the hole in being.

Hiding behind the rigid pseudo-structure known as self is more importantly an entity much more difficult to define—though it surely has *some* kind of identity—more akin to what Lacanians would term the something-one-desired-that-is-in-fact-no-longer-there. This kind of description holds the spotlight on the fact that our self-concepts are based on narcissistic images presented to us through the mirroring functions of parents, language, and the environment far more insistently than we like to imagine, and far earlier than we can imagine. By the time the human mind has established itself behind what it deems a stable sense of identity (let alone when conflict, perversion, and psychosis reign), it is quite difficult to become aware of the extent to which this "identity" has been predicated upon, and continues to respond to, the desires of the other. Throughout a lifetime, the individual keeps replacing with whatever one can the cut in the continuity of being, the very cut that introduced the infant to temporality, self, other, pleasure, and unpleasure.

Of importance is not only the fact that the self is elusive, in the sense of its being comprised as much of the other, and hence of what is not-I, as it is of what is recognizably own or self. Also important, but equally problematical, are certain subtleties inherent in the very mechanisms by which a self, like any internal mental representation, is formed. Every definition of self, implicitly, involves some form of negativization or nihilation, as opposed to mere, pathological "negation." This, too, is a complex domain of the contemporary literature, but the basic notion is as follows.

Whereas the common form "negation" is currently used solely to indicate concrete forms of denial and perceptual-emotional exile and excision (illustrated by psychotic denial, negative hallucinations, and blocking), the terms negativization and nihilation (or *neantisation*), rooted in the work of Kojeve (1939), Sartre (1943, pp. 3–45), and others, have been resurrected by Lacan (1955–1956) and Green (1999) to indicate certain healthy and ineluctable dimensions of mental functioning. Specifically, these authors highlight the ways in which the limiting systems of language, perceptual consciousness, and symbolization must interact in order (a) to "create" themselves, by a process of mutual detraction and contraction, and in order (b) continuously, throughout a lifetime, to defeat (or depress) the concrete multiplicity and polysemy of "raw, buzzing" experience in order to enable relatively discrete perceptions of self (as opposed to all that is not-self), other (as opposed to all that is self and non-other), objects (as opposed to all that is non-object), place (as opposed to all that is full), and time (as opposed to all that isn't-ceasing-to-unfold).

Without the capacity for negativization or denegation, as Freud was the first to clarify (see Green, 1999, pp. 1–25), the very "basic" ego functions of judgment and reality-testing itself would be impossible. Furthermore, without negativization, the fundamental human psychological process of *identification* would amount to nothing more than mere imitation. Different than mere ablation or foreclosure from the mind, negativization—as it sublates the concrete and transforms it into symbols while retaining the experience of the loss of the presymbolic state—gives rise to *representation.* Representation, unlike simple signs, houses desire and attempts to contain loss. To be sure, signs enable one to indicate everything that can be said about an object, but a representation is everything that can be experienced in the mind regarding relationship, which is itself always a consequence and personification of some kind of loss (Green, 2000, p. 59). Only representations etched from negativization lead to symbolic elaboration of loss; signs, on the other hand, tend to be unable to pick up the intrapsychic primary inscriptions of loss, instead turning concrete bits of memory over and over again in futile search for meaning. Indeed, it is only this healthy form of negativization that enables man to limit the sheer, pulsing, inchoate *possible*—and, in a sense, even "destroy" it—that, in turn, renders thought and mind possible in the first place (see Aragno, 1997; Bion, 1965, 1967, 1992). The deeper nothingness to which healthier negativization always points, and from which it is always attempting to escape, is thus an inherent aspect of being (Sartre, 1943, p. 28). This chasm is seductively hidden at that perceptual non-point that Freud brilliantly referred to as the navel of the dream (1900, p. 525).[34]

The Problematic Dimension of Memory

The prerequisite to any subjective sense of self is a trustworthy architecture of memory and memory functions. Herein lies the second problem: the dimension of *memory.*

Much of what is defined by stable self-perception presupposes psychological qualities known as object constancy and object permanence that are predicated upon memory. Memory science has come far since its early static-camera model, and has even surpassed Freud's revolutionary concepts of the double-layered, retroactively or deferred memory functioning, screen and repressed memories, though these psychoanalytic concepts are among the many that continue to be supported by empirical research. While memory may store a certain degree of photographic-type (i.e., "objective") sensory data at some level of the mind, memory is essentially a multi-tiered series of overlain and intermixed versions of reality, some more stable over time than others, most of which are not only initially stored in subjective form, but are also reconstructed or modified on a constant basis. Reviewing a large literature, Barclay and DeCooke conclude (1988, p. 91), "Accuracies and inaccuracies in autobiographical memories result from judgments about what could or should have happened in one's life, not necessarily from direct access to event representations *isomorphic* with the way in which events actually occurred." Memory, in other words, is generally distorted in favor of the preservation of the integrity of self-definition.

To add to the dilemma, the inaccuracies and "viruses" that confound the memory process can occur at several different experiential levels: at the moment of perception, encoding and storage, and even during unconscious retrieval and recall, owing to a variety of factors (Schachter, 1996). For example, while storage might seem to be an innocuous enough task in the long chain of memory and recall, storage is in fact highly sensitive to symbolic-cultural codes, most of which are of the implicit and non-declarative type, meaning that one barely becomes conscious of their impact and mode of operation (Donald, 1991). The influence of these symbolic storage codes is powerful enough to wield a cross-generational effect, as much recent research in the area of the Holocaust and its reconstruction repeatedly confirms (LaCapra, 1994; Langer, 1991; Sebald, 2001).

As is well known, some memories are repressed, exerting their influence in alternative patterns that actually make recall extremely difficult, owing partly to the perverse pleasure gained by certain primitive forms of repetition (as opposed to remembering). Other memories are encoded *ab initio* during preverbal phases of development, such that in a sense they can never be recalled, only reconstructed. And still other memories begin to "decay" owing to certain environmental pressures that fragment a memory into so many distinct pieces that it is rarely possible to endure whatever process might be involved in restoring some similitude of a valid "original" memory.

From a broadly accepted point of view, then, the very data of human mental experience at any given moment are *representational* data, which means that whenever one purports to talk about *psychic data* one is referring at all times to a mixture of affective impressions, early memories, and current experience (Shanon, 1993). To this, psychoanalytic theory adds the ever-present, powerful impact of the need for wish-fulfillment as well as defensive screen memories designed to mask emotionally problematic wish-fantasies (Freud, 1898). According to this view, in the absence of these elements, what we know as "mind" simply cannot take form, or ceases to exist.[35]

The Problematic Dimension of Time

The vicissitudes of self and memory intersect with a third problematic domain: the dimension of *time*.

The view of contemporary science is deeply committed to distinguishing the mathematical or ontological properties of time from its subjective, psychological, or existential properties, thereby fostering a multi-dimensional approach to time as opposed to retaining, say, easily and objectively measurable and quantifiable but possibly trivial characteristics of time and discarding the more elusive and subjective elements that may turn out, in fact, to be the critical and most substantial aspects of what time really is.

Fink puts the dilemma starkly, "The moment time is experienced and the moment it is designated never coincide," since the time one takes to observe any given event in time has already transformed the event (1995b, p. 64). Even if a physical dimension that can be identified by convention as "time" does exist in some objective sense, it would be a *perceived* time, and perceived time is another illusionary construct, largely intertwined in a complex cause-and-effect relationship with the equally imaginary representation of memory and self, an epiphenomenon that emerges against the horizons of the peculiar perceptual apparatus of the human mind. And, again, paradox awaits: the inner states of self, permeating one another, lying side by side, as it were, perceive of time as duration, as moments that contain one another (see Hartocollis, 1983, pp. 17–30). However, once time is externalized and perceived as part and parcel of the world of external objects-in-space—yet another psychological development that comes into focus only through negativization (i.e., no two objects can occupy the same place at the same time)—time is suddenly perceived asymmetrically, as discrete moments taking place *either* in the past, present, *or* future, but not as simultaneous or intermeshed. Time as tense (or perspective) is "outer" time, extended sequentially over space; time as duration is the deeper, unconscious sense of time in which the internal self tends to bathe (deeper still in the unconscious is utter timelessness and pure symmetrical experience).[36]

When mental health holds sway, and in the absence of major psychological conflicts, the human mind switches constantly, even poisefully between these radically different time perceptions. The perception of time ought to eventually acquire independence from desires and wishes, but this is not always the case—and it is certainly not the rule within the unconscious. It is obvious, then, that no sense of time would be possible without consciousness of self, and it is therefore not surprising that disturbances of the sense of self invariably are accompanied by distortions in time (see Arlow, 1986; Hartocollis, 1983).

All social time syntheses are themselves transient time constructions and they potentially disappear when they are no longer necessary, though deep within the labyrinthine alleys of the repressed, the human mind remains "aware" of a certain chronological pulse, and the transient, vulnerable sense that accompanies this pulse. "We never completely get rid of the feeling of fragility which goes through all the carefully erected temporal scaffolding of our lives," wrote Brockmeier (1996, p. 244). This feeling of fragility is significant not because of the experiential element of transience per se, but because, as Freud detected (1916a), the feeling of fragility itself *represents* the trace of lost internal psychological objects—which includes the loss of any perceived aspect of an object or even the loss of the love of the object—and tends to immediately initiate an unconscious process of mourning. Here, too, the Kleinian notion of post-depressive psychological achievements is critical because the sense of the flow of time—and a clear-headed sense of past and future—requires the negativization of the concrete and the instantiation of the symbolic, both of which come into maturity during this phase. Only once these conditions have been achieved can the all-important perception of flux emerge from the boundary surrounding the rhythmic interaction between the memory of the concrete in the light of the symbolic frame.[37] It is for this reason, too, that, despite the physicist's version of reality (which, as we shall soon see, many physicists no longer uphold in a simple-minded way) according to which time is irreversible, in effect time is determined by the perceptual interchange with specific intersubjective objects, and by significant psychological qualities of the unique historical "file" these objects carry.

Thus, on some level, *all perception of time is mourning*, and the capacity for a creative sense of the flow of time, and of returning within time, is proportionate to the capacity to surrender lost objects, relinquish mourning, and symbolize their absence. Without mourning, man becomes immured in a state of *aveilut yeshanah*.[38] From the psychoanalytic view of time, one cannot re-experience time on a *representationally* mature level if all one can do is reproduce unsymbolized internalizations suspended or frozen in time. In the latter instance, one merely reminisces pathologically or repeats the past in one of many semi-concrete ways (or attempts to undo the past in a manner that tends inevitably to recreate the past in its original traumatic form, barely modified [Freud, 1916b]). "Change in time," other than by way of noticing the movement of the hands of a clock (and sometimes individuals succeed in misperceiving and blocking off even this elementary perception!), would not even be a psychologically relevant possibility if it could not be the product of an actual mutation within representational structures of the mind, and this requires, as we shall shortly elaborate, not simply intellectual prowess or a rational outlook,[39] but rather the deep travail of negativization, symbolization, and mourning.[40] Even well within the normal range of mental experience, those perceptions of time that are of greatest significance to us—whether because they register for us what we possess or what we have lost—are extremely difficult to capture in written or spoken language, unless one is especially attuned to a variety of paraverbal (kinesic, sensation, mimetic) dimensions of time and self-experience, and adept at expressing these dimensions. In her beautiful prose (which not only conveys the point but illustrates it as well), Julia Kristeva states (1994, p. 167): "Whether we are lost in time, losing time, or losing our lives without discovering anything in death, we are made of the same substance as time because it defines the boundaries of our speech."

For Kristeva, what Marcel Proust was seeking in *Remembrance of Things Past* (which she correctly retranslates as *In Search of Lost Time*) is "the lost, invisible temple of *the sensory time of our subjective memories*" (p. 171). It takes all of Proust's skill, his neurotic sensitivities, and, ultimately, his life to appreciate that he will never actually locate any particular substance or entity, but rather will at best ensnare or embody the mental space crafted by lost time, and transform it into something personally meaningful. Proust, I would say, could be seen as having been involved in a prototype of *teshuvat ha-mishkal*. That is, the preparatory efforts at balancing the work of mourning against the use to which one can now put the spatiotemporal psychic representations that thereby become available has to be the prestage for further moral achievements in the domain of *teshuvah*.

I would like to end this subsection with the following thoughts. There are other ways of imagining why the arrest of immorality, or psychological conflict, by appeal to the direct inquisition of memory and desire, understood at face value, may be doomed. I have in mind the specific way in which the psychoanalyst W. R. Bion (1967) amplified the problem.

Bion sought to prepare the analyst for the treacheries of self, memory, and time by warning against undue efforts to anticipate the patient, or of becoming overly preoccupied at the beginning of a therapeutic session with an overabundance of "memory" and "desire." Bion's intuitive grasp of the latent paradox here was based on the fact that one finds within the mind numerous "memories" that are, in fact, no more than *projections into the past* of deep-seated, essentially unfulfillable wishes. In reality (whatever *that* is), no historical event per se can be located where one might expect one to be were one to trace the path of such "memories," because in these cases we are referring, in fact, only to the memory of a wish. Alternatively, an actual historical event may be located at the end of the path of such memories, but the individual is uninterested in reconnecting with the actual buried historical event; that is, has no interest in *remembering* it, only in acting-out (*Agieren*) and concrete repetition, while claiming ignorance of the relevant but hidden motive.

Mutatis mutandis, numerous human "desires" are, in fact, *projections into the future* of deeply repressed memories of things past, whose actual achievement or fulfillment has been traumatic, and rendered non-I or non-self. Despite every protestation to the contrary, and the evidence of intense frustration and lack that the individual desperately wishes to fill, such "desires" rarely bear fruit *because they are based on a traumatic fulfillment that took place a long time ago*, now brilliantly disguised as preoccupation with the future and a wish for the yet-to-be. Bion's concept of memory, therefore, is not the equivalent of recollection based on mourning, and his concept of desire underscores the antithesis of mature hope.

Evidently, the *senses* of past, present, and future are woefully difficult to define and secure within defensible boundaries. Accordingly, from the standpoint of the perception of time and its impact upon psychological experience, the traditional concept of linear time, and even the more advanced circular models of time, cannot be held to the constraints once considered obvious. I certainly am not saying that memory is useless and time perception totally unreliable, only that these cannot be relied upon as unimpeachable sources of hard, objective data. As an outcome, any praxis of *teshuvah, or psychotherapy*, which focuses in an insufficiently weighted manner upon memory and time, traverses into a blind spot that effaces an entire host of memory-influencing factors that surely need to be the object of our attention. On the one hand, it is certainly true that religious ritual (be it sacrifice, prayer, or the order of penance) does serve as that unique enclave where forgotten dimensions of time can be practiced—where variant and invariant spheres of time mix (Elchardus, 1993), and creative forms of repetition render the old new (Siggins, 1983). Nevertheless, suitable psychological processes are prerequisite to render ritual appealing to the individual mind, or else the unconscious value of ritual will lie inactivated even if the ritual is behaviorally practiced. If God is to be encountered through these psychological processes, the concept of *teshuvah* must be opened out, exposing the interstices where objects of faith and objects of mind can be integrated, *including the traces of whatever has been lost or rendered absent by virtue of the presence of these unique representations that had taken their stead.*

The Problematic Dimension of Language

The final dimension of impossibility to be discussed with reference to repentance is language itself.

Language is there from the start, long before it is comprehended, and its significance draws from the fact that it is a symbolic frame or structure to which even the parents of the infant owe allegiance, as their parents before them, and so on, thereby creating the beginning glimmer of a cut or wedge in the absolute, unmitigated pleasure of autistic or symbiotic oneness, a painful cleft, to be sure, but one that will ultimately inaugurate healthy differentiation and mature internalization. Inasmuch as language is predicated upon loss, then the articulate self as well is instantiated in loss from the start, but this articulateness only approaches meaning when it can be symbolically structured (Lacan, 1964, p. 319). Language is a substitute for the earliest libidinal relationship, which could not possibly be maintained forever in pristine form and could not at first possibly be spoken, and yet, paradoxically, the moment one adopts language, one is forever lost to some degree in a world of signifiers that lead to other signifiers but never directly back to an original object encounterable as it was prior to language.

Borch-Jacobsen summarizes the Lacanian thesis as follows (1991, p. 161): "If to become a subject the subject must speak himself, and speak himself to an other, it is because he can come to full consciousness only at the price of reflecting himself, of separating himself from himself in order to better (re)present himself 'before himself.'"

Language is thus inherently other, even when one has internalized one's own personal version of it. Through it, we not only communicate needs and thoughts; as one frames the mind in language, spatial and temporal dimensions are created with which we negotiate the relationships among minds. That which we perceive as "present" or "absent" is a perceptual experience that transpires in the mind, to be sure, but, more correctly, within the linguistic dimensions of which mind is comprised. Jacques Lacan worked these ideas through repeatedly throughout his *oeuvre*, always articulating language, object relations, and emotion (contrary to the popular misconception), especially desire. His view regarding the "treachery" of language comes through particularly clearly in statements such as this: "All the needs of speaking beings are contaminated by the fact of being involved in an other satisfaction" (1972–1973, p. 51). In an even more telling comment (1964, p. 207):

> The relation of the subject to the Other is entirely produced in a process of gap.
>
> Without this, *anything* could be there. . . . Psycho-analysis, then, reminds us that the facts of human psychology cannot be conceived in the absence of the function of the subject defined as the effect of the signifier. The signifier, producing itself in the field of the Other, makes manifest the subject of its signification. But it functions as a signifier only to reduce the subject in question to being no more than a signifier, to petrify the subject in the same movement in which it calls the subject to function, to speak, as subject.

Through formulations such as these Lacan developed his thesis that language by necessity alienates the individual from the early, highly pleasurable body-self, and is always restraining desire while at the same time providing for written and oral texts from which trapped desire is always impatiently attempting to escape and seek fulfillment. Without the overall maintenance of the symbolic frame that enables the slips, clefts, and gaps in language to safely reveal just how unaccidental and desire-oriented they are, human communication would be either purely random or an automaton-like mutual echoing of signifier to signifier. Only the larger symbolic transformation renders abject repetition, always full with itself, into an accidental existential encounter, which Lacan terms the *tyche*, graciously marred by the frustration of inevitable incompleteness, or, under other conditions, the obverse.

In essence, Lacan's stellar accomplishment lay in his exercising to the fullest the deconstruction of language initiated by De Saussure and his followers[41] as well as the multiple re-layering of mental representational inscription inaugurated by Freud, freeing the meaning of human speech from limits that had been inadvertently imposed on it by the theory of the arbitrary interrelationship of signifiers as well as from a theory that relied too heavily upon preordained latent fantasy structures. Instead, Lacan widened the scope of the meaning of discourse by arguing that the relationship among signifiers and symbols was to a large degree influenced by the nature of the way in which the individual sought to express desire, and to deal with the problem of the inescapable mirroring in human interrelationship; the structure of language, in turn, laden with the emotional discourse and desires of the Other, heavily directed the pathways of intersubjective relationship (see Van Buren, 2002).

The dimensions of loss and negativization in language are not wholly the preoccupation of exotic French psychoanalysts. The developmentalist Daniel N. Stern (1985) also underscored the sense of loss or "distancing of subjective experience" created by language and inherent in its use (i.e., to be speaking is to be regulating in one way or another the subjective sense of relationship with that regarding which or whom one speaks). More important, he notes that simple language (or, non-narrational modes of experience) allows for a chronic sense of "slippage," that we are somehow missing out on the fuller dimensions of experience, yet that is also precisely what generally impels an individual toward adopting advanced forms of speech, poetics, the use of metaphor and metonym, and other devices that enable one to recapture some of the deeper, more heavily affect-tinged sensory aspects of intersubjectivity, those which tend to bring to mind the original spheres of experience with the "impossible" primary objects. Language is thus not simply an expressive modality but a container.

If the hieratic liturgy and related linguistically moderated rituals of *teshuvah* are viewed in this light, then it is clear that in the act of making words with God, in the art of articulation, and through the pains and pleasures of enunciation and alliteration, one is already deeply involved, at one and the same time, in the processes of transgression and return (until one or the other takes primacy). That is, with words, and with speaking, comes a searching of the deepest elements of the human mind for that fundamental flaw or moment of severance from which emanates an agonizing cry that gradually took the concrete form of alienation, mistrust, apathy, and sin, but also potentially expresses the search for the original union.

Language, after Lacan, permits better perception of yet another aspect of human experience that is at once object-relational, semiotic, and faith-oriented. The otherness inherent in language redirects one's focus to the ways in which the self has been the basis for the other's desire. In the case of psychotics, this use of an individual through concrete, desymbolizing, and de-differentiating language slowly fragments and destroys the individual, replacing him physically by the psychotic desire of the disturbed parent or family network. In less primitive circumstances, the symbolic structure of language is maintained and loved in its own right, yet the subtleties of language—the tropes, slips, lacunae, tonalities, and the intertexts, and the un-said—all conspire to modify the relation-in-desire between two individuals in a variety of manners. Within the neurotic-to-normal range, we are generally too conscious of the flow of desire to ignore the pleasure or the pain entirely, and too unconscious of the symbolic hidden "real" (original) referents of desire to be able successfully to identify its source or aim. Language is always revealing a gap, a lack or inadequacy or incompleteness in my desire, and the pathways I have opened up for the other to fill that gap. Here lies the fork where one turns either toward the mutual embrace in good faith or to the theft of hearts and minds.

Only by accepting the interplay between the symbolic frame and the imaginary can one "play" with creating gaps, lack, and desire in an other: to symbolize entirely would eliminate desire while pure imagination would gradually unmoor one's grounding in physical reality (see Julien, 1994). Only this dialectical structure makes feasible rich time perceptions *with voids and vacuums*, spatial perceptions *with aporia*, and self perceptions *that acknowledge the inherent Otherness within* (be this "Other" language itself, the unconscious, the representations one has internalized into one's unconscious, and so forth).

This idea was so significant for Lacan that he once said that the pain experienced upon loss of a loved one is to be explained with the following formula: "I mourn the loss of the one for whom I was his lack!" (Might one hear "*Hareini kapparat mishkavo*" [Talmud, *Kiddushin* 31b]?). As Harari interprets this notion (2001, pp. 8–10), mourning is not about loss alone, it is about an awareness, an anxiety, that the survivor suddenly has of a hole or vaccum in the self that had formerly been filled by virtue of being the object of the deceased's desire, which itself contained traces of the incompleteness or lack that the deceased had experienced in those domains where his or her self had been a site of lack for an earlier generation's desire, and so forth. And in existentially meaningful relationship—by definition marked by the influence of a willingness for continuous refinding and *teshuvah*—there is also a willingness to allow the identifications and object representational fixations of the mind to "fall," revealing desire, even if the fall or "transgression" is ensconced by a sudden anxiety commensurate with the resulting vulnerability. For this type of anxiety to be at all manageable, and not result in psychosis, this lack must be negativized and turned into symbolic structures *containing* lack. The alternative is the experience of a hole in the mind.

Repetition, Loss, and Impossibility: Kierkegaard

The four domains of the impossible (self, memory, time, and language) coalesce within the troubling momentum of repetition.

The compulsion to repeat, as Freud observed (1932) with as much irony as Rabbi Soloveitchik (1974), is almost an inevitability of human existence. As we stated above, this factor was one of the reasons why Freud questioned the possibility of his entire psychotherapeutic enterprise. From an evolutionary point of view as well, "it is the nature of thought that human beings tend to be able to understand events only by virtue of the identities we imagine among them" (Kubler, 1962). Repetition is thus not indigenous solely to trauma but to all mental life. As Gilbert Rose put it (1980, pp. 185–86):

> The smallest frames of perception, thought, and memory include a microsecond by microsecond recapitulation of [the development of] psychic functioning from distant past to present, a rhythmic alternation between the blurring of boundaries and their redelineation. The melting-down and recasting goes on subliminally. In perception and thought, there are rapid, subliminal oscillations between global, scanning, condensing (primary process) operations, on the one hand, and discriminating, abstracting, focussing [*sic*] (secondary process) ones, on the other hand.

For Rose, then, the almost musical and yet repetitive shifting is a major and ineluctable characteristic of at least the undercurrent of all human mentation.

Is there an alternative to repetition, and in what ways is the re-turn of *teshuvah* fresh and authentic as opposed to compulsive and ineffectual? To the best of my knowledge, the problem of repetition was addressed by only four masters of psychology, each in his own way: Kierkegaard, Freud, Lacan, and Soloveitchik. To my mind, each could be understood to have been thinking simultaneously, directly or indirectly, about the impossibility of *teshuvah.* I will discuss the first three of these masters below, and return to Rabbi Soloveitchik in a later section.

Kierkegaard's little tract *Repetition* (composed in reaction to the dissolution of his engagement to his beloved Regina) (1843), featuring the curious subtitle, *An Essay in Experimental Psychology* (and the equally ironic pseudonym, Constantine Constantius), reaches out of personal despair into the heart of the challenge repetition throws down before repentance. On one hand, repetition is inevitable (p. 6):[42]

> If God Himself had not willed repetition, the world could never have come into existence. He would either have followed the light plans of hope, or He would have recalled it all and conserved it in recollection. This He did not do, and therefore the world endures, and it endures for the fact that it is a repetition.

However, if existence continuously repeats itself, and if time marches forward cyclically, in what manner can moral *change* be properly introduced? First, Kierkegaard creates the following elegant distinction (p. 4):

> Hope is a new garment, starched and stiff and glittering, yet one has never had it on, and hence one does not know how it will become one and how it fits. Recollection is a discarded garment, which beautiful as it may have been, does not fit for one has outgrown it. Repetition is an imperishable garment, which fits snugly and comfortably, neither too tight nor too loose.

Subsequently, Kierkegaard lays out three further significant distinctions (p. 34ff):

> Freedom first defined as pleasure, fears repetition, because repetition seems to possess a magic power to hold freedom captive once it contrives to get it under its influence; despite all of freedom's inventiveness, repetition chronically makes itself apparent. When a man in such a state of naive freedom falls into despair, a higher form of freedom makes its appearance.
>
> The second form of freedom is defined as "shrewdness." It acknowledges repetition, but in reality strives to convince itself that it welcomes something new that it discerns in repetition.

In psychoanalysis, this second form is a particular powerful form of resistance that undercuts the establishment of the transference neurosis (see Laplanche & Pontalis, 1967, pp. 455–63) that the analyst is patiently trying to nurture, since by the patient's emphasizing "novelty" and insisting that he is basically bringing forward something "new," "exclusive" to his contemporary relationship with the analyst, the transference dimension is ignored or stifled. Finally, Kierkegaard adds:

> This effort to delude repetition also inevitably fails, and despair sets in again.
>
> The third and highest form of freedom takes place in relation to itself. Having circumnavigated life' s experience, the individual now comprehends that the goal of freedom is to solicit, bring forth, and acknowledge repetition; and it, if anything, fears "change" insofar as this seductively challenges the true eternity of

> repetition. *Freedom now is repetition.* Mere recollection only affirms that all that is has been; repetition affirms that existence that has been *now becomes*!

With regard to this higher form, Kierkegaard no longer means repetition of physical nature, but spiritual or psychological repetition, an inward repetition, or return within the self. This form of repetition is latent, but is constantly drowned out by noise and distraction, the chatter of the drawing-room that drowns out a word that the listener knows his colleague wishes to say, or the banter of the patient who is attempting to "free associate" away from anything meaningful, usually in a pattern that repeats the modes of denial and dissociation that had always held sway in the past. But Kierkegaard was not a practicing psychoanalyst, and did not yet have the tools of transference analysis at his disposal, and thus is left simply to recommend as compassionately as he can that one philosophically embrace a higher form of repetition by recognizing that "change" is an illusion.[43]

Repetition, Loss, Impossibility: Freud

Freud's concerns about repetition carry us deep into the nub of the problem, primarily because of the relationship he eventually articulates between repetition and loss.

Freud took the clinical aspect of repetition very seriously, once he noticed that even while engaged in analytic therapy, the individual in treatment, who had earlier promised (possibly with every enthusiasm) to *say* everything that comes to mind with as little censorship as possible, does not (or, at least, not always), strictly "*remember* anything of what he has forgotten, but *acts* it out. He reproduces it not as a memory but as an action; he *repeats* it, without, of course, knowing that he is repeating it" (1914, p. 150). Furthermore (p. 151):

> As long as the patient is in the treatment he cannot escape from this compulsion to repeat; and in the end we understand that this is his way of remembering. What interests us most of all is naturally the relation of this compulsion to repeat to the transference and to resistance. We soon perceive that the transference is itself only a piece of repetition, and that the repetition is a transference of the forgotten past not only onto the doctor but also on to all the other aspects of the current situation.

This form of repeating is inherently related to the nature of the Unconscious, into which, so to speak, the analysis has dared to insert its hand. Elsewhere, Freud explains this inevitable initial "dynamic" (or anti-dynamic) of the transference (1912, pp. 107–8):

> The unconscious impulses do not want to be remembered in the way the [psychoanalytic] treatment desires them to be, but endeavor to reproduce themselves in accordance with the timelessness [*Zeitlosigkeit*] of the unconscious and its capacity for hallucinations. Just as happens in dreams, the patient repeats the products of the awakening of his unconscious impulses as contemporaneous [*Gegenwurigkeit*] and real; he seeks out to put his passions into action [*seine Leidenschaften aigeren*] without taking into account the real situation.

Thus, acting-out, as revealed by its atemporality, is contraposed to the temporal authenticity of true remembering (see also Laplance & Pontalis, 1967, pp. 4–6).

What can be accomplished in the face of this compulsion to repeat? Freud's answer is the cornerstone of psychoanalytic treatment (p. 154):

> The main instrument, however, for curbing the patient's compulsion to repeat and for turning it into a motive for remembering lies in the handling of the transference. We render the compulsion harmless, and indeed useful, by giving it the right to assert itself in a definite field. We admit it into the transference as a playground in which it is allowed to expand in almost complete freedom . . . provided only that the patient shows compliance enough to respect the necessary conditions of the analysis, we regularly succeed in giving all the symptoms of the illness a new transference meaning [*Ubertragungsbedeutung*] and in replacing his ordinary neurosis by a "transference-neurosis"[44] of which he can be cured by the therapeutic work. The transference thus creates an intermediate region between illness and real life through which the transition from the one to the other is made. The new condition has taken over all the features of the [original] illness; but it represents an artificial illness which is at every point accessible to our intervention. It is a piece of real experience, but one which has been made possible by especially favorable conditions, and it is of a provisional nature. From the repetitive reactions which are exhibited in the transference we are led along the familiar paths to the awakening of the memories, which appear without difficulty, as it were, after the resistance has been overcome.

Freud has introduced here the notion of a transitional or potential space, long before Winnicott (1953), in which some amount of in-mixing takes place between past and present, though it is clear that under general conditions we cannot readily differentiate acting from remembering. It is also important to point out that Freud, writing in German, defines the acting out that defeats remembering with the narrow term *reprodziert* as opposed to the relatively more malleable form of reliving, defined as *Wiedererleben.* And while pragmatically there is much that we gain by being able to insert an interpretation between the patient's attempts to cast as novel his neurotically repetitive concretization of the transference and the analyst's attempts to maintain the symbolic weave, if pre-patterned, of the transference, the obvious paradox remains. That is, if even during analysis, or *teshuvah*, the individual is nevertheless involved in a returning—or, a *repeat turning*—we must, in a sense, conclude that an individual is either always within transference analysis, or always in a state of *teshuvah*, or he may very well be dead!

All of the preceding would be far less valuable were it not for a unique little phrase that appears twice in Freud's work, giving expression to the major conceptual key to mitigating the impossibility of *teshuvah*. In the essay "Negation" (1925b, p. 237–38), Freud states:

> The antithesis between subjective and objective does not exist from the first. It only comes into being from the fact that thinking possesses the capacity to bring before the mind once more something that has *once* been perceived, by reproducing it as a presentation without the external object having still to be there. The first and immediate aim, therefore, of reality testing is, not to *find* an object in real perception which corresponds to the one presented, but to *refind* such an object, to convince oneself that it is still there.

And as he stated the matter in an earlier work (1905, p. 222), "The finding of an object is in fact a refinding of it" ["*Die objektfindung ist eigentlich eine weiderfindung.*"]. The full meaning of Freud's idea, according to most interpretations (e.g., Castoriadis, 1975), is that one really never *knows* the aboriginal love object in the first instance since it is encountered by definition during an essentially prelinguistic phase of life (though language is always present in some fashion). It is more proper to state that the "lost object" in fact never *was*; rather, it is only subsequently constituted-as-lost. That is, the first object—and, again, impossibly!—cannot be known since the object that first instantiates the psyche cannot itself also be perceived or represented, such that only the impact or trace of the primary object can attract the attention of the mind, and by the "time" the mind perceives it, the object-as-perceived is therefore already a retread, a substitute, a repetition.[45]

The Lacanian Contribution to Repetition and Impossibility

Taking this point one octave higher, the Lacanians have added that the original object is unknowable because, prior to differentiation, language, and symbolization, it could not possibly have been experienced any more as object than as self; and once the object and subject are constituted as somewhat distinct, it is no longer possible to experience a sustained (sane!) state similar to the primal state of self-objecthood. It is the gap, or the remainder between these two types of perceptions of the original object—the one, dim and imperceptible but necessary to presume; the second, recognizable but illusory—that we keep trying to re-find and re-approximate.[46]

Whatever linguistic traits might be dimly perceivable about the original loved object only lend to its seductive appeal, but are initially also nearly maddening, such that the aboriginal object per se is usually tucked away deep within the vaults of repression that are instated automatically once the symbolic mechanisms (negativization, representation, etc.) begin to shape the frame of the post-linguistic mind. Owing to the inherent perversity of the relationship between the sexual drives and the reality drives—in the sense that "reality" from the start accepts substitute after substitute for its original source of pleasure and survival, ultimately replacing much of primitive sensuality with language and other symbolic structures—the pure sexual satisfactions of the earliest moments of life are lost forever, part objects give way to whole objects, primary-process concrete presentations give way to secondary-process representations. *Refinding* is therefore the business of representation, and representationalizing. Thus, by elevating *refinding*, in his beautifully laconic statement, to the status of the sole comprehensible mental activity, Freud combined the ineluctability of repetition, on the one hand, with the hypothesis of the impossibility of recovering the initial losses upon which the sentient mind is based, on the other hand, despite the fact that the latter assertion essentially means that certain psychological experiences, though perhaps not strictly mentalizeable as such, *do not repeat.*

Lacan's linguistically oriented views summarized in the previous sections would seem to underscore a practical differential (or *nafka minah*, in talmudic terminology) regarding his and Freud's approach to repetition. Whereas Freud primarily saw repetition, or at least uninspired repetition, as an obstacle to remembering, Lacan excelled at envisioning repetition as a necessary and automatic process based on the drive for meaningful encounter and the power, and the limits of the symbol. For Freud, repetition was a form of acting out; for Lacan it was the proper impulse of the networking of signifiers. Freud seemed to have favored, as often as possible, a dismantling of repetition, whereas Lacan sought to blend repetition by altering the symbolic machinery that underwrites repetition, modifying the temporal conditions that portray a given behavior or fantasy as repetitious. However, having clarified these distinctions, my impression is that there are sufficient clues in Freud to permit one to view Lacan's theory as an elaboration of latent or germinal notions of Freud's, and to merge these two views into a single, complex one.

If this is the case, *then interpretation in language truly holds quite a big share of that which is arguably new and non-repetitious in human experience.* Andre

Green's Lacanian inspired yet novel analysis of time makes this point succinctly (2002, p. 122):

> Repetition knows no other outcome than to wait indefinitely for the recurrence of the same; even when it could content itself provisionally with the recurrence of what is similar. Hence the importance of the *object*, whom we never know one hundred percent fully or clearly, and of fantasy and of the imaginary which allow the drives to thrive while preventing utter sameness. . . . [p. 89] It is the conflict between the pleasure principle and the death instinct which prevent pure sameness, unmarked by difference, i.e., mere re-presentation. Pure reproduction occurs only when there has been no inscription capable of becoming the object of psychic work.

According to Green, then, the genius of Freud's comprehension of repetition is that although it is subject to the principle of "no change"—all evolution being erased by the return of what is merely repeated—it is an integral part of all drive functions because *going backwards*, if linked to the service of representationalizing and binding of psychic drives, is still a creative force (in contrast to mere "undoing"). Only if repetition exceeds certain limits does it defeat the life forces, such as when the compulsion to repeat abolishes repression itself, the latter being always in some way linked to an internal index of temporality. Hence the additional importance of Freud's notion of the double inscription of experience (*Neiderschrift*), the multi-tiered layering of endogenously produced wish-fantasies and "realistic" perceptions, and the principle of the retroactive conferral of meaning (*Nachtraglichkeit*) upon past events (see Denis, 1995; Freud, 1920; Green, 2002; Laplance & Pontalis, 1967, pp. 111–14, 474–75). Without the impact of these mechanisms, mental life could be satisfactorily evaluated on an essentially mathematical plane.[47]

The Halakhic-Midrashic Approach to Time, Repetition, and Loss

Good hypothetical models evince duplication. Are there other areas within the literature of the midrash and halakhah where one might discern a meeting between time, repetition, and negativization?

Stated simply, the greatest fear of the rabbis was repetition or relapse to sin (Jer. 3:22; Talmud, *Yoma*, 86b), or repetitive, and therefore meaningless, *teshuvah*. Gellman, in a moving personal essay, wrote that repentance that merely repeats is of no value; mere repetition, overwhelmed by precedent, is conformity, and hence inauthentic (1982, p. 249). In Soloveitchik's assessment (1974, pp. 162–66), the mere repetition of any psychological dynamic is a lifeless process (or non-process); repetition as "backsliding," of which Jeremiah speaks, a sin in its own right aside from any other transgression that might be described generally as backsliding.

Backsliding—which I would interpret as the propensity to be unanchored historically as a result of failure to have undergone the salubrious transformations into the post-depressive position—cannot be guaranteed by *teshuvah mi-yirah* alone, despite the creative impetus the latter is often capable of setting in motion. We must therefore characterize repentance from fear in a new way: as gradual and slow only *in the sense that its ultimate frequency is plodding and not sufficiently symbolized* (and, as such, it results in a lower level of "acquittal"). Repentance from love, on the other hand, which stems from the depths of the anguished soul, is more cataclysmic *in the sense that it is busy deconstructing the symmetrical granite of the mind at every step of the way.* Only this latter mode of repentance can result in the apparently sudden but resilient, intense transmogrification of personality that Maimonides had in mind when he referred to the type of *teshuvah* that results in immediate, "overnight" transformation from being despised by becoming the beloved of God (*M.T., Hilkhot Teshuvah,* 7:6). It is only at this stage that the individual can take up the fantasy of God Himself as witness to the individual's self-assuredness that he will never again repeat his transgression (*M.T., Hilkhot Teshuvah,* 2:2). That is, only the post-depressive achievements in *teshuvah* enable a mind to be psychologically capable of enjoying the paradox of a fantasy-wish for a desired state ("yet-to-be") that playfully evokes the memory of the pristine moments of existence when all that existed was God' s expectation that Adam not sin; a state that is not limited to any historical epoch or concrete object as much as it is bound to the true memory of the transgressive discovery of speech and the onset of the psychic work of regaining control of one's psychic momentum. This is the *eidut* of symbolization.

All the more remarkable, therefore, is the support our thesis can muster from the well-known attitude expressed by Rabbi Eliezer (*Pirkei Avot* 2:10; *ARN*, xxix; Talmud, *Shabbat* 153a; Eccl. *Rabbah* 1:7):

> Rabbi Eliezer ben Hurkanus taught: "Return [*Shuv*] one day prior to one's death." When his disciples asked him, "But does a man know the day of one's death?," he responded, "The more reason to repent everyday lest one die on the morrow [*mahar*], and [in this manner] all one's life is spent in *teshuvah*!" And even did Solomon in his wisdom state, "At all times let your garments be white." [Eccl. 9:8]

This reference holds its share of the keys to a broader comprehension of the crucial dilemma of the time factor, repetition, and *teshuvah*: the apparent futility of any attempt at reversal of that which belongs to the past. For, how, in the end, are we to balance the strange tension between all of the citations and practices that promulgate the efficacy and possibility of rapid, spontaneous repentance, the so-called *teshuvah shel shaah ahat*, on the one hand, and, on the other hand, the equal number of references to the extensiveness and complexity, and perhaps even impossibility, of permanent, reliable *teshuvah*, or that portray it as being, at best, a project for a lifetime? Where meet the *teshuvah* of mere *hirhur*, characterized by its overnight and even instantaneous nature, packing such cataclysmic ordnance in terms of real halakhic efficacy yet seemingly so contrary to logic, with the sense, expressed by the Psalmist, that the pain and time of one's iniquity are ineradicable (Ps. 51:5)?

On the face of things, Maimonides contributes to this problem. Maimonides' wording of a particular critical law in *Hilkhot Teshuvah* (7:6), made famous by Rabbi Soloveitchik's well-known treatment of the topic (1974), emphasizes the efficacy of sudden conviction, apparently despite the possible superficiality of this level of *teshuvah*:

> *Teshuvah* brings near those who are far. Yestereve [*emesh*] this individual was hateful before the Lord . . . and today [*ve-ha-yom*] he is beloved, comely, close, and a friend. How excellent is the attribute of *teshuvah*: yestereve [*emesh*] this individual was separated from God . . . and today [*ve-ha-yom*] he cleaves to the Holy Presence.

So we are left to speculate: Are there, perhaps, more complex dimensions to certain kinds of apparently instantaneous change that are not visible to the naked eye, and that in essence belie the simplicity of radical *teshuvah?* Alternatively, perhaps instantaneous repentance is no more and no less than what it seems to be, on the evident spatial or temporal plane, and yet it derives a unique efficacy from some *other* element, notwithstanding its brief duration and ephemeral content, or from some other latent structure that it symbolizes or represents?

Thus, I offer that at least part of the solution is to be extracted from Rabbi Eliezer's well-known ethicomoral recommendation, intentionally and not merely playfully set in full poetic irony, that one ought to repent *yom ehad*, one day—or perhaps even *shaah ahat!*—prior to one's death, lest one leave this life without repentance. To his students' rhetorical wonderment as to how one could be expected to surmise the penultimate moment for *teshuvah*, Rabbi Eliezer makes plain that one must therefore repent *every* day: *shema amut mahar, ve-nimtza kol yamav be-teshuvah* ("lest one die tomorrow, and thus all his days are [spent] in repentance"). *Rabbi Eliezer recommends the impossible because he appreciated that this is the only way to express to his students his own awareness that without "impossibility" mental life as we know it could not exist in the first place.* Some future-oriented form of "death" or deconstruction—which, owing to the constraints of the human mind, always seems to be projected toward tomorrow (*mahar*)[48]—must be brought into articulation with the contemporary moment, the specious present, the fleeting but critical time frame known as *shaah ahat.* Put differently, the human being's existential and ontological limitations, though truly unpredictable, can be brought to the fore in a very important sense by the constant exercise of the processes of psychological representationalization—that is, not by being morose or neurotically obsessed with melancholy sentiments and delusions of death, but by virtue of an awareness, or a willingness to become aware, that as one deconcretizes, symbolizes, and internalizes external experience one is simultaneously destroying, or bringing about the demise of its concrete, mortal, uni-dimensional spatial and temporal characteristics.

Rabbi Eliezer, in my analysis, comprehended that the brute facts of death and time put the lie to life without repression, for the simple reason that the human mind cannot experience everything and never experiences everything. This elision is not simply due to the fact that humans fear life as much as they fear death. The inability to take in all experience is inherent to the fundamental apparatus of attention, concentration, and symbolization upon which human sensibility is predicated. However, repression is but one method for internalizing the awareness of death, rendering it thereby somehow manageable (see Becker, 1973; Boothby, 1991); so are acting-out (Freud's *Agieren*),[49] mere reproduction, and repetition, only more concretely and less fruitfully. Repentance, as well, could serve in this capacity, but not unless we expand its meaning beyond "mere" repression of the past nor any kind of facile restoration or ersatz sublimation.

To that end, we have proposed that language itself, and signification, creates the critical mortifying effect. Rabbi Eliezer's attitude was that "*teshuvah* analysis," properly effected, requires the *practice of death,* the negatizivation of encrusted network of meaning. And in the process of negativizing the standard temporal frames of mind that maintain illusory distinctions between past, present, and future—in the ability to suspend oneself for a moment within the *shaah ahat*—the true temporal mode of the lost object becomes central.

I would also say that Rabbi Eliezer's mortification of *teshuvah* forces us to reconsider kabbalistic and related doctrines that tended to portray the "deathly," dark, and depressive elements of repentance as necessarily reflecting lower levels of *teshuvah.* In some sense this might be the case, but from the Kleinian point of view there can be no perfect *teshuvah* that has not partaken of the depressive and mournful aspects of experience, for the care and love such *teshuvah* ultimately requires are nourished by these "darker" dimensions. Similarly, these "darker," deconstructive aspects of mental functioning would not motivate as creatively toward mature repentance if they were not tempered by being bound to care and love.

In the thought of Rabbi Kook, for example, the fear of death and other dark impulses within the soul are distinctly marked as prestages, at best, and at worse, as imperfections in the grander accessions in repentance, potentially destructive flaws indicating lesser faith and distance from God: "As there is a deepening of repentance, so does fear of death decrease, till it ceases completely. Its place is taken by the condition of 'and she laugheth of the last day' [Prov. 31:25]" (*Orot* 11:3). Early we examined the death motif in Rabbi Soloveitchik's work, and noted that in the preponderance of this work his central contribution to the repentance literature is his emphasis on the essential link between sin, the awareness of death and finitude, mourning, and then redemption, and, though we may simply presume that Soloveitchik would never put aside the overriding halakhic teleology toward joyful love of God, his man of faith reaps more religious growth from a sense of the tragic, guilt, and loneliness.

Even these kinds of distinctions can probably also be reduced into merging perspectives, but such an effort would only be valid and worthwhile if it can be done without violence to the style and provenance of each writer (or the schools of thought they represent). Drawing liberally upon some of the concepts that characterize the writings of Rabbi Kook and Soloveitchik to purposely create a few significant antinomies, we may have arrived closer to understanding the subtleties that actually explain how, again citing Rabbi Kook (*Orot* 16:5):

> the soul perceives thereby the void of its bounded containment and is embittered over all that causes restraint of spirit, and weakness. This, in truth, is the basis of "repentance based upon love," which is appropriate for every sensitive soul; to accept this experienced bitterness in joy and gladness of heart; then this deepest of pains is transformed into a content of uppermost delight, wherein are revealed abounding pleasures of sanctity.

The Halakhic Concepts of Mahar/Emesh as Temporal Markers

The term *emesh* ("yestereve") appears in another context in Maimonides' code, though they have never benefited from comparison. In forbidding any inappropriate goading or teasing of converts or penitents in general about their past, Maimonides states, citing the Talmud (*Hilkhot Teshuvah* 7:8; also *Hilkhot Mekhirah* 14:11–12):

> Penitents tend to be downtrodden and exceedingly downtrodden and humble. And thus [it is sinful] if fools vilify them [by flaunting] their earlier deeds, saying to them: yestereve [*emesh*] you were wont to do thus and such; yestereve you were wont to say thus and such.

Here again is the familiar term *emesh*![50] It had been used as a badge of courage in The Laws of *Teshuvah* to underscore the laudatory, momentous change from the sinner of "yestereve" to the religious hero of the next morn.[51]

Apparently, *teshuvah* must represent a grappling with the yawning void between two halakhic, existential modes of *emesh*; both, paradoxically, impossibly, referring to the identical psychotemporal location. That is, despite referring to the same kinds of events that occurred "yestereve," one mode of *emesh* is prescribed, the other proscribed. The first *emesh* is subjectively historical, redefined by the acknowledgment of the testimony of the Other, capable of being refound and retold; the second *emesh* is objectively historical, but in a schizoid way, capable of being tattletaled endlessly and mindlessly. The first *emesh* enables articulation, permits the pleasure of polysemy and representation, and therefore deserves to be continuously spoken, while the second *emesh* obtunds symbolization, dictates repetition, freezes full speech, and therefore requires therapeutic refinement. If I am correct, the nature of the *emesh* one assigns to one's perceptions largely determines the timing and the quality of the *mahar* one can anticipate.

If there is a halakhic rationale for distinguishing between levels of *emesh*, is there also a source, besides logical inference, for distinguishing between halakhically relevant, qualitative levels of *mahar*, "tomorrow"? As always one looks to the biblical text to provide the first clue. In the passages attending the historical episode of the Exodus from Egypt and the halakhic institution of the festival of Passover that commemorates this great event, a familiar formula stands out in relief: *Ve-hayah ki yishalekha binkha mahar leimor* "And it shall be when your son inquires of you tomorrow saying" (Ex. 13:14).

Important for us in our current application is a statement of the midrashic *Mekhilta* (*Bo* 18) cited by Rashi in his commentary to the biblical passage:

> There is a *mahar* that is now—*yeish mahar shel akhshav*—and there is a *mahar* that applies to a further time [*ve-yeish mahar le-ahar zeman*] [e.g., Josh. 22:27].

Later on (Deut. 6:20), where the almost identical passage appears—minus, perhaps significantly, the conditional *ve-hayah*, "And it shall be"—Rashi seems to have nothing much to contribute besides alluding to his commentary in Exodus, and cites only half of the midrashic exegesis he had brought at that earlier point: "There is a *mahar* that applies to a further time."[52] Now the observation of the *Mekhilta* seems eminently mundane, and yet evidently Rashi perceived something that the text left unspoken.

In and of itself, the brute, concrete notion of a "yestereve" can be perceived in different ways: it can be a nonporous, unidirectional marker of the forgotten, neglected, or repressed past, an open wound permitting a chaotic intermingling of quasi-temporal impulsions, or a bivalent portal enabling integrated, syncopated movement among temporal dimensions. *Emesh* therefore requires symbolic weighting. In the earlier citation, Maimonides praises the *emesh* of creative synthesis, but in the preceding citation, the *emesh* resulting from contamination, symmetrization, and the failure to differentiate is singled out for censure. Halakhah denounces the identity confusion that results from any intentionally diabolical collapsing of the penitent's or the patient's history and memory. Granted the fleetingness of the overnight *teshuvah* transformation under consideration, it is the change in internal representational state that had made this initiation possible, thereby rendering *emesh* into a significant temporal span to be reckoned with. Under conflictual circumstances, represented by Maimonides' prohibition, where sheer repetition is afoot, this span is lost (see Eliade, 1949, p. 86; cf. also Loewald, 1973). As we know, archaic phylogenetic modes persist in individual ontology, and to the degree that such archaic modes of temporality persist in a given individual, *teshuvah* will be perceptually and representationally impossible because the past is effectively coterminous with the present. At best, dumb, or "clever" repetition will predominate. Brute repetition perseverates into defective or degenerate forms of *emesh* and *mahar*, the *mahar shel akhshav*, where *akhshav* is not simply the categorization of a subtype of *mahar*, but rather a continuation of the term, an amplification of the quality of *mahar*, a way of indicating a paradoxical mental state where past and present simply collapse into an indeterminate, inchoate mass—a "nowmorrow," as a psychotic might posit neologistically. No questioning can emerge from this state, no inquiry, and hence no transformation of personality.[53]

The *mahar she-le-ahar zeman*, on the other hand, represents "tomorrow" placed within a dialectic, where there is diachronic and synchronic movement, and it becomes possible to question, and compare different dimensions of time and their meaning. And perhaps because the term *ve-hayah* with which the passage in Exodus begins implies, according to rabbinic formula, an inevitability rather than the outcome of choice, Rashi sensed that there had been some problematical elision of the hard task of discriminating between two possible modes of time. However, when the Bible records for the second time in Deuteronomy, "*ki yishalekha binkha*," but now without the term *ve-hayah* that appeared in Exodus, Rashi's comment is not intended as mere repetition. It is in fact precision itself. By stating now only the phrase "*yeish mahar she-le-ahar zeman*," Rashi, it seems to me, is now definitively categorizing the existential inquiry posed in Deuteronomy as the result of choice, the transformation of rote, middling curiosity into the higher pitch of lack acknowledged and historical dislocation embraced, the outcome of the entry into the dialectic of repetition in halakhic categories of *zeman*.

Juxtaposed between his subjective interpretations of time, man is always caught in the vortex of a neurotic turning within time and an existential returning toward time, between an *emesh* and *mahar* bifurcated in an unnaturally polar manner, and an *emesh* and *mahar* that mutually influence each other, permitting *teshuvah* and the transcendence of the impossibility of time. Like Samuel, man must constantly refind and redefine his home from the rhythmic turning movement within the circuit.

Refinding Gaps in Symbolic Time

The need to refind the lost object, to establish the refinding process as inherently more pleasurable than the hope of actually ever recovering the original concrete object, is inherent to symbolization and language, and is the opposite of what it is to be psychotic.[54] We should pause to weave into one cloth what we have learned about time, language, and repetition.

In language, negativization of the concrete and abandonment of the immediately restorable and retrievable meet. The symbol results from separation—it transcends loss. Impossibly, as Aragno put the matter, "no true symbol can be formed without a prerequisite distancing from that to which it refers, and only that which becomes symbolized can be evoked or articulated mentally" (1997, p. 161). Representations are even more advanced "apparati" than symbols, and certainly than signs, as we noted above. The sense of psychic distance created by symbols, and the delay processes suitable for thinking and imagining that symbols negotiate and maintain, leaves an ineluctable psychic gap that is a necessary part of the human experience. Reality always defaults against the symbolic structure at some point, and thus there will always be gaps in the symbolic—culverts where time is arrested and desire revealed—but are rendered valuable by virtue of the meaning lent to them within the symbolic structure.

The view espoused here is shared by many, among whom Kenneth Wright has added a significant element (1991, p 241), and it, too, is a paradoxical and impossible one. In Wright's analysis, signs, metaphors, and ultimately symbols are not possible until the young child is able to fully relinquish—and, ultimately, mourn—the expectation of actually securing the concrete object, exchanging literal experience for the internalized pattern of his or her representations, and yet it is the symbol that enables reunion or links with the damaged or absent object. Unfortunately, this gap is usually not what one thinks of oneself as looking for when searching for some lost *thing* or some bygone *time*—or, as we shall soon examine in the context of halakhah, some *davar*—and yet the gap can be more important that all that surrounds it, certainly for specific periods of time. Again citing Aragno (p. 278):

> As long as experience is retained in the senses, of the senses, it is us and is therefore unconscious, not yet of the mind. We have not distanced ourselves from it or taken something to represent it and therefore have no means by which to make it comprehensible to ourselves or others. Only when we have, as it were, lost it, organically separated ourselves from the experience or object, and perceived it as other, beheld it, and given it a name by which to refer to it, only then can we become conscious of it.

And thus the use of speech means always accepting the partial satisfaction of desire.

The notion of the gap also has meaning for our sense of hope, and our sense of time, both of which need to be considered briefly in order to complete the scientific aspect of our analysis thus far. First, I refer to a recently translated work of Aulaugnier (1975, p. 116):

> A gap must remain between the I and its projects: what the I thinks it is must reveal a "less than," which is *always there* in relation to what it wishes to become. The future I and the present I must maintain a difference, an *x* that represents what would have to be added to the I for the two to coincide. That *x* must remain lacking: it represents the resumption of the trial in the register of identifications, and it recalls what this trial leaves intact: The narcissistic hope in a self-encounter, always deferred, between the I and its ideal that would permit the cessation of any identificatory quest. It is therefore a compromise that the I makes with time: it renounces making the future that locus to which the past might return; it accepts that fact, but it still hopes that one day that future will give it back possession of the dreamt-of past.

In Julia Kristeva's aching prose (1977, p. 184), "The love of God and for God resides in a gap: the broken space made explicit by sin on the one hand, the beyond on the other. The topography of a *symbolic* relation posits my otherness as impossible. Love, here, is only for the impossible."

Rabbi Soloveitchik approximated this perception of the gap as well when he publicized a fascinatingly novel insight of the Vilna Gaon regarding the quality of King David's repentance (1980, p. 263):

> After Natan reproved David for his sins, we read, "And David said to Natan, 'I have sinned against the Lord!' And Natan said unto David, 'The Lord hath put away thy sin; thou shalt not die'" (II Sam. 12:12–13). All David said was one word—*hatati*—"I have sinned," and for this he deserved pardon and forgive-

> ness?! Rabbi Elijah, the Gaon of Vilna, noted that in the Masoretic text of the Bible, after the phrase "I have sinned against the Lord," there is a gap, *a blank space.* David actually wanted to say more, to make a detailed confession, but he choked up with tears and could not utter another word. At that moment, David achieved complete self-negation, a sense of sheer worthlessness. He was so profoundly shaken that he could say no more. But Natan the prophet sensed the inner drama that was taking place in David's soul.[55]

The notion of an interval, arrest, hesitation, or gap plays a pivotal role in linking *teshuvah* and time. The reader will recall that in older writings on *teshuvah* the temporal element of sin was viewed essentially unalterable; if forgiveness is obtainable, this development was considered to devolve wholly from the forward-looking aspects of the transformation that has been effected (and from divine *hesed* or loving-kindness). For example, Rabbi Dessler wrote (1991, p. 83):

> The matter is that *teshuvah* is effective only [as an operative] in dimensions higher than time. In the dimensions within time, there can be no complete correction [*tikkun gamur*] for sin, for the time in which he sinned and the time in which he returned are forever divided.[56]

Most writers on *teshuvah* offer similar versions of this linear conception of time, but we have seen that it is essentially incomplete from the standpoint of contemporary views in psychophysics.

Soloveitchik on the Irreversiblity of Time

Rabbi Soloveitchik, however, offers a different possibility regarding time, first mentioned his early thesis *The Halakhic Mind* (1944, p. 162), informed by his mastery of the then regnant theories of time, particularly Bergson's. The influence of his view can be felt throughout his subsequent writings on *teshuvah*, though nowhere, to the best of my knowledge, does he write as specifically about the temporal element of repentance as he does in the material to be cited.

In his thesis, Soloveitchik combines the time concepts of the physicist and the psychologist. On one hand, he certainly appreciates the physical irreversibility of time—"Sin is not to be forgotten . . . on the contrary, sin has to be remembered. . . . The penitent does not wipe out the past nor tear the pages of sin from his memory" (1974, p. 255). But he is not satisfied with these notions, and instead opens up the vistas of repentance to more complex dimensions of time, in which the past is continuous and stable and does not pass or slip away through one's fingers; it is static, in some senses, but also capable of intermingling with the present. Soloveitchik argues (1944, p. 48–49):

> If one operates with static, spatialized, and reversible [retraceable] time then "becoming" is precluded; for a universe that can retrace its course and regress in the minus direction with the same ease (theoretically) as it progresses in the plus direction is not susceptible to change. It can always undo the cosmic occurrence of the past.

Within such a time framework, *teshuvah* would be unthinkable, and this, despite the previous sentence, because within such temporality there would be no *sense of past* against which to anchor forward or return movement. On the other hand, writes Soloveitchik, "Time flow which is continuous becoming renders reversibility impossible [since] change denotes a movement of superseding phases which cannot be reversed." Against this, Soloveitchik offers his religious awareness of time, ripe with impossibility:

> Religious time awareness is so paradoxical as to register both becoming and reversibility. As to becoming, the act of Creation introduces it metaphysically; and the religious norm with its associate postulate of freedom sponsors it ethically. Nevertheless, the reversibility of time and of the causal order is fundamental in religion, for otherwise the principle of conversion would be sheer nonsense. The act of reconstructing past psychical life, of changing the arrow of time from a forward to a retrospective direction, is the main premise of penitence. One must admit with Kierkegaard that repetition is a basic religious category. The *homo religiosus*, oscillating between sin and remorse, flight from and return to God, frequently explores not only the living traces of a bygone past retained in memory, but a living "past" which is consummated in his emergent time-consciousness. . . . The paradox of a directed yet reversible time concept remains. Neither physicalism nor psychologism throws any light upon a dichotomy which can be understood only against the typical religious cognitive background. It must be understood that these time concepts are not mere fantasy. They are inherent in the religious consciousness which apprehends time in its own fashion.

It is a minor but not unimportant vexation when, every so often, Soloveitchik commits his own form of tautology: for, stating that religious minds (even *serious* religious minds) conceive of time in a certain manner hardly proves that these perceptions are *eo ipso* not fantasy Soloveitchik has outlined the horizon of change within the then current models of temporality that might make the paradoxical temporality quality of *teshuvah* more potable to the modern mind. However, despite his perspicacity, he is essentially offering no more than a very sophisticated version of the pan-halakhistic view (referred to above in the name of the Sage of Navaradok) that if halakhah demands *teshuvah*, one must assume perforce that it is possible, for he has not articulated within what container or through what psychological mechanism are the temporal vicissitudes of *teshuvah* to be carried out.

But we now know that we can add at least three elements to the picture that go further than ever before toward clarifying the specific psychological dimensions that *contain* the otherwise impossible or paradoxical temporal paradox described by Kierkegaard, Freud, Soloveitchik, and Lacan. First, anxiety always signals the lack of the symbolized lack of the other, and the sudden implosion of the concrete Real into the imaginary or symbolic register. Hence, it is not an epiphenomenon of *teshuvah* but essential to it and cannot be absent if *teshuvah* is taking place, or else all that must be "present" by virtue of anxiety will be missing from the task of refinding. Second, transference emerges whenever humans enter into an intensive relationship rich with illusion under the aegis of a relatively stable surrounding symbolic framework. The transference dimension of relationship must be focal for *teshuvah* to be aimed at its most appropriate and relevant objects, even when these seem closely identified with the subject, and vice versa. Third, the negativizing and symbolizing capacities of language and speech (the mind itself cannot perform these functions without a language) are not only a handy but expendable tool for "doing" *teshuvah*—as I have tried to persuade, in that sense *teshuvah* is impossible!—but perhaps the most essential property of the phenomenon of *teshuvah.* That is, as we shall elaborate in the concluding sections, to speak properly is to be refinding. Refinding, *re-turning* even when one cannot possibly return, is the proper momentum of *teshuvah.*

Time, the Gap, Hesitation, and Desire

We can now link the preceding discussion of time, particularly Rabbi Soloveitchik's view, with the notion of the gap and desire.

I mentioned above two views regarding the phenomenon of Samuel's *teshuvah* to his home: Rabbi Soloveitchik's, emphasizing a metonymic, circular movement within the context of a physically true homecoming, and the Talmud's, seemingly emphasizing a metaphoric return divorced from any physically actual homecoming. I believe that we need not impose a choice between these two expositions.

In effect, true internalization always transoms between dual directions, involving processes of moving away from early symbiosis toward maximum differentiation and separation, and the reverse, of introjection and projection, of symmetrization (i.e., the use of the symmetric mode of logic) and asymmetrization (the asymmetric mode of logic), closer at one time to the subjectively experienced "self" component in self-experience and closer at other times to the subjectively experienced "other" component in self-experience. A kind of commutation or mutual enfolding of polarities, as opposed to primitive splitting or even commonsensical but superficial "integration," is what lies at stake. Bearing in mind the truism "You can't go back home," we must demur that, in fact, one *can*—and so trusts the psychoanalytic literature as well as the masters of *teshuvah*—provided that one is always "destroying" one's home as a concrete, embedded, rigid mental presence, playing with shades of internalization and memory so as to relocate "home" in a psychologically more meaningful manner within a symbolic geographic frame that enables existential circulation and repetition. One cannot return home, but one can, and must be always *re-turning* to home, *refinding* it wherever necessary in all that must be characterized as "homely," canny, and known as opposed to *unheimlich*, uncanny, and alien. This latter form of "return" home can only occur as one simultaneously returns to a *representation* of home, not a concrete home, that has been psychically symbolized, including the working-through of the painful loss of the concrete, literal, brute real home, in favor of a personal reconstruction that, while technically or photographically inaccurate, enables an always enriching sense of recollection and rootedness.

According to this revised view, if the prophet Samuel excelled in his pastoral role it is because the original matrix—the aboriginal maternal person/object—physically located in Ramah was capable of giving way to its own deconstruction and replacement, allowing itself to be "refound" repeatedly under any suitable internal or external set of circumstances.

Where does the gap fit into the temporal impossibility/possibility of *teshuvah?* Here, too, we can begin by recalling Rabbi Soloveitchik's nugget regarding King David's repentance. The halting interval, the pause, the arrest in time is always a moment of transgression; gaps convey the transgressive effect of desire, whether for sin, literally construed, or in service of development, such as the transgressive impact of the first signifier on the chaotic miasma of symbiotic existence. Boundaries and time always interrelate—sometimes dilating one the other and other times coarctating one the other. Nothing carries this relationship better than language; nothing makes this relationship more evident than the transference neurosis during psychoanalysis . . . and the encounter of *teshuvah.*

The very concept of temporal intervals and gaps is intrinsically related to the patterns of presence and absence that infiltrate language. When the neurotic, or the sinner,[57] is near to the conscious realization that his perceptions are about to change (one can only observe this under the microscope of psychoanalysis), a hesitation takes place—of variable duration and "weight" —

and for that moment one becomes aware that the other (God, the analyst, an internal ideal), by virtue of having not hesitated, by virtue of being present in a unique way waiting for him at the end of his hesitation, appears radically different than one had perceived the other only a moment before. At this point, a deeply subjective panic sets in, even when, as is sometimes the case, the other is not different than the individual in precisely the way the individual imagines him to so be. The individual becomes convinced that if he hesitates much longer, he is lost, possibly forever.[58] *This awareness is the sole difference between all of the moments prior to the hesitation and following it*; it is this that characterizes those unique watersheds of change in psychoanalysis.

And while this may seem like a simple enough observation, it is remarkably difficult, and in some sense impossible, for humans to be conscious of such hesitations. This is so not because any special logic is needed to comprehend it, but because up until that point the individual has not been engaged in refinding but only in finding, in the effort (always in some manner perverse) to return literally to some concrete representation of that which he truly lacks. And all humans, at root, are always truly lacking the original object: mother, father, God, or their prelinguistic *jouissance* and near delirious experience of symbiotic selfobject-hood. The instantiation of language decrees this so. Installed deep within the mind is the immemorable memory of the first awareness of time, the first gasp or hesitation, at which point the individual wondered (if that is the correct term) whether a single breath's more hesitation—"Either I go with this seductive, amazing novel Word or Gaze or I revert now while I still can to that which preceded it?"—might be his last![59]

Critical is that this hesitation opens up the fundamental capacity to reconstruct the past. If one were to lend words to the hesitation, it might take the form known grammatically as the future anterior (or future perfect); for example, "By the time you return from Jerusalem to visit me in Bet-Shemesh, I will have already left for Tel-Aviv," or "He could not have known when he got on to the bus that it was not to ever arrive at its destination."[60] This is not simply a matter of speech (so to speak!): it is the grammatical-propositional basis, psychologically speaking, for moving on beyond trauma or sin. It is also the meaning of Freud's principle of deferred action or *Nachtraglichkeit*: an event, as it occurred in the past, and even to the degree that we experienced what we believed to be its motive, cannot have a complete determinative status—even though it already "happened"—until the mind fully grasps the event. Determination is always largely retroactive (*le-mafrei'a*). Only the "fall" or hesitation reveals this, because it will always contain within itself (*present* by virtue of having been negativized and symbolized) the memory of that *second* moment of mental life when we discovered that the event *just prior* had been given meaning after the fact, and that the object just now comprehended as present was utterly unidentified only a second ago even though it was by no means *absent.*

The same type of hesitation, hopefully following suitable additional developments, will also mark the termination of a psychoanalytic treatment. Freud was no fool, and he did not think that psychoanalysis does not have a point at which it stops. The question that perplexed him in "Analysis Terminable and Interminable" (1937) concerned the *proper* end of analysis; specifically, it revolved around the fact that, on one hand, the analytic associative process is as unending and timeless as the unconscious itself, and, on the other hand, an analysis in many ways *begins* only in the light of that which will determine its end. Clinically, this dilemma is manifest in the type of patient whose analysis seems to drag on in a difficult, almost interminable manner for years, and in a paradoxical sense seems never to have taken off—even though it certainly officially "began"—because the patient harbors a corrupted concept of time, has insufficiently mourned, has frozen the object they would like to believe they are seeking to find, and has a minimal sense of the future and expresses an empty sense of hope.

It is for this reason that Freud considered it axiomatic that a psychoanalytic treatment must begin with no declared termination range or date, and perhaps must be conceived of as having no end at all. Thus, in his essay, Freud ironically states that an analysis ends when the patient ceases to come to speak to the analyst! More deeply, in order to work effectively with the significations of birth, unmirroring, and castration latent in any cessation *requires* that the frame of analysis stand outside of the possibility of its own termination, enveloping it, and not simply "stopped" by it, thereby enabling the influence of a terminated analysis to continue even beyond the chronological date of its end. Adding only the Lacanian twist (see Jaanus, 1996), one might say that a patient deserves to cease his formal analysis when he has come to understand that subjecthood *is* castration, negativization, symbolic presence, and retroactive determination, and can live relatively comfortably and even creatively with this. The patient finally understands that neither God not the analyst can be used as the clue to what the self is because a truly symbolized Other is no longer identified as the alter ego nor confused with the ideal ego. This is not merely a satisfactory realization with which the analysis terminates: it is a true gift. The patient has refound his home.

Summarizing the above, and responding to one of the key riddles announced at the outset, we can say that the power of *teshuvah* is by definition experienced retroactively, and only as language makes possible, when we find that it will have been possible for it to have been said of us—by the Knower of Mysteries, by God, by the grammatical foundation of the unconscious—that we shall sin no more! In so saying, I am taking the position that Maimonides' inclusion of this unique criterion of satisfactory repentance is to be understood, in its greatest significance, as a natural consequence of the engagement of hesitation and of adopting the attitude of refinding one's lost object, an attitude that by definition will mean a recursive immersion in the symbolic appreciation of retroactive meaning. Maimonides was thoroughly on target to state only here in the Laws of Repentance that by definition (recall Wright's comment above) entry into the circuit of symbolic *teshuvah* is a direct result of the cessation of the expectation of the return of, or to the original concrete object of demand, and of all efforts to "find" it. At this juncture, a certain sense of future potentialities becomes inherent in how the past has now been reunderstood. At this moment, in the gap between last night (*emesh*), today (*hayom*), and tomorrow (*mahar*), the link between signifiers and objects of desire that carry these temporalities has become sufficiently more tight as to permit a new increment in moral clarity. In this light, how better to appreciate Rav Kook's poetry, "Every thought of repentance unites the entire past with the future; thereby is the future exalted in the ascent of the will of repentance motivated by love" (*Orot* 7:2).

Talmudic Case Studies in *Teshuvah*

The following might be considered case studies that epitomize in one way or another the role of time, memory, the moment of hesitation and the changes of perception that comprise *teshuvah*. Common to all of them, in one sense or another, is that they are paradigmatic instances of so-called instantaneous repentance (cf. Zevin, 1944), *teshuvah be-shaah ahat*, but in the most profound sense.

From the standpoint of normative ethics, it has been customary to qualify that the Talmud was not seeking to argue that such cataclysmic *teshuvah* transformations necessarily represent the highest form of repentance. For the question will always be whether or not such a transformation was the result of a sufficiently deep alteration of internal psychological conditions as to enable the change achieved to be sustained. Nevertheless, in their radicalness and suddenness—or, in one of the incidents, the radical *lack* of suddenness—these episodes support the idea that psychic transformation begins in the gap, in the moment the individual has formulated a linguistic expression of sufficient weight as to capture a life's misguided wanderings, negativizing that which was concrete, immovable, irreversible—even a divine prognostication—and changing it by symbolization into an element that can be assimilated and integrated into human identity. Against this view, these episodes become self-explanatory.[61]

The first encounter is with Eliezer ben Durdaya, a sage with an unlikely, but honestly reported life-long struggle with sexual lust (Talmud, *Avodah Zarah* 17a). Complying with a habit that earned him the reputation of having frequented every known prostitute, he was prepared on one occasion to undertake a perilous journey and great financial expenditure to test the skills of an especially well-publicized courtesan. *In flagrante*, he passed wind, and the surprised prostitute was inspired to declare: "Just as the wind does not return to its point of origin, so will Eliezer ben Durdaya never be accepted in repentance!" Stricken, ben Durdaya beseeched the mountains, heavens, sun, moon, and constellations to intercede on his behalf. Each was unable. He then cried out, *Ein ha-davar taluy ela bi* ("Then the matter is dependent solely upon me!"). He put his head between his knees and cried until he expired. At that moment, a heavenly voice declared: "Rabbi Eliezer ben Durdaya is invited to the World to Come!" When Rabbi Yehudah the Prince heard this, he, too, exclaimed, "There are those who acquire their share in many years and those who acquire their world in an instant! Moreover, they are titled, 'Master.'"

The second episode concerns the famous Shimon ben Lakish, known as Resh Lakish, who as a young man had devoted his uncommon prowess to banditry and later became a gladiator (Talmud, *Bava Metzia* 84a; *Gittin* 47a). In the pertinent circumstances, Rabbi Yohanan was traveling and found himself unable to cross over a deep river. Resh Lakish spied him and leaped clear across the river to Rabbi Yohanan's side. Amazed, Rabbi Yohanan exclaimed, "Your strength [is suitable] for Torah!" Resh Lakish retorted, "Your beauty [is suitable] for women!" To which Rabbi Yohanan replied, "If you would return [in *teshuvah*], I would marry you to my sister, who is more beautiful than I." Resh Lakish accepted the offer, whereupon he turned to jump back over to the other bank of the river to fetch his articles, but was unable to do so. While the linguistic formulations in this story are succinct, they are laden with erotic undertones, which were apparently sublimated through the interchange itself. The Talmud, in fact, relates that Resh Lakish subsequently either divorced this wife, or declined to marry her at all, lest it be said that he repented simply for ulterior motives.[62]

In each of these instances, the role of the unpredictable and spontaneous arrest or hesitation is clear, as well as the sense in which each penitent discovers that return cannot be literal, but may be powerfully virtual, and that the sign of a satisfying transformation is the symbolization of loss within a higher framework of meaning.

The next two episodes also feature hesitation, and repetition. Significantly, they illustrate, respectively, the successful and the failed negotiation of mourning and memory in the effort to achieve *teshuvah.* In the first episode, impossibility is contained; in the second, it tears the mind to bits.

The first tells of the reaction of Rav's students to his death (Talmud, *Berakhot* 42b). As the Talmud relates, while returning from the funeral, the students invited each other to sit and eat bread together. When the time came for reciting the blessing over the bread, doubts about custom suddenly emerged and a discussion ensued regarding the halakhic particulars. The students were surprised to find that they were unsure of the law. Rabbi Ada bar Ahavah stood up and reversed his cloak where it had been previously torn as a sign of mourning (*keriah*), and rent it a second time on the same place. Rashi (s.v. *ahadar*) comments, "he tore on the same place to demonstrate mourning fresh as the day of death since they had need to rule but were unable to rule."[63] The student then exclaimed, "Rav's spirit has passed and we do not know the blessings over bread!" At that point an elder passed by and they engaged him in study over the proper ruling: inviting each other to sit and eat has the same status as formally reclining together, and one member may bless for all.

While strictly speaking this is not a story about *teshuvah,* it does contain the powerful motifs of return, death, remonstration, mourning, and restoration. It is fascinating to see an almost made-to-order illustration of the compulsion to repeat—tearing the already rent garment on the already torn place—and its transformation into re-creative repetition by virtue of symbolization and a linguistic framework for emotion. As well, there is no attempt here to change the fact of Rav's death, and the grief over his death is not suppressed. Instead, against a backdrop suggestive of the return of repressed oral, cannibalistic desires, the students modify their own identification with the absent master, rendering him and his teachings "present" in the fuller sense of the term. It was not biological death that they were assessing; this had already been marked by the first tearing. Now came the anxious pain that indicated the "presence" of a lost object that cannot be restored, and the potential danger of lost teachings remaining lost (by virtue of being identified with their author). The students intuitively *refound* their master Rav through speaking, and now, impossibly, he could be absented. Through language and symbols they were able to acknowledge their desire, and make sense of the anxiety-evoking fullness (i.e., food) where formerly had been the site of lack they represented for Rav. This they indicated by symbolizing the lack of lack, by the expedient of a symbolic "second" *keriah* and a return to the register of law. Only at that point in the episode does a revenant-like elder appear who enables the closure of the transformation.[64]

The second episode is the tragic and oft-discussed final moments of Elisha ben Abuyah, known by the sobriquet *Aher,* the Other (Talmud, *Hagigah*

15a–16a; Jer. Talmud, *Hagigah* 2:1, *Sanhedrin* 102b, *Kiddushin* 40b). Among other significances, this case is particularly interesting because of divided opinion as to whether or not Elisha actually repented in the end. My account here summarizes the several variants of the story. One *Shabbat*, Rabbi Meir accompanied his now apostate master, Elisha, who was on horseback. Rabbi Meir began to discuss some of the homilies he had delivered in the academy. He then expounded on the particular text, "Gold and glass cannot equal it [the words of the Torah]; neither shall vessels of gold be its exchange" (Job 28:17). Elisha rejected his student's homiletic and exclaimed,

> By God! Akiba, your teacher, would not interpret it thus. The passage teaches that just as vessels of glass and gold, if broken, can be repaired, so, too, can a scholar who becomes corrupt be returned to his former stature.

Rabbi Meir seized the opportunity and cried, "Then return *you*!" Elisha refused, saying, "Once I was riding past the academy [variant: The Holy of Holies in the Temple] on Yom Kippur which coincides with Shabbat [a blatant double desecration], and I heard a voice declare [variant: "I heard from behind the Veil"]: "Return, O errant sons" [Jer. 3:14], "Return to me and I will return to you" [Mal. 3:7]. All may return except *Aher*, for he knew My strength and rebelled against Me!" When the two scholars arrived at the *tehum Shabbat* (the *Shabbat* traveling limit, beyond which point it is forbidden to traverse), Elisha warned Rabbi Meir, "Return!" Rabbi Meir again beseeched, "*Return you!*" Elisha again demurred, "Have I not already told you: I heard from behind the Veil, 'All may return except *Aher*!'"

At one point, Elisha taunts some passing schoolboys to recite the first passage that comes to their lips, and interprets each as damning him. Ironically, the last child cited an especially encouraging passage, but owing to the child's stutter, Elisha believed that he had heard yet another, final indication that his fate was sealed.

When Elisha lay dying, Rabbi Meir begged him to repent. Elisha again taunted, "Now they will accept my repentance!?" Citing verse [Ps. 90:3], Rabbi Meir encouraged him, "Even until his spirit is broken!" At that moment, Elisha ben Abuyah cried, and expired. Rabbi Meir proclaimed happily, "It would appear [*nidmeh*] that my master departed in repentance!" And so maintain many of the commentaries, although his final atonement was held in abeyance throughout a protracted punishment.[65]

Eliezer ben Durdaya ultimately accepted responsibility for a *guilty* self, and, in a state of remorse and contrition, realized that random, promiscuous, anonymous "containers" would never be able to satisfactorily envelop him, or fill the maw or lack left by some apparently intrapsychically unavailable primary object. Need cannot out-shout desire, save by the compulsive, exhausting repetition of the need-act. But this is self-depletion. And yet despite the *a priori* declaration—the winds of time are not reversible!—ben Durdaya in the end presumably refound his lost object, his blocked-off domain of self, and reintegrated it into an experience of *teshuvah.*

The matter is more complex with Elisha ben Abuyah. Analyzed not simply as a story but as a myth wherein each character represents a part of self (which holds for all of these cases), Elisha is corrosive, self-vilifying, and wanders dangerously across boundaries, all of which suggests an *ashamed,* imploded self. He can no longer accurately assess, judge, and evaluate objective linguistic patterns, and projects his own hesitant, stuttering, agonal palpitation into the external world that appears to reject him. It is a dramatic psychophysiological arrest and death. The potential containers for such pain appear to him as veiled, the plaintive cry to return seems so distant and cold. Had it not been for the unswerving attention and pressure from a devoted, desiring other, one is left with grave doubts as to whether Elisha would ever have become contrite.

Helen Keller and the Well-House

I offer two final episodes. Neither is recorded in any Midrash or Talmud, but each is a worthy example of all of the aforementioned issues, and can be correctly considered an incident of *teshuvah be-shaah ahat.*

The first case is provided by the autobiography of Helen Keller (1905, pp. 23–25), who lost her senses of sight, speech, and hearing following severe illness in early childhood. She was gently pushed along by her dedicated teacher, Anne Sullivan, who had been trying, with more frustration than success, to break through Helen' s sensory barriers and by now brittle emotional carapace. Earlier in the day to be described, Sullivan had been trying to get Helen to understand that "w-a-t-e-r" meant *water* and "m-u-g" meant *mug,* but Helen kept interchanging the two. Later, Sullivan had given Helen a new doll. She placed the new doll in Helen's lap, and also Helen's big old rag doll, and tried to bring Helen to comprehend that "d-o-l-l," as she signed into Helen's hand, applied to the new doll and Helen's old rag doll. With growing impatience and frustration, Helen seized the new doll and smashed it on the floor. As Helen recalled,

> I was keenly delighted when I felt the fragments of the broken doll at my feet. Neither sorrow nor regret followed my passionate outburst. I had not loved the doll. In the still, dark world in which I lived there was no strong sentiment of tenderness. I felt my teacher sweep the fragments to one side of the hearth and I had a sense of satisfaction that the cause of my discomfort was removed.

At this point, both Keller and Sullivan went outside to romp about. The next development is crucial:

> We walked down the path to the well-house, attracted by the fragrance of the honeysuckle with which it was covered. Someone was drawing water and my teacher placed my hand under the spout. As the cool stream gushed over one hand she spelled into the other the word *water*, first slowly, then rapidly. I stood still, my whole attention fixed upon the motions of her fingers. Suddenly I felt a misty consciousness of something forgotten—a thrill of returning thought; and somehow the mystery of language was revealed to me. I knew then that "w-a-t-e-r" meant the wonderful cool something that was flowing over my hand. That living world awakened my soul, gave it light, hope, joy, set it free!

The next phase is redolent of true *hirhurei teshuvah*:

> I left the well-house eager to learn. Everything had a name, and each name gave birth to a new thought. As we returned to the house every object which I touched seemed to quiver with life. That was because I saw everything with the strange, new sight that had come to me. On entering the door I remembered the doll I had broken. I felt my way to the hearth and picked up the pieces. I tried in vain to put them together. Then my eyes filled with tears; for I realized what I had done, and for the first time I felt repentance and sorrow.

In this remarkable confession, language kindles consciousness, and angry demands ignite desire. Desire for more objects, more self, more representations, and more historical space, wherein the clock of emotional, psychologically relevant time—a true remembrance of lost time—can be kick-started into life. At the well-house, the signifiers of "w-a-t-e-r" chained rapture and delight to anger, rage, remorse, and mourning, leading Helen to recover a different, and far more important doll than the one she broke earlier that day. One might think that she could only possibly locate the fragments of the broken doll, but that would be only trivially correct, and entirely wrong from the point of view elaborated here. The moment of hesitation—experienced as hesitation and not dumb restraint—brought on by the finally successful transgressive act of language enabled Helen to *refind* a doll at once the same and different than the one she had broken. More important, she was able by this measure to discover more of her own self.

Helen came to realize the meaning of the word "water" in a flash of insight, and one could certainly say that there was a simultaneous impression of the return of a lost object, in its refound form. What had been missing within her own mind was now given form and dimension. Though impossible, Helen discovers the signifier r-e-p-e-n-t-a-n-c-e!

Donald Winnicott and the Repaired Doll

I chanced upon another account strikingly similar to Keller's repentance, but only up to the point of even more remarkable differences, in the autobiographical notes of psychoanalyst Donald W. Winnicott.

Describing a childhood memory of a particular geographic slope of special personal significance in his childhood family's yard—"fraught with history," as Winnicott puts it—the analyst recollected the following (1990, p. 9):

> It was on that slope that I took my own private croquet mallet and I bashed flat the nose of the wax doll that belonged to my sisters and that had become a source of irritation in my life because it was over that doll that my father used to tease me. She was called Rose. He taunted me by parodying some popular song. . . . I was perhaps somewhat relieved when my father took a series of matches and, warming up the wax nose enough, remoulded it so that the face once more become a face. This clear demonstration of the restitution and reparation act certainly made an impression on me, and perhaps made me able to accept the fact that I myself, dear innocent child, had actually become violent directly with a doll, but indirectly with my good-tempered father who was just then entering my conscious life.

This testimony is different in significant ways from Helen Keller's. Obviously, unlike Keller, Winnicott is not describing the transition from chaotic speechlessness to speech. And his discovery is not about the elemental instantiation of mindedness, but rather about a more mature Oedipal-type recognition. However, like Keller, Winnicott has annotated the fact that it is language that temporalizes and nominates emotion, thereby making retrospection possible, which in turn enables reparation and restoration of what was at first unknown and incapable of being found.

In this light, more crucial distinctions between the two cases can be discerned, but only if we agree to not approach this distinction literally, but as representative of typologies. Keller cannot yet repair the doll, nor does this even seem to occur to her, so enthralled is she by the sheer rhapsody of emerging from isolation and imaginary phantasms into symbolism. Since, as everyone recalls, she enjoyed normal sensory experience until her tragic illness wrecked its havoc, the "instant" at the well-house was a true form of refinding the lost dimension—and not the concrete return to a specific lost object nor a circular repetition of lost experience—*be-shaah ahat.* As a consequence, a same but by no means identical mental representation of her parents, her teacher, and self are refound; a self that includes a heightened sense of otherness within: the inner depths of languaged mind. Within this new state of mind, Keller locates a new doll. Not a physically repaired doll, of course, nor even a physically new one, and not an obsessive fantasy representation of perfect unbreakable doll-self . . . but a *negativized fragment of a doll* that can be brought close and embraced owing to remorse and atonement, enabling the self to be re-experienced as somewhat more whole and relocated within a flowing history. Winnicott, on the other hand, has the benefit of obtaining a literally repaired doll, and, as he emphasizes, the new representational experience of a tolerant father, benevolently containing his son's Oedipal rage and overexcitement. "New," because this specific internalization within language of the discrete event at hand on that given date and place had never occurred before and can never be identically repeated or undone. But also *not* new, but renewed, or refound, since this episode would have transpired and been resolved differently had it not taken place within external and intrapsychic environments well disposed toward language, symbolization, and containment, and so on back within the previous generations.

At first blush, one might think that the fact that Winnicott has the good fortune to receive a repaired doll is a regressive or infantile solution (it certainly *would* be if his father always rushed to undo every bit of damage wrought by his son!). Whereas Helen Keller appears resigned to loss and brokenness, and uses language to mourn and return wholly within the ambience of symbolic structures, Winnicott seems to get to have his cake and eat it as well, pacified by his father's rush to facilitate concrete repair. Perhaps he might even have experienced his rage and masculine protest as neutralized, his breach too quickly "whitened as snow."

But I would say that Winnicott nevertheless was able to accept the phenomenon of repair, and to make psychological use of the "new" doll and not hurl it away in a fit of narcissistic injury owing to his having internalized that while time *is* irreversible, breakage, loss, and death are *tolerable* holes in a symbolic structure that has already taught one that linguistic-representational holes, lacks, and absences can be borne and actually contribute. Thanks to the expedient of the "merely" repaired, refound doll—as opposed to a banished old one replaced immediately by a brand new one—young Winnicott gained a "new" bit of father, a new dimension of otherness. And perhaps, owing to this salubrious development within him, the communities of psychotherapists, pediatricians, artists and authors, and suffering patients gained a Winnicott whose "return in faith" before the Other (doll, father, self, and language) enabled him to become the innovative psychoanalytic thinker he was.

Lack and Desire in the Case of Job

The difference between Keller's and Winnicott's doll self-objects in their unique intersubjective contexts and the qualities of their corresponding moments of loss, restitution, and repentance bring to mind one final incidence of *teshuvah.*

If the reader can bear a wide jump back in time, recall that poor Job, akin to Winnicott's circumstances, would, at first blush, also seem to have been granted brand new replacement "dolls" in return for his troubles and humility—the return of even more wealth, twice the number of male children, and daughters even more beautiful than the ones that were originally taken from him (Job 42).[66] The issue is a complex one, and has troubled many.

However, I believe that the main principles outlined in my essay are quite well respected in the case of Job, on condition that we adopt the view, compatible to the view of some of the rabbis as well, that the Book of Job is a *mashal,* a parable or morality tale, or, in my way of understanding this, a mythic narrative representation of an otherwise incomprehensible piece of psychic structure.[67] In this view, the outline of specific sequences of events (and other qualitative factors) within the narrative may reveal significant latent levels or complexes of psychological and moral development. In the case of Job—again, by analogy—one gets a very accurate glimpse of the movement of the infant mind from primitive oral and skin fixations, characteristic of the earliest representational-enveloping structures of the budding mind,[68] through mirroring (i.e., the back-and-forth dialogue with his friends and, ultimately, God), toward recognition of loss within himself by virtue of the acquisition of words with which to speak. This characterization is signaled by such important textual evidence as the peculiar facts that (a) Job had always dreaded that the fullness of his life was illusionary (cf. Job 1:5 and 3:25), and (b) at no time throughout the narrative does he express longing for his allegedly missing children.

Reconsider, accordingly, the exact sequence of events that unfold in the famous epilogue. First, we notice the key phenomenon of *shaah ahat:* Job humbly *hesitates*—literally placing his hand over his mouth—before the initial transgressive "call" to respond (40:1–5), much as he sought such response, and only begins to speak when this can be accepted as his own initiative and he is not pressed to reply (42:2–5). As Buber wrote, eloquently joining the primitive and the mature metaphors of the book, "Job is the first clothing of a human quest in form of speech" (1942, p. 52). Now Job begins to successfully articulate his acknowledgment of the paradoxical closeness and incomprehensibility that God was lovingly made available to him (1–6) that alone satisfied his singular request, consistently miscomprehended by each of his friends: Will God speak to me? Does my lack or absence *matter* to God?

The crucial element of change within Job is his articulation of the fact that he has developed a novel personal interpretation of God's relationship to man different from that which he had merely absorbed through tradition. No, Job did not literally find a "new" God, or a new self. Rather, by relinquishing the search for the same old thing, and by refusing the standardized explanations designed to render Providence *possible*, Job discovers that he has *refound* an entirely unique experience, *the processes of negativization and symbolization themselves.* This, I believe, is testified to by the famously vexing intransitive phrase (42:6) *Al kein emas ve-nihamti al afar va-efer*—"Therefore I abhor, and repent; seeing I am dust and ashes."[69] It has never been clear what exactly Job "abhorred" or recanted: presumably his earlier theses and ravings. However, it is quite feasible, and I think probable, that Job's use of the term *emas* (with a possible linguistic connection to the term meaning "to melt away") indicates a vaporization of the rigid self-structures that until this point served as a bulwark or enclave that made it impossible for Job to sense the lack within the structure of the Other and within himself. His reference to "repentance" or symbolically reparative *nehamah* strongly suggests this qualitative transformation.

Subsequently, another development occurs as the internal psychic fragments represented by the companions coalesce into what is now a comforting internalized structure. Only following this advent are we told (42:10): "And the Lord changed the fortune of Job [lit, returned his captivity (*shav et shevut*)], when he prayed for his friends." In other words, God, or Job's superego representations, had enfolded symbolically upon his acceptance of the impossibility of concrete return and "finding." I understand this to indicate that the "refinding" process had progressed as a result of further empathic dialogue within the *internal* forgiving mental structures of Job's personality, following which an even further advance occurs. As the text portrays it, there is now an in-gathering of all of Job's former relations, or his psychic "components," which, though belonging to an historically earlier period of his development, had now matured along with the overall transformation within, and we are informed of an additional *nehamah* (11). It is only after all of these developments that the text indicates the return of his former material wealth and children. Or, in the grammar of the future anterior of *teshuvah*: "By the time Job comprehends that he has reworked his terrors, it will have become clear that he had been dreaming, and that his children were safe and sound."

Summing up what has transpired, it seems to me that the psychic structure offered to us analogically in the form of trials of Job and his component "fellow" structures has matured from an extremely primitive state toward mature recognition of the inevitability of lack and the unslackable nature of desire. He finally appreciates that *lack* is the characteristic of utmost relevance in the universe, and *evil* only secondarily.[70] What is returned to Job, the literal description of the text notwithstanding, is not any brute, material "Thing" (*Das Ding*), or a bit of *efer* or *davar*, but rather a new relationship with as yet unexperienced dimensions of the other and of the self in interaction with it, a refound structure—(*ve-khol yode'av le-fanim* [42:11]). Such change, like that of Helen Keller and Donald Winnicott, *mutatis mutandis*, is the result of psychological *nehamah* on the highest level, the capacity for symbolic representationalization of not only what has been lost and mourned—since, owing to language, *everything* truly knowable has its lost and need-be-mourned-for dimension—but also of the *relation* of the self *to* that which has been lost and mourned.

Teshuvah Mei-Ahavah and *Teshuvah Mi-Yirah* Distinguished

I will now interpret the distinction between the primacy of *teshuvah mei-ahavah* over *teshuvah mi-yirah*, continuing with the approach to which I alluded above. In so doing, I will aim for an interpretation of the distinction in a manner that does not coerce one simultaneously to lose sight of the praiseworthy aspects of the immediate and direct transformation from "yestereve" to "today," which Soloveitchik among others considers the epiphany of loving, deeply redemptive repentance, over and against our tendency to view such a shocking transition as the least productive and less existentially valuable level of personality change.

I will begin this section by citing a rather different approach to two levels of *teshuvah* expressed by Rabbi Soloveitchik, as moving poetically as it is instructive. The first he calls "repentance which carries no resolve for the future," beyond mere restraint in the here and now owing to the immediacy of the sense of shame and disgust with oneself. This form of *teshuvah* is spontaneous and flooding, and is preoccupied with the sensation of sin itself and with tormented emotions of perversity and withdrawal. On the other hand, "repentance with carries resolution for the future" is based on a full intellectual understanding of sin, derived from careful reflection, from guilt rather than from shame, less concerned with transforming one emotion into another than with a daring decision to sin no more (pp. 204–6). Interestingly, after comparing the two types of repentance, and offering evidence that Maimonides seems to have favored the intellectual type as "perfect repentance," Rabbi Soloveitchik himself demurs, "Which of them is greater I do not know. . . . I think that the first kind of repentance, which comes spontaneously and emotionally, is also 'perfect repentance'" (p. 206).

What is significant about this sense of future that Soloveitchik places in such a pivotal position? In part, it is this element that may have been expressed by Maimonides' special demand that the penitent individual be able to call upon God to testify, as it were, that the individual has achieved sufficient growth as to be able to project his self-transformation into the future. But we can amplify this point, and our larger interest, by recalling that we spoke earlier in terms of specific changes that take place in the movement from the early, paranoid-schizoid position to the more advanced depressive position, as delineated by the Kleinian school of object relations.

A synthesis between the Kleinian transformations outlined above and the corresponding changes in time sense emerge very clearly in the work of Andre Green in Paris and Thomas H. Ogden in California. Green (2000, p. 15) gives expression thus to the link between the flow of time and what we have been referring to as the achievements of the post-depressive phase of development:

> I have the feeling that psychoanalysis has to do with history and with the historical nature of man, not only in terms of recovering the past, but also of making the present and being *in* the present as a historical being, *through* a process during which progressing toward the object mobilizes all the transformations of the past that are relevant to the present situation, and finds new room for the expectations that have not found a way to link themselves with the transformations, either because of the anxieties associated with them, or because of the impossibility of avoiding destruction because of the fear that arises in the mind of damaging the objects toward which one's desires are directed.

Similarly, Ogden (1990b, p. 71) elaborates the emergence of *historicity* after the advent of the depressive position, which takes hold as the infant begins to learn to relinquish omnipotent defenses, enabling itself to be angry at the loved object without fearing its own or the object's destruction, and at the same time, by being able to perceive change in the object without recasting it as an entirely, concretely new object with each passing mood or emotional state. Ogden's view is that (p. 71):

> Although one's interpretations of the past are evolving (and therefore history is continuously evolving and changing), the past is understood to be immutable. This knowledge brings with it the sadness that one's past will never be all that one had wished. . . . At the same time, the rootedness in time also brings a depth and stability to one's experience of self. . . . The past is never lost completely in a depressive mode, in that one can repeat something of the experience with the original object in a relationship with a new object.

Accordingly, I would view *teshuvah mi-yirah*, on the main, as a simple, one or two dimensional, primarily not-yet-fully-historicized phenomenon, a testament to the best that the personality can offer from within an internal mental context most similar to that described by the paranoid-schizoid position. That one occasionally witnesses explosive and sudden forces in association with *teshuvah mi-yirah* is in no way contradicted by this categorization; if anything, it is very likely that as any behavioral constellation yaws closer to the symmetrical, presymbolic pole of the spectrum, emotional tensions are less representational in quality and hence less capable of being contained and channeled. I thus think that the first type of repentance isolated by Soloveitchik could only be deemed "perfect" in the sense that all primitive, presymbolic, prehistoric mental states seem perfect—that is, their apparent perfection inheres in a glaringly pristine, logic-tight, and often autistic-like self-sufficiency. The emotional set of spontaneous *teshuvah*, focused primarily on the immediate, in a sense protects itself with its veneer of perfectness, but must in fact *not* be presumed perfect.

Nevertheless, *teshuvah mi-yirah* is indubitably a breach in the wall; it marks the registration of a hesitation in time, allowing for the symbolization of repetition, and it stands as a bulwark against the miasma of constant mental flow.

The uniqueness of *teshuvah mei-ahavah*, on the other hand, must lie in its being an achievement predicated upon the dissolution of concrete perception, mourning whatever emotional losses this modification entails (e.g., the diminution in megalomaniacal influence over the introject-level, part representations that typify the early emotional phases of development), and the advance toward increased levels of symbolization. In Kleinian terms, *teshuvah mei-ahavah* is a post-depressive phenomenon; it is the product of reparation and caring that accompany, even if painfully, the mitigation of the seductive "perfection" of the previous phases of emotional development.

Reassessing *Teshuvah*

We have traveled very far from any traditional conception of repentance as the return of the human being to *redintegratio in statum pristinum.* Rather, the apex of *teshuvah* is achieved when the individual is able to stand outside of the cyclicity of events, where the seductive modes of repetition and chance can be reinterpreted symbolically by virtue of the fact that *teshuvah* is a language process steeped in impossibility, loss, negativization, and mourning. The solution to the dilemma of disowning the past versus living in the present through the past; of *teshuvah mi-yirah* versus *teshuvah mei-ahavah*; of returning to oneself versus returning to God, of the instantaneous dimension of *teshuvah* versus its lifetime dynamic draws to resolution by recognizing that *teshuvah* is, in fact, a nexus or "third" place where these sharp dichotomies lose their illusory distinctiveness. This place *is* otherness and it is *of* the Other—whether this means God Himself (*hinneh makom itti* [Ex. 33:21]), or every individual's unconscious, or the representational space that literally exists nowhere save in some hyperlocale shared by two or more intersubjective participants who can experience it when their relationship has achieved sufficient existential-symbolic structure. It is extremely difficult to locate, but it is the only place where the penitent stands a chance of refinding himself. In the nexus point of *teshuvah*, past and present, fear, anxiety, and love, and temporal simultaneity and discontinuity all cross. And it takes the role of this third place by virtue of its being a linguistic exchange between the subject and the Other, against the background of a heightened awareness of what basic existential nothingness or emptiness lies at the root of the language vehicle *when language is (mis)identified as the route to the objective truths of the illusory ego. Teshuvah* thus construed is a type of psychic envelope for the most subtle of internal contradictions within the self. Among these contradictions are the fact that the self is, in fact, in many ways as much "other" as it is self, and that it is an expression of the ongoing conflict between the asymmetrical versus symmetrical logics of consciousness and unconsciousness, and because it is a cauldron for the constant interplay between the paranoid-schizoid and depressive position.

Since subjectivity is a process sustained by speech in the symbolic register, and since the ego, in the contemporary view, is an image maintained by the imaginary, the gap between "subject" and ego is always analogous to the discrepancy between the registers themselves. Abandoning all conventional assumptions, I insisted in this essay that "Who—or which—components of the so-called self is actually doing *teshuvah*?" and "Is *teshuvah* even possible?" are the most legitimate questions for study. And my reply has been, to reiterate but always in a new way: "*Teshuvah* is impossible because *nobody* is doing it!" "Nobody" is the only one to effect a truly relevant and remarkable *teshuvah* in the sense that the very agencies, structures, and mechanisms that are likely to effect repentance are themselves enveloped within alienating, imaginary structures of language, and heavily modeled upon identifications that imbricate the other (the loved object, the unconscious, language itself as other, the subjective self as other to the ego), and otherness itself, deeply within almost everything that could be considered self. Pangs of conscience alone are not the essential desideratum in repentance. Only when the humbled, internally divided, alienated self has reached a sense of its utter emptiness as it confronts the silent Other—in whose Presence the dwarf-like quality of our lifelong task of representationalization becomes unavoidable—only when it appreciates that the linguistic enclaves of the ego essentially house a "no-one," can it be said that *teshuvah* has reached its zenith.

That said, I obviously harbor some faith in the usefulness and ultimate value of the concept of *teshuvah*. For, on one hand, a *teshuvah* project that fails to reckon with the role of language and symbolic transformation—internal and shared—as the focal points of change risks not fully differentiating itself sufficiently from the very misperceptions and *meconnaissance* that result in sin. Too much "doing" when doing *teshuvah* tends to result in an overemphasis on mimesis (unreflective identification) at the expense of a more symbolized, narrational *diegesis* (reflective identification),[71] accompanied or augmented by emotional or mimetic nuances. On the other hand, when an individual is willing to risk the temporary increase in alienation and decentering that a full suspension and scrutiny of self will cause, as well as the emptiness that is experienced when acknowledging the unfathomable depths of desire that are otherwise false-bottomed and masked by demands and "needs," the individual then moves into position to perceive more of the true nakedness of the self and the other, and the gap between the imaginary and the symbolic registers becomes so much less distressful. Every human joy, pain, aggressiveness, and creativity is rooted to this gap—to the acknowledgment of it or the failure to do so—and to the temporal moment in which the gap emerged from the miasma of symbiotic undifferentiatedness (prior to which it could not even be perceived, though it is surely felt) into that which repetitively takes form in the imaginary (the neurotic chasing of stereotypic holes and missing letters), and eventually, for those willing to make the effort, submits itself to meaning within a formal structure (the scientific or halakhic use of negativization).

Put differently—and why not risk an overextended metaphor?—*teshuvah* is a returning image, like that reflected back by the mirror. What "returns" to the penitent is perhaps not so much an "object" per se but a slightly more clear perception of that which had become inherently other in the self, an awareness of that bit of self that had been subsumed or lost owing to internalization of the other, or had been experienced as diminished owing to the distractions of the hyper-specific, entrapping form the self had acquired owing to the limits of the structures of language. It is a symbolic attitude toward mirroring that ultimately releases from repetition the souls of Moses hidden in the cleft, Job humbled by the whirlwind, and the shepherd-*nazir*, told of by the Talmud, whose spontaneous disillusionment elicits a unique hesitation and admiration from the High Priest Shimon ha-Tzaddik, the *nazir*'s skeptical inquisitor, and an exclamation of *jouissance* that breaks a certain lifelong ritual scruple of the high priest (Talmud, *Nedarim* 9b, *Nazir* 4b; cf. Spero [1994]). Each is enabled to surrender their demands and, by acknowledging lack and alienation, experience a bit more of their unspeakable desire for God and God's desire for the speaking being. This is the state most suitable to a religious experience of God—in what is shared as well as in what is necessarily, impossibly lacking—shorn to as great a degree possible from our own psychotic, split-off, and imaginary projections and misrepresentations. Throwing oneself before the ultimate otherness of one's identity, and of the mind itself—perhaps in prayer at first, but ultimately surrendering even some of the very conventions of speech—produces loneliness and alienation. Yet this loneliness yields the most blinding light imaginable; the alienation it exposes is the greatest possible bond to one's fellow man.

Final Reflections

In summary, *teshuvah* is not possible; it is not possibility, but rather impossibility. It cannot *possibly* "be done"; it is to be struggled at through *im*possibility. The intention and goal of *teshuvah* is that it embrace all of the central and intrinsic paradoxes of self, memory, and time and invites the yearning human being to grasp unabashedly the concrete experience of impossibility—the impossibility of pure selfhood, the impossibility of exact restitution of lost objects in memory, the impossibility of pure reversal of time—and contain their intrapsychic influence by symbolization, negativization, and linguistic representation.

It has also been expressed, more than occasionally, by contemporary psychoanalytically oriented writers favorably predisposed toward religious experience overall, that the final goal of that which is subsumed under the term repentance is to move beyond mere superego, and closer to some internalized structures that allow for a less guilt-ridden, more empathic, forgiving representation of God, which in turn permits deeper transformations within the self. The sentiment is, of course, wonderful, but, as we have seen, if it is at all possible, requires more than mere "self-knowledge." Knowledge of oneself does not restore lost selfhood, as long as knowledge of self is "objective" or extrinsic. For the kind of change valued by Jewish philosophy, the modes of knowing and living-in-time must also be existential-halakhic: not recollecting, cataloguing memory upon memory,[72] sin upon sin, nor even simply repairing or attempting to literally restore the past. The knowledge one seeks has been acquired through mourning and loss, and the representations it makes possible are redolent of this quality. Knowledge of self thus acquired persists in a near-constant state of unfolding, refolding, and repair through symbolic repetition, a refinding of the lostness of things alongside the evidence of an ever-available, ever-possible well-spring of new elements and dimensions of experience since-lived—like the dimension of Job's relationships characterized as "from before"—that had not been sufficiently mentalized, hence not conceptualized, and hence not brought into alignment with ethical perspectives.

I have attempted throughout the essay to keep the lessons drawn from clinical reality in plain sight. And, thus, I emphasized that trauma is reworkable—and sin repairable—not because one literally changes the past or locates the identical item, object, feeling, or memory that was lost. Instead, trauma, conflict, and moral immaturity are reworkable because one can come to accept the fact that no event can ever be known completely or apprehended by the mind in all of its dimensions, owing to the limitations of language, the perceptual apparatus, and the wiles of unconscious fantasy, and the interminability of desire. Any experience can be known and representationalized in an infinite number of ways—even though we tend to prefer certain ways more than others. This is all the more the case for the great range of human psychological experience that has been veiled by repression, splitting, psychotic foreclosure, and other defense operations.

Teshuvah, then, at its most significant, does not achieve its zenith in the moment of concrete repair of a specific iniquity or trait, nor even within the movement from soul-felt disparity with God to the returning embrace, though these developments are certainly desirable to man and God. I have tried to recast *teshuvah* less as a discrete, event-bound event and more as a complex psychic movement comprised of linguistic, temporal, and representational elements—constant and vibrant under healthy conditions, nurturing all of the mind's psychological dynamics, and frozen in states of crisis, bad faith, and breakdown, pulling to stagnation or arrest all of the mind's psychological dynamics. The true movement from stalemate, panic, despair, apathy, melancholy, and self-misrecognition in the face of the perceived tragic impossibility of *teshuvah* toward a creative, flourishing envelopment within its paradoxes requires a reentry into language, a reconsideration of the losses and lacks it masks symbolically, and courageous struggle with the type of refinding it potentiates. In that moment, one feels strange and alien within one's own language; in that moment, one has the chance of recalling and apprehending a bit of that ancient search for the primary Object who preceded the allure of speech—which cannot be found but can be refound.

The denouement of the drama of *teshuvah,* like the dispersal of the transference during analysis, lies in the reinstatement of the object whose absence—implicit in the anxiety of sin as well as in the use of language as a force of change—had been hidden, or had been eclipsed by the strident claim that there was something *else* we lacked and to whose attainment we turned our energies obsessively and repetitiously. Strictly speaking, *teshuvah* neither violates the irreversibility of time nor does it "unbreak" that which was "broken." It does not locate that which cannot be found. In *teshuvah,* the overall internal representational space, crafted by language in sin and retooled during repentance, suddenly reshuffles the relationship between self-representations and their objects, which formerly were sustained by one level of interaction, degree of internalization, tolerance of similitude, and freedom of movement, and henceforth another. From the ethical down to the neural-cognitive dimension, all psychic representations are maps, with spatial, temporal, emotional, and intersubjective contents and qualities—that are always in process, never *a priori* (see Edelman, 1990; Buonomano and Merzenich, 1999). It is precisely for this reason that one literally "returns" to an identical representation only in instances of fear, or pathology; in the process of more mature repentance, one instead comes to a more radical reorganization of their quality and content.

The brilliant cultural anthropologist Ernest Becker did not ignore the concept of repentance (1964, p. 250):

> When we respect the object, we lose in a sense our power over it. But we gain something in the loss: respect of the object relinquishes power over it only to gain the harvest of the object's privacy. Our world is this enriched by the uniqueness of the loved object. Forgiveness absolves the object from any connection with its own past action. This means that with forgiveness we harvest a potentially new object, in the form of new spontaneity.

If we permit ourselves to paraphrase Becker as referring to the *internalized*, mental representational object—for surely that's the dimension that counts—it becomes clear that the change in repentance comes from a new relationship with that which was always there to be refound. "'The Lord, the Lord'—I am the Lord before man sins, and I am the Lord after man sins and repents" (Talmud, *Rosh ha-Shanah* 17b). As object is refound, self is restored, consciousness alters, memory reshapes, gaps are restored, time refolds, and fresh wisps of desire shuttle freely among the uncharted, newly available space in the mind. Indeed, Becker sensed the irony that one advances morally and psychologically by virtue of becoming somewhat *less* familiar with one's all-too-well known objects; a certain modesty appears as one acknowledges the degree to which we cannot fashion the other in our own image, and in the light of our insistence that it take the role of that which really desire but cannot fully acquire.

When we began, language was indicted as a major target of suspicion in the impossibility of *teshuvah*. How to entrust *teshuvah* to her care, we asked, when she would lead *teshuvah* into uncharted depths of the mind while not even having the compassion to provide sufficient consciousness to allow the speaking individual half the chance to be able to recognize the misplaced, displaced, inverted part-objects that disguise his desire? In the end, there is no retracting our suspicions, and one must indeed consider that the rabbinic mind seems to have comprehended this impossibility. For, in the final analysis, the lacerating cynicism implied by the rabbinic concept *devarim she-ba-lev einam devarim* ("Words of the heart are not words") is sustained: matters of the heart and mind are incredibly difficult to ever present in linguistic form, for the "language of the heart" is symmetrical, and we speak of it only as a legal fiction in order to enable a minimum of consensual reference, whereas *language*, the language of the *languaged*, is asymmetrical. However, the rabbis also held forth the possibility that under the special circumstances of true repentance of love—*teshuvah mei-ahavah*—a person of faith could acknowledge that his language is failing him, *and from that unburdening sense of destitution* gradually begin to accept the fact that the anxiety and shame his language releases steer him toward a better identification of the false objects that have distracted his desire, and toward more authentic representations of self and other, including the Other, that become more clear in the wake of ever broader moral symbolization and internalization.

Teshuvah is the meeting ground for the God of language and his speaking creations. Language, memory, time, and self are the vehicles of choice for repentance because *it is precisely these things that* God *wants of* man: *they* are the representations of the range of human success, or failure, with the hard and unending work of accommodating to the lost hedonism of an initial monadic mindlessness that only subjectivity and speech can ensure. "In the language that one transgressed; in that language shall one repent" (*Pirkei de-Rabbi Eliezer* to Ex. 90). Some of the rabbis used this statement to enforce the mechanics of *teshuvat ha-mishkal*. The far greater importance of the aphorism is to make plain that although acts of sin and repentance are diametrically opposed, they both operate out of the same rules, properties, and structures. Both are efforts to address the alienation that results from the struggle with the ultimate impossibility of desire; that which is called *teshuvah* represents the willingness to open oneself up to a language more permeated by the presence of God, even if this means marking it with absences where, formerly, smaller gods resided.

In my opinion, this essay does not paint a wholly frustrating portrait of *teshuvah*, nor does it demand nothing more than resignation to the grim reality that the self is more elusive than we ever imagined, time too slippery, repetition too compulsive, and the conditions of language overwhelmingly antagonistic to the wish to find something very real placed into our open palms. On the contrary, a language-oriented approach to *teshuvah*, though sobering, underscores exactly what it is: a divinely inspired gift, the ultimate expression of the *ruah memallela* breathed into our souls,[73] a symbolic edifice that permits man to grasp a small, and sometimes even a large portion of the impossibilities that are inherent in our mortal, language-bound, other-dependent minds. And thus we are enjoined: *Kehu immakhem devarim ve-shuvu el ha-Shem*—"Take with you *words* and return unto the Lord" (Hos. 14:3). Happily, the quality of our discourse has not been left entirely to chance or absolute impossibility. There is more freedom and randomness in language and time than ever before imagined, and yet even the defiles of signification obey certain limits and frames of the larger superstructure of symbolic logic. We have been "instructed in the manner of speaking the thirteen signifiers of God's name,"[74] by an Other whose desire for us so readily escapes His own lips (Jer. 31:19):

> Is Ephraim a darling son unto Me?
> is he a child that is dandled?
> for as often as I speak of him,
> I do earnestly remember him still;
> Therefore my heart yearns for him,
> I will surely have compassion upon him,
> says the Lord.

Notes

1. That is to say, as a writer with faith in an involved God, I could never regard, for example, psychological or neurological factors, or the rules of the most primitive level of semiotic structuring, as the least common denominator of language, or the unconscious, since I must always leave room for the position and influence of God as a denominator sui generis. And while I can readily accept the distinction between God as noumena (as it were) and what I have called divine object representations in the mind, and the fact that both coexist (as it were) and that a human *needs* both (i.e., God would want it that way, too), I must still maintain, as a religious believer, that the former could never be exhaustively explained in terms of the latter (see Bomford, 1999; Jones, 1991; Rizzuto, 1979). At the end of the day, the believer needs room for at least *some* divine residuum in personality (though, by definition, it could not be coeval with the material aspects of personality) that is *not* autochthonous, and that is *necessarily* of God (cf. Grotstein, 1997). Kierkegaard once tried to resolve this problem by speaking in terms of a dialectic between the finite, immediate self and what he termed the "projected image of self" with God-like features believed on faith to derive from exposure to and actual relationship with the Spirit of God (1849, pp. 162–67; see Coles, 1971).

2. Despite every temptation, all review of the impact of post-modernity upon the status of psychoanalytic interpretation and the definition of "reality" and "knowability" must be sought elsewhere, such as in the ample discussions by Balter, 1996; Carnochan, 2001, chaps. 6–7; Leary, 1994; Spero, 1997; Stolorow, Orange, & Atwood, 2001. Mitchell (1998) represents a postmodern revision so sweeping as to give cause to doubt whether his version of psychotherapy can be considered psychoanalytic, strictly speaking. The redefinition of the boundaries of self (or subject) have reached the point where recently Rucker and Lombardi (1998) insist upon the term "subject relations" instead of the traditional conceptualization of "object relations" in order to better emphasize the near constant interchange between state of embeddedness and separation, states of self in which self-other differentiation is very loose and ambiguous and other states of self in which self-other differentiation is highly demarcated.

3. One can question whether the application of psychoanalytical science to religion truly represents an advance in grappling with the dimension of the impossible, as I have begun to delineate it in the text. For, to remain "scientifically" psychoanalytic, true to its terms, which respects the limits of time, space, logic, and so on, would seem to leave the domain of the *unknowable* unknown, and the dimension of the *impossible* impossible. On the other hand, to transcend its scientific limits, psychoanalysis would itself become a form of mysticism or spirituality (see Leavy, 1988; Stein, 1999).

4. I am perforce leaving aside an entire set of significant and legitimate preliminary questions, such as "What *is* the self?" and, more pertinently, "Does Jewish thought, ancient, medieval, or contemporary, even have a well-developed concept or philosophy of self?" Clearly these are appropriate issues in their own right, and our discussion is not independent of them, but they are constituents of a field so vast that they must be reserved for another time. As well, questioning the very existence of a Judaic concept or philosophy of self may strike the reader as counterintuitive, as one might quickly point to the very existence of *teshuvah* as evidence of the self, and hence defend the need for a concept of self. However, until a lot more work is done regarding the problematics of *teshuvah* as I shall define them, this expedient intuition simply begs the question. See Subert Spero's (2004) recent assessment of this problem.

5. Ariel, 1998; Bialik and Rabinsky, 1962; Cohen, 1989; Montefiore, 1904; Schechter, 1909; Steinsaltz, 1988; Urbach, 1971.

6. Adler, 1999, pp. 496–674; Behr, 1981; Cohen, 1989; Don, 1975, 1985; Goldberg, 1976, 1982; Gottlieb, 1975; Gruenwald, 1985; Heschel, 1950, pp. 143–47; Katz, 1965; Metzger, 1968; Pachter, 1985; Schweid, 1985; Shokek, 1985; Spero, 1977, 1980, chap. 6; Steinsaltz, 1988; Ury, 1970.

7. This is developed by Rav Kook in *Orot,* but see also Ramban at Gen. 6:11 and Lev. 26:5.

8. Compare Rav Soloveitchik's comment with Proust's, "For a man cannot change, that is to say become another person, while continuing to obey the dictates of the self that he has ceased to be" (1913, p. 410).

9. Rabbi Shimon ben Lakish, himself the product of a repentance process of stunning and perhaps complex proportion, is the author of this distinction (Talmud, *Yoma* 86b):

> Resh Lakish said: "Great is repentance, for because of it, premeditated sins are accounted as errors," as it is written, "Return, O Israel, unto the Lord, your God, for you have stumbled in your iniquity" [Hosea 14:2]. "Iniquity" is premeditated, and yet [the prophet] refers to it as "stumbling?"—But is that really so? Has not Resh Lakish said that repentance is great that premeditated sins are accounted as though they were [full-fledged] meritorious acts, as it is written: "And when the wicked turn from his wickedness and does what is lawful and right, he shall live thereby" [Ezek. 33:19]? There is no contradiction: the [latter] refers to a case of [repentance] motivated by love; the [original statement] [refers to repentance] which derives from fear.

10. Note also that insofar as the process of mourning has been deemed by writers such as Rabbi Soloveitchik (1975) to be essential to the completeness of the repentance process (see Ex. 33:4, and cf. Job 42:10, 11, where *nehamah* once again *follows* the

terms *ve-shav et shevut*), one can sense additional reason to presume that the dimension of *nehamah* must be unpacked in its fullest psychological dimensions.

At the same time, the fact that the terms *naham, va-yinnahem* or *nehamah* oftentimes follows the word *shuv* in the biblical text does not automatically help us to decide under which circumstances *nehamah* refers to consolation and the related intrapsychic dynamics of reparation, or to a more rational, executive type of reconsideration, akin to a philosophical decision to rescind a formal commitment. I suggest Jon. 3:9–10 as a good example of just how unclear the matter can be:

> Who knows whether God will not turn and repent [*yashuv ve-niham*],
> And turn away [*ve-shav*] from His fierce anger, that we not perish?
> And God saw their works, that they turned from their evil ways;
> And God repented [*va-yinnahem*] of the evil, which He said He would do unto them;
> and He did it not.

In any event, I suggest that we not draw too fine a line here. For practical purposes, we can state that in general there are *nehamah*-like processes that lead up to *teshuvah* and then there other *nehamah*-like processes that result from or run parallel with the advancing stages of *teshuvah*. This distinction gains support in Samson Raphael Hirsch's grammatical commentary. Elaborating upon the word *va-yinnehem* (at Gen. 6:6, to which I will return below), Hirsch argues that the forms *hinnahem* and *yinnahem*, as opposed to the reflexive form *yitnahem*, indicates the influence of external forces upon the individual's decision or mental set. Thus there are probably certain qualitative levels of *nehamah* that are correlated with certain qualitative levels of repentance, *largely dependent upon whether the motivation, or locus of control, is exogenous and coerced or endogenous, independent, and reflexive.* Interestingly, Hirsch refers to the appearance of the reflexive term *yitnahem* in Deut. 32:36 and Ps. 135:14 and states, "These have a deep implication: [the divine promise here is that] even if [Israel] does not return in full repentance, and [thereby] do not bring about a change in the divine decree, God is destined to nevertheless bring the Redemption *of His own free will*."

Rashi also seems to have appreciated different qualities within the concept of *nehamah*. In his commentary on Gen. 6:6—*vayinnahem Adonai ki asah et ha-adam, va-yitatzev el libbo*—Rashi highlights the divine ambivalence inherent in the passage. On one hand, God was "consoled" that the rebellion was not of a more egregious kind, perhaps of a sort that would have required of God that He destroy the entire cosmos—yet He also "recanted" of the original joy that brought Him to the decision to create humankind. In the end, Rashi emphasizes a mature, panhistorical (post-depressive) view whereby God defends the wish to create man, despite His awareness of human foibles, "for during the time of joy, there is joy, and during the time of mourning there is mourning."

Somewhat more problematic, but no less fascinating, would be the use of the term *mitnahem* in Gen. 27:42, where Rivkah, wise to Eisav's designs against Jacob, advises Jacob to flee, informing him: *Hinei Esav ahikha mitnahem lekha le-horgekha*, "behold your brother Esav is *comforting himself* [in order], to kill you" [trans.: Soncino]. The commentary Rashi interprets this odd usage with its full ambivalence (i.e., allowing the reader to chose either of two alternatives or to consider their combinatory effect on the mind): In the simple explanation, the term refers to pure bereavement (and consolation?) over the loss of the birthright to Jacob, which provides the motive for Esav's murderous rage, or it refers, in two possible modes, to a psychological reversal of internal feelings toward Jacob himself: "he is divesting himself of his brotherly feelings for you, to render himself estranged from [consoled of] you" or "he is acting as if bereaved of you, as if the cup of consolation has already passed before him" [i.e., as if you were already dead in his mind]. In either of the latter two interpretations, it would not seem that this would have been an uncomplicated case of "consolation" but more an example of the aggravated need to ablate or foreclose the representation of Jacob from his mind prematurely, more similar to the use of the term in Gen. 6:6 (there may also be a link in Hebrew between *nehamah* and *nekamah* ["revenge"], as in Isaiah, 1:24, further supporting the notion of the need for reining of aggressive energy in the process of loss, mourning, repair, and consolation).

11. To be sure, R. Abraham de Boton (*Lehem Mishneh, ad loc.*, 2:2) translates the operative verb *ve-ya'id* ("will testify") transitively, i.e., that the proper level of *teshuvah* has been achieved when the individual feels prepared to invite God to testify as to his resoluteness, rather than the direct actualization of divine testimony. This is a significant modification, but does not alter entirely the riddle of why this factor merits mention at all.

12. Bloch is among the many who attempt to render comprehensible the philosophical dilemma implicit in the notion of God's allowing change and in reversing His judgment regarding the sinner. He ingeniously focuses on the statement "Thus saith the Lord, behold I return [*hineni shav*]" (Jer. 30:18), cited by the *Pesikta de-Rav Kahana* (chap. 44), which depicts God as "repenting" in order to make *teshuvah* appealing to His incredulous creatures, which Bloch interprets as meaning primarily that God "repented" of His earlier decision to fashion a world lacking *teshuvah*, and subsequently installed *teshuvah*. Most importantly, Kierkegaard also characterized repentance, and forgiveness, as "an offense against reason," on two counts: first, it presupposes that the past can be annulled; second, it rests upon the assumption that the eternal can enter into time (1843, p. 21).

13. At Num. 14:36, the *Meshekh Hokhmah* indicates that the "healing" (*erpa*) model of *teshuvah* is restricted to the future and cannot modify the past, save by effecting *teshuvat ha-mishkal*. In Deuteronomy, in his commentary to the prophet Hosea, 14:2 (Copperman ed., pp. 308–11), Rabbi Meir Simhah makes several important distinctions bearing upon the conditions under which repentance achieves its retroactive effects. Stunningly, at least to my mind, he reinterprets the tradition (Gen. *Rabbah* 84:18; Num. *Rabbah* 13:18) that Reuben was the first to initiate repentance—"*ve-attah patahta bi -teshuvah tehillah*"—(notwithstanding the tradition that Adam eventually repented of his sin in Eden) to mean that Reuben was willing to analyze the root determinants, the *hathalah* (drawing from and embellishing the midrashic term *tehillah*), of his sin.

14. Gen. *Rabbah* 1:5; Ruth, no. 469; Talmud, *Pesahim* 54a, *Nedarim* 55a.

15. Maharal, *Tiferet Yisrael*, no. 48.

16. Isa. 45:6; Ezek. 33:11; Talmud, *Yoma* 38b; see *Sefer ha-Ikkarim*, 2:3.

17. The efficacy of the famous *teshuvah* of *hirhur be-libbo* is actually only "nearly full halakhic efficacy," since the Talmud's original ruling, as codified in *M. T., Hilkhot Ishut* viii:4, and in *Tur Sh. A., E. H.*, xxxviii:31–32 (see also Karo, *Bet Yosef*, ad loc), is reshaped by a significant qualification. According to these codifiers, and *per* the ruling of Rabbenu Asher, the marriage in this case would only be a doubtful marriage (*safek kiddushin*) or *kiddushin le-humrah*—rendering the woman forbidden to other men but not permitted to her husband, and their relationship thus requiring either additional substantiation or else nullification of the marriage—since it is only predicated upon the *possibility* (*shema*) that the individual had in fact fully transformed his personality. The matter is further discussed by Epstein, *Arukh ha-Shulhan, E. H.*, xxxviii:95–96. Thus, in order for such a condition of betrothal to be fully binding, the individual would need to also state audibly that he recanted his sinful ways, in which case halakhah would accept his conditional statement *al menat she-ani tzaddik*, despite the fact that his internal transformation cannot be directly observed, and without further test.

18. The "cost" of Horowitz's approach to repentance, which was distinctly different than the approach adopted by his mentor, Rabbi Yisrael Salanter, is that his vision of "perfect repentance" demands nothing less than total self surrender (via "consent"), and rather flies in the face of all that we know about human conflict and the inaccessibility of the deepest reaches of the psyche. Again, my point is not simply to discount such views, but rather to emphasize that their extremeness results in the failure to incorporate the paradoxes of human memory and self-experience (to be discussed) into the *teshuvah* experience, and also promote a lack of appreciation for the role of language itself and the phenomenon of *teshuvah* in containing these paradoxes.

19. Winnicott authored famous paradoxes about the transitional object and transitional phenomenon (1953), expressed most notably by his declaration that the transitional object "has to be found there in order for the child to create it." Christopher Bollas (1979) has recently added to this range of experience a phenomenon he refers to as the transformational object, defined as presences which influence behavior as soon as the individual is within their shadow, but cannot in themselves be known cognitively. The influence of these early experiences—in some sense not yet quite *mental*—is powerful, and they impact directly upon human behavior through affective signals or body memories (see Mitrani, 1995; Spezzano, 1993), and tend to be responsible for some of the most primitive forms of communication during the therapeutic hour, conveying strange forms of "psychological impact," for lack of a better word, deep into the mind of the therapist where he or she may actually be able to formulate out of such "impacts" interpretations of great and lasting value for the patient.

20. Balter (1999) recently reconsidered anew the dilemma of the unknowable or impossible dimension in psychoanalysis, and many of his thoughts would apply to what I consider the impossibilities of *teshuvah*, if expanded fully. The relevant characteristics of Balter's thesis are: (a) any given mental state is by definition always a compromise formation of some kind, and bears multiple functions; (b) unconscious mental processes are inherently unobservable directly and are highly dependent upon the influence of adjuvant operational and representational properties of the conscious mind (i.e., something is always lost in the "translation"); (c) as with any frame, the

psychoanalytic frame by definition excludes certain extra-analytic determinants, and cannot possibly account for all of them in any psychoanalytic explanation; and (d) there are innumerable unanalyzeable effects of the observer's own interaction and theoretical persuasion.

21. I've borrowed this manner of expression from Lacan, who clarified matters by clearly distinguishing between transference insistence, which he viewed as symbolic repetition (expressed in full speech), and transference resistance, which he viewed as imaginary projection (expressed in empty speech) (see 1954–1955, pp. 87–90; 1964 [1977], pp. 33, 69, 146, 149). For further discussion of the existential psychoanalytic reconsideration of transference as repetition-compulsion, and as different, see Muller & Richardson (1988, pp. 55–98); Fink (1995, pp. 153–72); Nobus (2000, pp. 117–22).

22. *Direkte und unmittelbare* are the words Freud himself used to express this prehistoric identification (1923, p. 31). The significance of this primordial, unknowable-yet-incapable-of-being-ignored, almost objectless—and hence "impossible" and paradoxical type of identification (since, by definition, one presumably always identifies with an object!), is carefully discussed by Kristeva (1983, pp. 243–48). The "impossible" profession, and the "impossible" *teshuvah,* cannot bring about full knowledge of the object of this early identification, and yet it serves, as interpreted in Kristeva's chiseled prose, as the "magnet of identification constitutive of identity and the condition for that [eventual] unification which ensures the advent of a subject for an object" (p. 247).

I should point out that I do not mean to suggest that the "impossibilities" of psychoanalysis or repentance are strictly limited to the realm of rationality, or the processes of thought and cognition. Emotional or affect-oriented impossibilities are involved in these endeavors as well. Thus, for example, the consubstantiality of love and hate (both are forms of binding, despite the apparently opposed aims of each sentiment!), or the "narcissistic" incorporation of the object (e.g., the primitive totemic devouring of the father, which quite effectively and unilaterally absents him, being at the same time the mode of *being* him! [see Borch-Jacobsen (1991, pp. 12, 32) who discusses these and other emotional paradoxes]), would play major roles in a full consideration of *teshuvah.*

23. See Aviad, 1983; Bet-Hallahmi, 1991; Buchbinder, Bilu & Witztum, 1997; Danzger, 1989; Diller, 1980; Exline & Baumeister, 2000; Ganz & Harrison, 1978; Greenberg & Witztum, 2001; Knoblauch, 1997; Mester & Klein, 1981, 1986; Ostow, 1995; Randour, 1993; Rice, 1999; Shaffir, 1991; Spero, 1986, 1995; Ullman, 1989.

24. See Maharsha to Talmud, *Rosh ha-Shanah,* 30a, s.v. *man*; Maharal, *Derekh ha-Hayyim* to *Avot* 1:2, s.v., *ve-hinnei hitba'er; Peirush ha-Gra* to Isa. 1:2.

25. As Maimonides states, "Say not that repentance applies only to transgressions of action such as perversity, robbery, and theft. Rather, just as one must repent of these [*la-shuv mei-eilu*] so must he search within his evil traits [*dei'ot ha-raot*] and to repent of anger, and terror, and envy, and derisiveness [or, cynicism]" (*M.T., Hilkhot Teshuvah* 7:3).

26. See also Joel 2:14, Jonah 3:9; and related sources which suggest a transformative process (cf. Talmud, *Sanhedrin* 108a to Gen. 6:6; most interesting is the transformational role given to the imagination in the case of the punishment of exile described

by Meiri, *Beit ha-Behirah* to Talmud, *Makkot* 11a, s.v. *nigmar* !). The term *nihamtani,* Ruth 2:13 nicely brings forward the idea that she is consoled by the fact that she has also been transformed from simple handmaiden to forebear of the Davidic line of kings. Maimonides (*M.T., Hilkhot Teshuvah* 2:2) includes this important term only *at the culmination* of the previous two operations of "abandoning" the sin [*azivat ha-het*] and "resolution" in one's heart [*ve-yigmor be-libbo*]. And since, at least as Maimonides orders the proceedings, the critically important phase of actual verbal articulation [*mitvaddeh bi-devarim*] presupposes internal resolution (*Hilkhot Teshuvah* 2:3 [although some *Rishonim* argue that these prior operations do not themselves need to be articulated (cf. Meiri, *Hibbur ha-Teshuvah,* p. 199; *Kiryat Sefer,* intro. to *Hilkhot Teshuvah*)]), one may state that the process of *nehamah* or consolation, or reconciliation, is the consequence and fulfillment of the repentance process. Regarding the role of *hirhur* , note also Chazon Ish's comment [*O.H., Moed,* no. 29] that *hirhur* does not mean mere "thoughts" ["*hirhurim*"] but rather the "intentions" or "representations" of the mind ["*kavvanat ha-lev ve-hainu tziyyurei ha-devarim*"]. A similar point can be made by considering the rabbinic-midrashic interpretation of the term *devarim* in Gen. 48:1 (*va-yehi aharei ha-devarim ha-eilleh* ["And it was following these things"]), and several other passages (see *Torah Sheleimah* to Gen. 48:1, no. 1 [vol. 2, p. 1737]), as a reference to *hirhurei devarim* or "inner reflections." Here again I would say that the inner *hirhur* or representational process is that which enables concrete perception of *devarim* (*das Ding-an-sich,* the "thing-itself") to be transformed into *dibburim* or words.

When more fully interpreted according to the approach to be outlined here, it will hopefully become more clear that in order for the *halakhic* ontological experience of *nehamah* to be complete, there will need to take place *psychological* ontological experiences akin to (using for this example Melanie Klein's terminology) the movement from the fragmentative paranoid-schizoid position toward the reparative mode of the so-called depressive position, with the latter's emphasis on the development of existential guilt, caring, and symbolized forms of desire.

27. Regarding the Judaic-existential approach to suffering and pain in general, with tangential reference to the atonement and repentance processes, see Carmy (1999), Fox (1987), Lichtenstein (1999), and Spero (1999). None of the authors of these papers, myself included, question the root impossibility of *teshuvah* per se, except in so far as they imply a proportionate relationship between the depth of an individual' s psychological insight and the net quality of repentance and atonement.

28. During different periods, tendencies may be discerned among religious authors to emphasize either the former or latter portrait (see Dan, 1975, 1985; Gruenwald, 1985; Katz, 1964; Shokek, 1985). Nevertheless, it is quite evident that almost all writers on the topic of repentance concur—owing in the main to the supreme weight of biblical and talmudic texts and traditions upon which all Jewish philosophy must ultimately be grounded—that repentance admits of several forms, phases or stages; the "lower" or more basic, primitive, or even regressive ones characterized chiefly by the dimensions of terror, abject loss, fear, shame, and pain, and the "higher" or more complex, advanced, or creative ones tending to be characterized chiefly by the dimensions of satisfaction, joyous longing, and a sense of pleasure. In the paragraphs following, I

offer an alternative conceptualization of the relationship between these two types of repentance.

29. These two positions, and possibly a third known as the autistic-contiguous position which precedes but more correctly underwrites them (Ogden, 1990*b*), do not move one into the other in a unilateral manner, signifying the total disappearance of the earlier mode. Rather, they interleaf with each other throughout life, lending depth and color toward the normal and creative edges of the spectrum, but also maintaining pockets of rigidity and potential "black holes" toward the psychopathological edge of the spectrum. Furthermore, it is conceived that there are subdivisions with each Kleinian position, emphasizing the special disposition toward the latter part of the first stage that favors the next development (e.g., a subtle movement from grossly pathological fragmentations at the beginning of the paranoid-schizoid stage toward more "normalized" splitting in its latter half), but also the potential for cessation in growth, and the creation of unhealthy cusps, impactions, bulwarks, or retreats within which the ego encapsulates itself (see Steiner, 1992). Each of these fine points of developmental theory might have implications for the theory and practice of *teshuvah.*

30. A brief bit of background for the uninitiated. The general assumption is that symbols (and symbolization) reflect the highest mental representational achievement possible. However, some authors speak of symbols that are a lesser level of abstraction than, for example, metaphors; symbols, goes the argument, while of the imaginary, tend not to be imagined; i.e., symbols tend to lean toward the concrete and "to be embedded in the flesh of desire" (Wright, 1991, p. 165, 170–72; see also Langer, 1942). While clinically one does often encounter this *type* of symbol, which may indeed be more akin to a concrete entity than a wholly verbal metaphor, we generally do not restrict the concept of symbol, or the symbolic and symbolization, to this unique sense. More useful is Edelson's delineation of three categories (1975, p. 82): (1) unconscious "symbols," or protosymbols and symbolic equations which are essentially highly concrete, personal signs or indices for some other personal internal or external state or object (cutting one's skin in place of directing these action-motives toward a clearly differentiation original object), (2) presentational symbols, such as in dreams, music, or art, which recall an abstract facsimile of an earlier object, with no anticipation of its immediate return or response, and (3) fully representational symbols, including the ranges of metaphor, where there is no longer any easily identifiable allusion to imagery that promises action or concrete return of the original pleasurable mental experience upon which the symbol may be based (e.g., a mathematical formula or a complex propositional schema). My references to symbolic structures will refer to the third category, unless otherwise specified. Freud' s own comments on negativization are best represented by hypotheses such as this: "With the help of the symbol of negation, thinking frees itself from the restrictions of repression and thus enriches itself with material that is now available for proper thinking" (1925, p. 236 [see also Parsons, 1999]).

31. For an even more striking example—which shall occupy a major portion of the forthcoming analysis of repentance—let us follow the hint in the text that the psychotherapeutic endeavor as a whole has its halakhic analogue in the paradigm of retrieving lost objects or *hashavat aveidah.* If this is so, then the various particularities

of the laws of *hashavat aveidah* should be found also pertinent to psychotherapeutic explicanda. For example, Deut. 22:1 states, "You shall not observe the ox of your friend, or his sheep, wandering and ignore them [*ve-hitallamta mehem*]; verily, you shall return them [*hashev teshivem*] to your friend." In passage 3 of the same chapter, the wording under an almost identical application is formulated in the negative, "and so shall you do regarding all lost objects that your friend has lost, and you have found them, you shall *not be able* to ignore these [*lo tukhal le-hitallem*]." The Talmud (*Baba Metziah* 30a) deduces from these variations of the injunctive *hitallem* that, in fact, under some circumstances one is actually *obliged* to ignore the lost object, foregoing the *mitzvah* of its retrieval and return: such as when a member of the priestly class notices the lost object in a cemetery into which domain he may not enter lest he become ritually defiled, or, even more interestingly, when the business of retrieving the lost object might involve excessive effort or even indignity to the erstwhile retriever (see also *Torah Temimah,* no. 5, to Deut. 22:1, comparing it to Ex. 23:5). I shall have more to say about this shortly, but for now what I wish to highlight is that, according to the halakhic metapsychology, one is led to anticipate that psychotherapy as well (that is, in the language of my model, psychotherapy-as-*hashavat aveidah*) acknowledges a *clinical* circumstance analogous to that described by halakhah. A psychoanalyst, for instance, might, under certain circumstances, consider it technically appropriate to temporarily and methodologically "ignore" an evidently "lost object"—i.e., some bit of memory, an idea, or an affect—that lies abandoned or misplaced, perhaps experienced as if "located" within the mind of the psychotherapist far, far from its proper "home" in the inner world of the patient, pending a more suitable situation, clinically speaking, for retrieving and return the "lost object." The psychoanalyst might do this, say, when, on the evidence, he believes the patient is as yet incapable of, or deeply unwilling to appreciate the return of the lost object, or might even deny that it is lost, or disown his object, and even, if necessary, dismantle his own psychic apparatus in order to create a more impermeable bulwark against intrusive locators of lost objects!

32. The *polel* form *shovev*—which some hold to be a distinct root—connoting "turning" occurs frequently (e.g., "*nafshi yeshovev*" in Ps. 23:3; cf. Jer. 8:5; Isa. 49:5, 58:12; Ezek. 38:8, 39:27; Mic. 2:4), but its reflexive form, the term *le-hishtovev,* which sounds like what we have in mind, contemporarily connotes "to act mischievous." As we shall discuss in the text, S. R. Hirsch (1878) discusses the reflexive properties of *le-hitvaddot* and *le-hitpallel* in Lev. 5:5 (p. 141) and Gen. 20:7 (p. 347), 40:11 (p. 649). As a wild venture, I wonder if the adjectival form *shuvah* or "[a state of] restfulness" that appears singularly in Isa. 30:16 ("*be-shuvah va-nahat tivvashe'un*") might not be the legitimate noun understructure which permits our term *teshuvah* to qualify as some sort of latent or quasi-reflexive dynamic.

Continuing with the problematic term *teshuvah,* it is interesting to note that Samson Raphael Hirsch interprets the related acts of *vidduy* and *tefillah*—which, as noted, do take introflexive form in Hebrew grammar—as the risky endeavor of imposing a new spiritual or moral element into a currently disorganized, heterogeneous state, unifying it into something more homogenous (at Gen. 20:7 [p. 347], 48:11 [p. 649]; Lev. 5:5 [p. 141]). In a remark regarding repentance, but rightfully suited to all three aforementioned concepts, Hirsch states:

> But the step toward this *teshuvah*-resolve is, making clear to oneself without any self-deception, exactly where, and in what, one has sinned. As long as one allows one' s mind to be bemused with apologetic self-deception, there is no thought of really pure conscientiousness for the future.

The preceding comment amplifies thoughts that can be found in germinal form prior to the time of Hirsch's writing as well as after.

33. Recall Hegel's teaching: self-consciousness is inherently reflective; the minimum sine qua non for subjective consciousness (i.e., not necessarily for the rudimentary neurological dawning of consciousness) is another self—always a matter of a "return from otherness" (1823, pp. 104–05). The embeddedness of self and object are actually a secondary development, subsequent to a prior embeddedness: that of the germinal ego and its primordial body-skin precursors qua originary "representations" *in* the ego as well as *of which* the ego is built. After reviewing Freud's description of the beginnings of the ego and of the first self-representation, Snelling asks (2000, p. 111): "But is it not a circularity to state that the ego derives its unity from a bodily representation of itself, and that a unified representation of the body is derived *from* the representation of that body-like unity of the ego?" The contemporary object-relational reply, which Snelling endorses, is No!, since ego and object, perceptually speaking, arise simultaneously (i.e, they are inaugurated by the same originary psychic event of loss of unconscious omnipotence in the moment of the instantiation of language). Thus, ego *is* object-relating; "the primordial intentional relation is the relation to internal objects in fantasy" (see also Borch-Jacobsen, 1991; Castoriadis, 1975).

34. While I am sure kabbalistic concepts such as *tzimtzum* and *ein sof* are also apposite here, I consider that it is even more significant that it is also implied in the halakhic conceptualization of *eino min ha-middah* encountered in certain instances (Talmud, *Megillah* 10b, *Menahot* 99a to Ex. 34:1; Jer. Talmud, *Rosh ha-Shanah* 1:1). Another categorical rabbinic statement that expresses most creatively an awareness of the presence inherent in absence emerges in the context of the biblical biographical comment regarding Esther, "*ein lah av va-em*" ([Esther 2:7] "she had no father and mother"): "Said Rabbi Levi: In every place where [the term] *ein lah* ["she had not"] appears, [it indicates via exegesis] that [eventually] she *had* [*havei lah*]" (Gen Rabbah 38:14, Eccl. Rabbah 1:26, 52; cf. Gen. 11:30, Jer. 30:17, I Sam. 1:20; also cf. Talmud, *Hullin* 65a, s.v. *af al pi*). Substance is thereby lent to a lack via symbolization. The literal intent of the exegesis is that Mordechai became Esther's parent, but the wider meaning of the category is essentially analogous to the classical Freudian (and ultimately Hegelian) working principle *negativa non percippi est affirmatio manifesta*, and to the notion that behind every erasure lies the trace of presence. Finally, Maharal (*Netzah Yisrael*, chaps. 26, 30, 39) who states that all being must be preceded by absence (*kol havayah yukdam lah he'ader . . .*) and discusses the historical implications of this law (see also Abravanel, Gen. 16:12, s.v. *ve-im nomar*).

35. As Aragno puts it, "The mind comes about through its own function; to 'represent' is the process of signifying meaning. To speak of 'mental' is to speak of representation—and to speak of representation is to speak of meaning. Mind and meaning are one" (p. 118; see also esp. Castoriadis [1975]).

36. For the discussion of asymmetrical logic—i.e., the tendency toward the "logical" principle of exclusion, that a thing cannot be the equivalent of its opposite (A cannot equal -A)—and symmetrical logic (A can equal -A), and its relationship, respectively, to conscious, secondary-process thought, temporality and spatiality, and unconscious, primary-process thought, atemporality and atopia, on the other hand, see the remarkable work of Matte-Blanco (1975, 1988; cf. Rayner, 1995). Bomford (1999) recently applies Matte-Blanco's work to a definition of the characteristics of God. Obviously, these are gradations, and demarcate points on a continuum, for in reality one works with both modes of logic throughout life, except under unique circumstances (which may be ultra-creative or simply psychopathological). Relating this notion to Klein's twin positions, one would note that the paranoid-schizoid position would tend to overemphasize sharply asymmetrical defenses ("You've changed your sweater, Doc, so I assume you do not care to hear me drone on today!") because of the insecurely repressed maw of symmetrical, symbiotic, poorly differentiated primitive mental states, whereas the depressive mode would indicate a relatively tranquil relationship between both forms of logic. Obviously, as worked out in greater detail by Matte-Blanco and Rayner, the beauty of sexual passion, the realm of the aesthetics, and the play of metaphor, metonym, and the host of other linguistic tropes is that these dimensions, each in its own way and provenance, allows us to "flirt" with approximations of the dizzying pleasure of symmetry, which in its absolute form evokes psychosis, or total mindlessness, and is cordoned off or "forbidden" by the aboriginal repressive barriers upon which sentience is predicated. That is to say, one needs to tilt toward symmetry, which always means taking on board that much more of the unrestrained qualities and characteristics of the primary unconscious. Yet without the basic restraining envelope of consciousness, and the characteristics of asymmetrical logic and other ego functions upon which consciousness—indeed, mind itself—is predicated, the value of the symmetry could not be contained nor appreciated. In a sense, one could say that pure symmetrization, at least as we might imagine it, is yet another of those phenomena that our mind can only regard retroactively as a "lost object," never really in one's possession, never to be fully retrieved, but always alluring.

37. In what remains an unparalleled intuition about the sense of flow of time, yet to be empirically proven or disproven, Freud (1925) conjectured that the sense of the pulse of time stems from the rhythmic "opening and closing" of the shutters of perceptual consciousness in tune with the episodic cathecting of consciousness with libido and the withdrawal of libido: "It is as though the unconscious stretches out feelers, through the medium of the system *Pcpt.-Cs.*, towards the external world and hastily withdraws them as soon as they have sampled the excitations coming from it . . . I further had the suspicion that this discontinuous method of functioning of the system *Pcpt.-Cs.* lies at the bottom of the origin of the concept of time" (p. 231).

38. The link of repetition and mourning is perhaps best clinched with the principle: *Ein puranut baah al Yisrael she-ein bah ketzat mi-piron avon ha-egel* (Talmud, *Sanhedrin* 102; Rashi to Ex. 32:34). Some of my original thoughts regarding the concept of *aveilut yeshanah*, mourning, and the psychological capacity for true *suffering*—as opposed to lingering in pain—were originally presented in Spero (1999). Julia Kristeva as well, in her searching essay *Black Sun* (1987), describes patients who "recall"

great amounts of data—not in the service of remembering but in order to refuse to symbolize loss, not to mourn the lost object but rather to idealize the "deceased object" (p. 46). With Kristeva, one might say that individuals engaged in constant, obsessional efforts at *teshuvah* are likely not working-through the relevant necessary losses, and instead are constantly evoking something traumatic in order to deny negativization and further symbolization. Compare with Radden (1987) who speaks of a deeply objectless form of melancholia. Ultimately, however, the minds of those undergoing all forms of mourning and melancholia, even the psychoses, are not absolutely objectless, and this is to say nothing about the indwelling presence of the divine object which maintains residence even should an individual reveal no evidence of a divine object *representation* in the psychological sense of the term (see Spero, 1992, 1998), though this is very difficult to conceptualize clearly. Even *divine* object representations can be overshadowed, eclipsed, repressed, and fragmented because representations are "mere" psychological phenomena of the mortal mind.

39. Rabbi Soloveitchik (1974, p. 224) often emphasizes the rational solution, and even when etching the emotional factors (e.g., pp. 101, 260, 279, 280–86), does not make clear exactly how such penitential change transpires psychologically. He may legitimately have not seen it as his task to touch to the deeper levels I am indicating here, but these levels are clearly the linchpin of the whole enterprise. One thing is for sure, as our review must surely have shown by now: There is a great expanse between the *peshat* of otherwise inviting passages such as "And his heart will understand, and he will return, and I will heal him" (Isa. 6:10) in praise of intellection (*binah*) (see Talmud, *Megillah,* 17b), and the deeper psychodynamic substrate that underwrite the processes of symbolization and mourning that enable such understanding. It is not the other way around.

40. Perhaps the tragic flaw of one who deludes oneself with the formula *eheta veashuv* ("I shall sin and *subsequently* repent!"), enumerated among those who become potentially refractory to repentance, is that the presupposition of the capacity to wait, delay, and renegotiate internal identifications across even small spans of time is perversely reversed in such persons. Though they speak as if they are possessed of temporal tenses, and of the capacity for waiting, they are actually trapped in a repetition-compulsion ("I shall sin *and sin again*"); but not in the simple sense. That is, they do not simply repeat their crime or sin; rather, they repetitiously inhabit the null "dead" non-time zone from within which perspective (if one could call it that) their "time" is one, extended, enduring act of bad faith or sin. This is not true, existential anticipation is but a mechanical self-replication which gives the illusion of being to refer to a tomorrow or other beyond (see Schiffer, 1978). Healthy waiting or "heavy time," on the other hand, generally invites self-reflection and self-examination, leading further to anxiety, guilt or shame which can then be put to moral purposes. Hence the penitent quality of this scene from Beckett's *Waiting for Godot* (1954), as interpreted by Clausius (1998):

> Vladimir: Suppose we repented?
> Estragon: Repented what?
> Vladimir: Oh . . . (He reflects.) We wouldn't have to go into details.
> Estragon: Our being born?

41. Hawkes (1977) reviews these developments.

42. While certainly not a text one would find among the general Musar section of the library, C. S. Lewis' *Screwtape Letters* (1961) provide brilliant insights into human nature, couched in the language of a sort of "nineteen letters" of guidance shared between the wise demon Screwtape, experienced in the foibles and weaknesses of mankind, and his young nephew, Wormwood, a beginner at the art of temptation.

43. Later writers were also to observe that there are gradients and shades of repetition, some more constructive than others. Loewald (1971) distinguished between *passive* repetition, closer to Freud's concept of "reproduction," under the sway of unconscious motivation, and active repetition, or "re-creation," largely deconstructive in nature. In a recent essay, Rothenberg (1997) made a serious comparison of Loewald' s later papers (1988), the work of Thomas Ogden (1990a) on post-depressive historicity, and Rabbi Soloveitchik's espousal of a religious consciousness which by virtue of faith enables one to speak of reversible time. However, the observer still does not know just what it is that makes it feasible to speak of a halakhically efficacious mode of repetition.

44. For a basic definition of "transference neurosis," see Laplanche & Pontalis (1967, pp. 462–64), who emphasize its dual role in Freud's purview as both linchpin of the treatment—inasmuch as it brings the latent, repressed infantile neurosis into the analysis—as well as a major expression of the potentially suffocating compulsion to repeat (note that Freud introduces the term in "Remembering, Repeating, and Working-Through" [1914]).

45. This principle also plays a role in the possibility of reconstructing trauma, in that one is not simply attempting to remember the known data of the past—the latter, as more than one psychoanalyst has observed noncynically, is generally *not* the significant matter! —But often to discover that much of what was involved in the impact of trauma was never fully psychologically experienced and processed in the first place until the advent of therapy. In this vein as well, *teshuvah* is not best understood as returning to muck about in the swamps of known data but more so as the rediscovery, or refinding, of that which was never known. According to trauma theory, it is the process of negativizing and forgetting—paradoxical forms of storage nevertheless, as opposed to, say, abject denial and psychotic blocking out—which enables traumatic events to ever possibly be known (Caruth, 1995, p. 7; see also Lyotard, 1988). According to Cavell (2000, p. 94), Freud is asking: "How can one use the past in such a way that it helps one to notice, rather than obscure, what is peculiar about this object [or need] in the present?" The answer: through negativization, symbolization, and mourning.

46. See Fink, 1995 *a*, p. 94; Lacan, 1946, 1964; Nobus, 2000, p. 83; Peraldi, 1988; Roudinesco, 1 993, pp. 386–98; Spero, 1993.

47. For the sake of augmenting our halakhic metapsychological frame, note that a biblical anticipation of the role of the lost object, analogous to Freud's, can be derived from a reinterpretation of the concept of *hashavat aveidah* (return of lost property). The proof text is familiar enough: "'And you shall return it to him' [*va-hasheivoto lo*]—but the gemstone is the following rabbinic interpretation "[This means:] Even himself you shall return to him [*af et atzmo meishiv lo*]" (*Sifrei, Piska* 13 to Deut. 22:2). The

Talmud, *Sanhedrin* 73a draws out a slightly different instruction: "This includes [returning] loss of his body [or his limb] [*le-rabbot aveidat gufo*]," or, as Rashi amplifies (ad loc.), " *hashev et gufo le-atzmo,* " and elsewhere (Talmud, *Bava Kamma* 81a), "If he is lost in his way one is obligated to raise him [up upon the right path]." Latent here is an important relationship between *hashavat aveidah,* the return of lost functions, properties, or objects, on the one hand, and certain existential elements of the processes of psychotherapy, on the other hand. This relationship will be spelled out in the text.

Bible scholar Bryna J. Levy has suggested (personal communication, October 2002) applying the concept I have delineated here to the biblical portrait of the sailors' inability to "return" to shore in the story of Yonah (Jon. 1:13): *va-yahteru ha-anashim le-hashiv el ha-yabbashah ve-lo yakholu,* an idea that I would purchase more wholeheartedly had the relevant passage been stated of Yonah himself, the protagonist of the story. However, within the larger context of the story—especially if one is willing to view each of the components and characters of the episode primarily as symbolic representations of Yonah' s own internal mental agencies (as in a dream [cf. *Commentary on the Book of Yonah* by Ibn Caspi, inter alia])—Levy's suggestion may work. She points out that the *hiphil* construction of the verb *le-hashiv* permits two exegetical possibilities: the sailors attempted unsuccessfully to turn the *boat,* or they attempted unsuccessfully "to turn" Yonah! In the mythic perspective, the floundering boat would represent the rapidly failing psychic container—*ve-ha-oniyyah hishevah le-hishaver* (Jon. 1:4) —for Yonah' s internal (but also externalized and enacted) conflict. In this conflict, one self-component is (blissfully?) dissociated in the ship's hold, content to abandon itself to the murky depths; the other self-component, emblematized by the conspicuously God-fearing sailors, tries to rouse itself to confess, repair, and return to shore. As we have said, and will say again in varying forms throughout the remainder of the essay, such literal return must be unsuccessful, for it is *impossible.* The myth-story reflects this as well. However, as Yonah matures both psychologically as well as morally—following a reparative episode within the container of the mother's/fish's belly—he begins to comprehend that there is a sense in which he *can* re-turn to the city of Nineveh (Jon. 4:2 depicts Yonah expressing that he had already foreseen the *teshuvah* of Nineveh).

48. As well, one might say: death, or negativization, is the "tomorrowing" of the specious present.

49. I noted above that acting-out is directly related to the element of time. This being the case, the maturation through *teshuvah* to full consciousness of the transition between *emesh* to *mahar* is the essential guarantor that the influence of the indigenous timelessness of the unconscious is being channeled into temporal pathways by language, further indicating that true remembering is in operation rather than mere repetition or passage-into-action. This is emblematized by the appearance of the Knower of Mysteries, the Symbol of symbols beyond and proceeding which there is no further signification.

50. Cf. variants at *Geirim* 4:1; *Baba Metzia* 58b; Rashi, Lev. 19:33. Interestingly, in *M. T., Hilkhot. Mekhirah,* xiv:13, Maimonides cites this ruling but deletes the term *emesh.*

51. The term *emesh* appears in two other biblical contexts: In Gen. 31:24 and 29, the context clearly emphasizes the value of *emesh* as a dream or perhaps a transitional state, and in Gen. 19:34 the context heavily implies a stuporous state, redolent of intoxication (see the Midrash), dangerous partial denial, and the blinding allure of illicit incestuous wish-fulfillment.

52. *Mahar* also plays a determinative role in Ex. 17:9, "Go and wage war with Amalek tomorrow [*mahar*], I will be standing at the peak." The talmudic sage Isi ben Yehudah enumerated this *mahar* clause as among the 5 biblical "passages" whose meaning is indeterminate owing to a key axis word which can be shifted into two equally meaningful positions in the passage; i.e., "wage war with Amalek tomorrow" and "wage war with Amalek [immediately]—and tomorrow I will be standing at the peak" (see Talmud, *Yevamot* 52a-b, Jer . Talmud, *Yevamot* 2:7; see also Gen. Rabbah 80:6). Most germane to our topic is that the *Mekhilta* cites Ex. 17:9 as an example of *mahar* [*shel*] *akhshav*, which, historical implications notwithstanding, suggests that the very collapse of the temporal dimension emblematized by *mahar=akhshav* is precisely what renders the term too weak to cut clearly into undifferentiated circumstances.

53. My interpretation of *mahar shel akhshav* as a relatively defective temporal state receives indirect support from a seemingly unrelated midrashic interpretation. Following Adam's sin in the Garden of Eden, the Bible records what is perhaps the first psychodiagnostic prediction, made by God Himself: "And the Lord God said—Behold! The man is become as one of us, to know good and evil; *and now, lest he put forth his hand,* and eat, and live forever" (Gen. 3:22), as a result of which evaluation Adam is exiled from the Garden. It was an appropriate prediction, I once argued (1986, chap. 15, "Emergence of and slight from self—The experience of shame in Genesis 3"), because the near-crippling anxieties (as opposed to a more mature sense of guilt!) evoked by the sudden flooding of self-consciousness following the transgression would quite probably have navigated man to adopt the most immediate mode of denial of death possible: incoporating concretely the fruit of the Tree of Life. The concise Hebrew terms for the section of the passage I emphasized—*ve-attah, pen yishlah yado*—alerted the rabbinical exegetical sensitivities for two reasons: (a) the rabbis considered the reference to the elements of time (*ve-attah*) and apparent conditionality (*pen*; "lest") inconsistent for a description of divine musings, and (b) the term "*pen*" usually denotes a proscription (the equivalent of "shall not!" [see Talmud, *Eruvin* 96a, and glosses]), in which case the text could be seen to state, quite problematically, "and now, he shall *not* put forth his hand." Thus, Rabbi Abba bar Kahanah offered a reinterpretation (for readability's sake I have combined the two main sources in Gen. Rabbah 21:6 and Num. Rabbah 13:3):

> This [text] teaches that God opened [to him] the door of *teshuvah*. "And now" [*ve-attah*]—this [term] always [refers] to *teshuvah* , as it is written, "And now [*ve-attah*], O Israel, what does the Lord, your God, ask of you, but that you fear Him . . . [Deut. 10:12]. And the Writ also states "Lest . . ." [*pen*]—this [term] always [means] "No!" [Therefore interpret the sequence of the passage as follows:] God said to man, "Even now [*afilu akhshav*] do *teshuvah* and I will accept you!" And the man said, "No!" [*pen!*] "I cannot!" [*i efshi* "Impossible!"] And [only then] the Lord said, "He *will* put out his hand . . ."

In line with the thinking presented in the text, I would say that we have here another illustration of the poverty, and ultimately the failure of impulsive repentance—the *teshuvah* of *akhshav* [*attah*], of the specious present, of pseudo-objective consciousness—even when it is under the benevolent counsel by God Himself. While it is difficult even to conjecture about primordial Adam's personal psychology (even as a mythic persona) without some form of evidence, I truly doubt that he had any conception of time (witness his panic at the initial transformation of day into night [Talmud, *Avodah Zarah* 8a]). And assuming that he had no concept of time, I presume that he perceived the invitation to "return" as somehow terrifying, as if possibly a seduction to be recycled into nonexistence, or any number of additional fantasy possibilities that would not have been entertained as mere abstractions by an as yet asymbolic creature. Indeed, for knowing no better, Adam may have been the first to profess the "impossibility" (the *i-efsharut*) of *teshuvah*!

54. While everyone is familiar with what psychoanalysts presumably say about motivated forgetting, and the hidden meaning of losing things, it is much less well known that Freud considered the finding process—or shall one say, the *refinding* process!—No less pregnant with potential meaning. Citing his then colleague Otto Rank, Freud states (1901, p. 208), "It is of interest when he writes later that often not only losing objects but also finding them appears to be [psychologically] determined. It is obvious that in cases of losing, the object is already provided; in cases of finding, it has first to be looked for."

55. I would compare the Vilna Gaon' s example with the case of the masoretic blank inserted starkly in the middle of Gen. 35:22: "and Reuben went and lay with Bilhah, the concubine of his father, and Jacob heard [?] and the sons of Jacob were twelve." Despite the consensus rejection of the literal interpretation of the passage (though not unanimous; cf. *Torah Sheleimah, ad loc.*, nos. 93–94), the rabbis acknowledged that Reuben had exhibited *some* essential impropriety in this incident, and suggested that the blank space represents the suspension of Reuben's fate owing to his sincere repentance, followed by the announcement of the moral equality of all of Jacob's twelve sons. Another case of "presence" in relief against absence.

56. Nevertheless, *teshuvah* is essentially unlimitable and always possible. The possibility of *teshuvah* can still be spoken of coherently, states Rabbi Dessler in full concurrence with one of the oldest of theological doctrines regarding repentance, because *teshuvah* is above all else a rudimentary change dynamic built in to the very creation and order of the universe (see Talmud, *Pesahim* 54a).

57. I mean this in the sense authorized by the halakhic metapsychology, and not so as to superficially identify every neurotic as a sinner and every sinner as deeply neurotic, though my point lies closer to that which the two share in common and not with that by virtue of which they differ.

58. See Fink' s excellent analyses (1995*a*, pp. 62–65, 1995 *b*, 1996) of Lacan's (1946) use of the so-called prisoners' dilemma riddle to illustrate this point. For Lacan, the key to the resolution of the riddle, for at least one lucky prisoner, will be nothing more nor less than his pausing at critical moments to reflect while other prisoners, at least at that given moment, seem not to be, and perhaps by definition are not capable of reflecting on that which the lucky prisoner is reflecting.

59. Regarding the enigmatic and seductive first signifiers, consider this anticipation in the Talmud (*Kiddushin* 30b): "Rabbi taught, 'It is known and obvious to He who declared that the world exist that a son honors his mother more than his father since she cajoles him with words [*she-meshaddalto bi -devarim*], which Rashi interprets, "she seduces him [*mefattah oto*]."

60. Deleuze referred to this dimension as "recursive futurity" (1994), which if we apply it to *teshuvah* would mean that its coming to pass envelopes a series of prior events, in an intense revisiting of the moments leading to its own emergence, all the way up the chains of expression, thereby defining the meaning of the future.

61. Three other incidents are of great interest—Ketiah bar Shalom (Talmud, *Avodah Zarah* 10b [cf. Deut. Rabbah 2:24]), the student and the *tzitzit* (ritual fringes) (Talmud, *Menahot* 44a; *Eliyyahu Zuta*, 22), and the *kalztanere* (corrup. of *questionarius*, executioner) of Rabbi Hananiah ben Teradeyon (*Avodah Zarah* 18a)—but the principles are essentially the same. Regarding the distinctions between them, see Ariel (1998).

62. In fact, the sublimation process was an extended one, and it revealed its strains: Resh Lakesh would review his studies 40 times before entering Rabbi Yohanan's lecture hall (Talmud, *Taanit* 8a), when studying with Rabbi Yohanan, would counter each teaching with 24 independent queries (*Bava Metzia* 84a, *Ketubot* 54b), and was characterized as "an uprooter of mountains who then grinds them together" (Talmud, *Sanhedrin* 24a).

63. *Ahadar kar'eh le-ahoreh ve-kara keriah aharina.* Rashi, s.v. *ahadar.* "The rent that he had torn during the funeral for his master he folded back over, and folded [forward] that which was behind in order to tear another rent, to demonstrate a new mourning akin to the day of [his master's] death."

64. Revenant indeed: according to Shmuel ibn Tibbon, "*Sabba*" always refers to Elijah! From a strictly Freudian-Kleinian point of view, I could further elaborate the story and claim that Rav represented the Hegelian archetype of *Der Tod* who is resurrected in a sense by his disciples (as would a father by his sons or the Master by his rebellious slaves) and lives on eternally in their guilt-ridden memories because of the fact that, in his absence, he represents *their* death, or their own unrepresentable death. Perhaps the double performance of the *keriah* effectively transmutated such guilt into something meretricious.

65. Rabbi Soloveitchik (1974) holds that Elisha should have anticipated that he would be allowed to repent. I recently noticed that Rabbi Issac Halevi Horowitz, in *Shaar ha-Otiyyot*, vol. 1, p. 130, opines that the *bat kol*, a heavenly echo, did not seal his fate, and cites precedent from Talmud, *Sanhedrin* 102a.

66. See *Tos.*, s.v. *ki*, Talmud, *Bava Batra* 116a, who notes that judging from the text alone there is no suggestion that Job lost all of his financial wealth or his properties; another of the many subtle hints (see note 67) that Job is not to be viewed as a literal, historical event—and maybe is not even a tragic one in the main. Rather, it is a mythic text, suitable (if not expressly intended) for analysis as one would analyze the fragments of a dream.

67. The rabbinic view I have in mind states that Job is a parable and not an actual historical event (Talmud, *Bav a Batra*, 14b, 16b), and this view was accepted by Maimonides (*Guide*, III: 28, 29), though other expositors upheld the more literalist ap-

proach (which, nevertheless, has had no better luck in placing Job or its author). I have recently attempted to address this dilemma by way of an alternative interpretation of the entire transaction between God and Job, entitled "The psychological redemption of Job: Beyond mirroring toward recognition of the desire of the Other," to be published in 2005 the *Proceedings of the International Conference on the Biblical Personality:* [as yet untitled], eds. Glas, Gerrit, Verhagen, Peter, & van der Pragg, Herman (New York: Kluwer Academic). The psychoanalytic methodological approach I adopt, leaving aside the content of my larger thesis, requires viewing all of the actors in the story as representations of different aspects and components of the author's personality (some fragmentary and "part" and others more completely internalized or "whole") engaged in various internal relations as the story unfolds (see Freud, 1908[1907], 1916, p. 307, 1919, p. 232n). In this approach, the image of God and to some degree Satan may represent superego introjects of varying qualities, and the four companions may represent different levels of self-observational functioning. This kind of approach authorizes one to pay special attention to the quality of interaction among these fragments, the meaning of repeated statements, appearances, and textual absences, and so forth. In the aforementioned essay I focus in particular upon the implications of this approach for the rabbinic view that the story of Job was a parable, in terms of the hypothesis that Job expresses Moses' desires and conflicts latent at around the time of his dialogue with God in the cleft in the rock.

68. Skin, flesh, and clothes metaphors are glaringly predominant in this text compared to any other text in the Bible, and references to the mouth and to speech far exceed any other book except Psalms, where the contexts are multiple (see Spero, *op. cit.*, 2005).

69. Cf. I Sam. 15:29; II Sam, 13:39.

70. See Lefort & Lefort (1980) regarding the infant' s transition from experiencing the world and the body-self as holed toward the recognition of the holed other, the other as signification system which alone enables the self to symbolize its holes into sentient structures. The preponderance of metaphors and metonyms in the Book of Job referring to skin, coverings, mouth, and words (*millim*) are part of the evidence I would muster (see note 85 *supra*) to prove just how early, psychologically speaking, must be the anchoring point of the develo pmental drama depicted therein.

71. "Diegesis" refers to Plato' s famous differentiation (*The Republic* , 392c–98b) between three forms of poetic enunciation: the poet speaks in his own name without seeking to make us believe that it is someone other than he who is speaking (*haple diegesis* or pure narrative); the poet speaks under the name of an other with whom he is identified (*mimesis*); and mixtures of the two.

72. Meares (2000) characterized the relative impoverishment of concretized "cataloguing" as opposed to qualitatively superior, resonating remembering: the language of the experience is linear (non-metaphoric), event dominated, unphilosophical or reflective, caught in the present, dominated by episodic memory, boring, and nonconsolitory.

73. *Targum Onkelos* to Gen. 2:7. Compare with Eccl. 8:11, "*Asher ein naasah pitgam maaseh ha-ra meheirah*" [lit., "Because sentence against an evil work is not executed speedily, the heart of the sons of men is fully set in them to do evil"], which suggests

that experience that fails to be linked to words, or in our terms here, representationalized, lends toward acting-out and nonsymbolized expression. The fundamental role of language in the basic instantiation of the mind can also be "heard" in the rabbinic mind if one is willing to indulge me a risky reading of the following talmudic derivation. In its introduction to the chapters dealing with the leper and other states of ritual uncleanness, the Bible states (Lev. 13:2): *Adam ki yihyeh* . . . "When a person shall experience . . ." (or "shall *be*"), referring to the specific defiling flow or exudate described as ritually unclean. The Talmud (*Horayot* 10a) glosses this passage in a very arcane manner: "[When shall a person experience?]—*min ha-dibbur va-eilakh*—from the Word onward!" The nominal interpretation of this gloss holds that the rabbis were attempting to clarify that the entire institution of ritual defilement and purification takes force only following the revelation of *The Word at Sinai.* However, as Epstein points out, this would seem to not have merited mention since, according to tradition, as part of the phenomenal impact of the Revelation, all of the Israelites attending the Revelation were cured of any illnesses they might have had, such that there could not conceivably have been any individual to whom these laws would have applied until the subsequent years of sojourn through the desert (*Torah Temimah, ad loc,* no. 2). It would thus be peculiar to suddenly link this specific area of the commandments to the historical events of Sinai. In any event, I would like to read the text together with its gloss as follows: "An individual, or the mind, *becomes* from the word, from speech onward!"

74.

> "'and God descended in a cloud and took His position with him there, and He called out the Name of God' (Ex. 34:5): Had the Writ not stated so, one could not have dared to say it—The [passage] teaches that God wrapped himself [in a *Tallit* or prayer-shawl] as would a *sheliah tzibbur* and showed Moses the order of prayer [*seder ha-tefillah*] and said to him: Whenever Israel sins they shall perform this order and I shall pardon them"

(Talmud, *Rosh ha-Shanah* 17b; cf. also Rashi to Ex. 33:19). The order or *seder* of prayer is to cast before man certain divine instrumental invariants and to invite man to exercise his language capabilities as well and create a web of signification.

References

Author's Note: All citations of Freud's works are from *The Standard Edition of the Complete Psychological Works of Sigmund Freud*, ed. and trans. J. Strachey. XXIV vols. London: Hogarth and The Institute of Psycho-Analysis, 1953–1974, subsequently referred to as *S. E.*

Adler, Samuel A. (1999). "*Teshuvah.*" Pp. 496–674 in *Aspaklaria.* vol. XXIX. Jerusalem: Aspaklaria Publishing Co.

Anzieu, Annie. (1977). "L'heure de la Repetition." *Nouv. Rev. Psychanaly.* 15:163–75.

Aragno, Anne. (1997). *Symbolization.* Madison, CT.: International University Press.

Ariel, Yigal. (1998). *Erpah Meshuvotekhem: Ha-Teshuvah ba-Mikra u-ve-Aggadah.* Hispin, Israel: Midreshet ha-Golan.

Arlow, Jacob. (1986). "Psychoanalysis and Time." *J. Amer. Psychoanal. Assoc.* 34:507–28.

Aulagnier, Piera. (1975). *The Violence of Interpretation.* Trans. A. Sheridan. London: Routledge.

Aviad, Janet. (1983). *Return to Judaism.* Chicago: University of Chicago Press.

Balter, Leon. (1999). "Constant Mental Change and Unknowability in Psychoanalysis." Pp. 93–129 in *The Psychoanalytic Study of the Child,* vol. XLIV, edited by A. J. Solnit, P. B. Neubauer, S. Abrams, and A. S. Dowling. New Haven, CT: Yale University Press.

Barclay, Craig, and DeCooke, Peggy. (1988). "Ordinary Everyday Memories: Some of the Things of Which Selves Are Made." Pp. 91–125 in *Remembering Reconsidered,* edited by U. Neisser and E. Winograd. Cambridge, MA: Cambridge University Press.

Becker, Ernest. (1964). *Revolution in Psychiatry.* New York: Free Press.

———. (1973). *The Denial of Death.* New York: Free Press.

Beckett, Samuel. (1954). *Waiting for Godot.* New York: Grove Press.

Behr, M. (1980). *"Al Maasei Kapparah Shel Baalei Teshuvah Be-Sifrut Hazal." Zion* 46:159–81.

Bet-Hallahmi, Benjamin. (1991). "Back to the Fold: The Return to Judaism." Pp. 153–72 in *Tradition, Innovation, and Conflict,* edited by Z. Sobel and B. Bet-Hallahmi. Albany, NY: SUNY Press.

Bialik, Chaim N., and Rabinsky, Y. H. (1962). *Sefer ha-Aggadah.* Tel-Aviv: Devir.

Bion, Wilfred. R. (1963). *Elements of Psycho-Analysis.* London: Maresfield, 1989.

———. (1965). *Transformations.* London: Maresfield/Karnac.

———. (1967). *Second Thoughts.* London: Maresfield/Karnac.

———, ed. (1992). *Cogitations.* London: Karnac.

Bleger, Jose. (1967). "Psycho-Analysis of the Psycho-Analytic Frame." *Int. J. Psycho-Anal.* 48:511–19.

Bloch, Yosef Eliyahu. (1976). *Shiurei Daat.* Vol. 3. Jerusalem: Feldheim.

Bollas, Christopher. (1979). "The Transformational Object." *Int. J. Psycho-Anal.* 60:97–107.

Bomford, Rodney. (1999). *The Symmetry of God.* New York: Free Association.

Boothby, Richard. (1991). *Death and Desire.* London: Routledge.

Borch-Jacobsen, Mikkel. (1991). *The Emotional Tie.* Trans. D. Brick. Stanford, CA: Stanford University Press, 1993.

Bowlby, John. (1973). *Separation: Anxiety and Anger.* New York: Basic Books.

Brockmeier, Jack. (1996). "Anthropomorphic Operations of Time: Chronology, Activity, Language, and Space." Pp. 239–50 in *Dimensions of Time and Life: The Study of Time VIII,* edited by J. Fraser, and M. P. Soulsby. Madison, CT: International University Press.

Bromberg, Philip M. (1994). "Speak, That I May See You!" Pp. 241–66 in *Standing in the Spaces.* Hillsdale, NJ: Analytic Press, 1998.

Buber, Martin. (1942). "A God Who Hides His Face." Pp. 56–65 in *The Dimensions of Job,* edited by N. Glatzer. New York: Schocken, 1969.

Buchbinder, Tuvia, Bilu, Yoram, and Witztum, Eliezer. (1993). "Ethnic Background and Antecedents of Religious Conversion among Israeli Jewish Outpatients." *Psychol. Rep.* 81:1187–202.

Buonomano, Dean, V., and Merzenich, Michael. (1999). " A Neural Network Model of Temporal Code Generation and Position Invariant Pattern Recognition." *Neural Computation* 11:103–16.

Camus, Albert. (1946). *The Stranger.* Trans. A. Bower. New York: Knopf, 1965.

Caper, Robert. (1999). *A Mind of One's Own.* London: Routledge.

———. (2000). *Immaterial Facts.* London: Routledge.

Carmy, Shalom. (1999). "Tell Them I've Had a Good Enough Life." Pp. 97–153 in *Jewish Perspectives on the Experience of Suffering*, edited by S. Carmy. Northvale, NJ: Aronson.

Carnochan, Peter G. M. (2001). *Looking for Ground.* Hillsdale, NJ: The Analytic Press.

Caruth, Cathy. (1995). *Trauma: Explorations in Memory.* Baltimore: Johns Hopkins Press.

Cassirer, Ernst. (1953). *Philosophy and Symbolic Forms.* Vol. I: *Language.* Trans. R. Manheim. New Haven, CT: Yale University Press.

Castoriadis, Cornelius. (1975). *The Imaginary Institution of Society.* Trans. K. Blamey. Cambridge, MA: The MIT Press.

Cavell, Marcia. (2000). Keeping Time: "Freud on the Temporality of the Mind." Pp. 85–100 in *Th e Analytic Freud: Philosophy and Psychoanalysis*, edited by M. P. Levine. London: Routledge.

Clausius, Claudia. (1998). "Waiting and Deconstruction: Between Order and Chaos in Postmodern Drama." Pp. 63–78 in *Time, Order, and Chaos: The Study of Time IX*, edited by J. Fraser, M. P. Soulsby, and A. Argyros. Madison, CT: International University Press.

Cohen, Yosef. (1988). *Sefer ha-Teshuvah.* Jerusalem: Harry Fischel Institute.

Cole, J. Preston. (1971). *The Problematic Self in Kierkegaard and Freud.* New Haven, CT: Yale University Press.

Danzger, Moshe H. (1989). *Returning to Judaism.* New Haven, CT: Yale University Press.

Deleuze, Gilles. (1994). *Difference and Repetition.* Trans. P. Patton. New York: Columbia University Press.

Denis, Allan. (1995). "Temporality and Modes of Language." *Int. J. Psycho-Anal.* 76:1109–19.

Dessler, Eliyahu E. (1991). *Mikhtav me-Eliyyahu.* Ed. Aryeh Carmell. Jerusalem: Sifreati.

Diller, Jerry. (1980). "Identity Rejection and Reawakening in the Jewish Context." *J. Psychol. and Judaism* 5:38–47.

Don, Yosef. (1975). *Sifrut ha-Musar ve-ha-Derush.* Jerusalem: Keter.

———. (1985). "A Note on the History of *Teshuvah* among Ashkenazi Chasidim." Pp. 271–82 in *The World of Rav Kuk's Thought*, edited by B. Ish Shalom and S. Rosenberg, translated by S. Carmy and B. Casper. Jerusalem: Avi Chai.

Donald, Merlen. (1991). *Origins of the Modern Mind.* Cambridge, MA: Harvard University Press.

Edelman, Gerald. (1990). *The Remembered Present: A Biological Theory of Consciousness.* NY: New York University Press.
Edelson, Marshall. (1975). *Language and Interpretation in Psychoanalysis.* Chicago: University of Chicago Press.
Elchardus, Mark. (1993). "The Revolutionary Transition and the Modern Problem of Process." Pp. 87–107 in *Time and Process: Interdisciplinary Issues—The Study of Time VII*, edited by J. Fraser and L. Rowell. Madison, CT: International University Press.
Eliade, Mircea. (1949). *Cosmos and History: The Myth of the Eternal Return.* New York: Harper, 1959.
Emde, Robert N., and Oppenheim, David. (1995). "Shame, Guilt, and the Oedipal Drama: Developmental Considerations Concerning Morality and the Referencing of Critical Others." Pp. 413–36 in *Self-Conscious Emotions*, edited by J. Tangney and K. Fischer. New York: Guilford.
Exline, Julie, and Baumeister, Roy F. (2000). "Expressing Forgiveness and Repentance." Pp. 133–55 in *Forgiveness: Theory, Research, and Practice*, edited by M. McCullough, K. Pargament, and C. E. Thoresen. New York: Guilford.
Fairfield, Susan. (2001). "Analyzing Multiplicity: A Postmodern Perspective on Some Current Psychoanalytic Theories of Subjectivity." *Psychoanal. Dial.* 11:221–51.
Fink, Bruce. (1995a). *The Lacanian Subject.* Princeton, NJ: Princeton University Press.
———. (1995b). "The Real Cause of Repetition." Pp. 223–29 in *Reading Seminar XI: Lacan's Four Fundamental Concepts of Psychoanalysis*, edited by R. Feldstein, B. Fink, and M Jaanus. Albany, NY: SUNY Press.
———. (1996). "Logical Time and the Precipitation of Subjectivity." Pp. 356–86 in *Reading Seminars I and II: Lacan's Return to Freud*, edited by R. Feldman, B. Fink, and M. Jaanus. Albany, NY: SUNY Press.
Fox, David. (1987). "Suffering and Atonement as a Psycho-Judaic Construct." *J. of Psychology and Judaism* 11:91–102.
Freud, Sigmund. (1898). "Screen Memories." Pp. 301–22 in *S. E.*, vol. III.
———. (1900). *The Interpretation of Dreams.* In *S. E.*, vol. V.
———. (1901). *The Psychopathology of Everyday Life.* In *S. E.*, vol. VI.
———. (1905). *Three Essays on the Theory of Sexuality.* Pp. 135–243 in *S. E.*, vol. VII.
———. (1908 [1907]). "Creative Writers and Day Dreaming." Pp. 142–53 in *S. E.*, vol. IX.
———. (1912). "The Dynamics of the Transference." Pp. 99–108 in *S. E.*, vol. XII.
———. (1914). "Remembering, Repeating, and Working-Through (Further Recommendations on the Technique of Psychoanalysis, II)." Pp. 145–56 in *S. E.*, vol. XII.
———. (1916a). "Some Character-Types Met With in Psycho-Analytic Work." Pp. 309–33 in *S. E.*, vol. XIV.
———. (1916b). "On Transience." Pp. 305–7 in *S. E.*, vol. XIV.
———. (1916–1917 [1915]). "Mourning and Melancholia." Pp. 243–58 in *S. E.*, vol. XIV.
———. (1917). *Introductory Lectures on Psycho-Analysis.* Part III: General Theory of the Neuroses. Pp. 243–448 in *S. E.*, vol. XVI.
———. (1919). "The Uncanny." Pp. 219–56 in *S. E.*, vol. XVII.
———. (1920). *Beyond the Pleasure Principle.* Pp. 7–64 in *S. E.*, vol. XVIII.
———. (1923). *The Ego and the Id.* Pp. 12–59 in *S. E.*, vol. XIX.

———. (1925a [1924]). "A Note upon the 'Mystic Writing Pad.'" Pp. 227–32 in *S. E.*, vol. XIX.

———. (1925b). "Negation." Pp. 232–39 in *S. E.*, vol. XIX.

———. (1937). "Analysis Terminable and Interminable." Pp. 211–53 in *S. E.*, vol. XXIII.

Friedman, Lawrence. (1994). "Overview: Knowledge and Authority in the Psychoanalytic Relationship." *Psychoanal. Quart.* 65:254–60.

Fromm, Erich. (1966). *You Shall Be as Gods.* New Haven, CT: Yale University Press.

Gellman, Yehudah. (1982). "Teshuvah and Authenticity." *Tradition* 20:249–53.

Glanz, David, and Harrison, Michael T. (1978). "Varieties of Identity Transformation." *Jewish J. of Sociology* 20:129–41.

Goldberg, Hillel. (1976). "Toward an Understanding of Rabbi Israel Salanter." *Tradition* 16:83–119.

———. (1982). *Israel Salanter: Text, Structure, Idea.* New York: Ktav.

Goldberg, Peter. (1990). "Actively Seeking the Holding Environment." Pp. 271–99 in *Master Clinicians Treating the Regressed Patient*, edited by L. B. Boyer and P. Giovachinni. Northvale, NJ: Aronson.

Gotfryd, Arnie, Branover, Herman, and Lipskar, Shalom, eds. (1990). *Fusion.* Jerusalem: Feldheim.

Gottlieb, Mel. (1975). "Israel Salanter and Therapeutic Values." *Tradition* 15:112–29.

Green, Andre. (1971). "The Unbinding Process." Pp. 331–59 in *On Private Madness*, translated by L. Duisit. Madison, CT: International University Press, 1986.

———. (1973). "The Double and the Absent." Pp. 311–30 in *On Private Madness.* Trans. J. F. Houis. Madison, CT: International University Press, 1986.

———. (1999). *The Work of the Negative.* Trans. A Weller. London: Free Association.

———. (2000). *Andre Green at the* Squiggle *Foundation.* Ed. J. Abram. London: Karnac.

———. (2002). *Time in Psychoanalysis.* Trans. A. Weller. New York: Free Association.

Grosman, Meir. (1986). *Al Het ve-ha-Teshuvah.* Jerusalem: Ahavah.

Grotstein, James S. (1997a) "Autochthony (Self-Creation) and Alterity (Cocreation): Psychic Reality in Counterpoint." Pp. 37–57 in *Who Is the Dreamer Who Dreams the Dream*? Hillsdale, NJ: Analytic Press, 2000.

———. (1997b). "A Fearful Symmetry and the Calipers of the Infinite Geometer." Pp. 59–81 in *Who Is the Dreamer Who Dreams the Dream?* Hillsdale, NJ: The Analytic Press, 2000.

Gruenwald, Itamar. (1985). "The Concept of *Teshuvah* in the Teachings of Maimonides and Rav Kuk." Pp. 283–304 in *The World of Rav Kuk's Thought*, edited by B. Ish Shalom and S. Rosenberg, translated by S. Carmy and B. Casper. Jerusalem: Avi Chai, 1985.

Harari, Roberto. (2001). *Lacan's Seminar on Anxiety.* New York: Karnac.

Hartocollis, Peter. (1983). *Time and Timelessness.* New York: International University Press.

Hawkes, Thomas. (1977). *Structuralism and Semiotics.* Berkeley, CA: University of California Press.

Hegel, G. W. Fredrich. (1810–1814). *The Phenomenology of the Mind.* Trans. J. B. Baillie. London: Allen and Unwin, 1949.

———. (1825). *The Phenomenology of the Spirit.* Trans. A. V. Miller. Oxford: Oxford University Press, 1977.

Heschel, Abraham Joshua. (1950). *Torah min ha-Shamayim ba-Aspaklaria shel ha-Dorot.* Vol. I: *Teshuvah ve-Kapparah.* London: Heinemann.

Hirsch, Samson [ben] Raphael. (1878). *The Pentateuch.* Trans. I. Levy. England: Judaica, 1976.

Horowitz, Yosef Y. (1950). *Madregat ha-Adam.* Jerusalem: Orliner, 1960.

Jaanus, Marie. (1996). "'Civilization of Hatred': The Other in the Imaginary." Pp. 323–55 in *Reading Seminars I and II: Lacan's Return to Freud,* edited by R. Feldman, B. Fink, and M. Jaanus. Albany, NY: SUNY Press.

Jones, Ernest. (1928). *Psycho-Analysis.* London: Ernest Benn.

Jones, James W. (1991). *Contemporary Psychoanalysis and Religion.* New Haven, CT: Yale University Press.

Julien, Philippe. (1994). *Jacques Lacan's Return to Freud.* Trans. D. Simiu. New York: New York University Press.

Katz, Dov. (1974). *Tenuat ha-Mussar.* Jerusalem: Brody-Katz.

Keller, Helen. (1903). *The Story of My Life.* 2nd ed. New York: Grosset and Dunlap, 1956.

Kierkegaard, Soren. (1843). *Repetition.* Trans. W. Lowrie. Princeton, NJ: Princeton University Press, 1946.

———. (1849). *The Sickness unto Death.* Trans. W. Lowrie. Princeton, NJ: Princeton University Press, 1954.

———. (1850). *Concluding Unscientific Postscript to the Philosophical Fragments.* Trans. D. Swenson. Princeton, NJ: Princeton University Press, 1944.

Klein, Melanie. (1935). "A Contribution to the Psychogenesis of Manic-Depressive States." Pp. 262–89 in *Love, Guilt, and Reparation, and Other Works 1928–1945.* London: Virago.

———. (1940). "Mourning and Its Relation to Manic-Depressive States." Pp. 344–69 in *Love, Guilt, and Reparation, and Other Works 1928–1945.* London: Virago.

———. (1946). "Notes on Some Schizoid Mechanisms." Pp. 1–24 in *Envy, Gratitude, and Other Works 1946–1963.* London: Virago.

Knoblauch, Steven H. (1997). "The Patient Who Was Touched by and Knew Nothing about God." Pp. 43–56 in *Soul on the Couch: Spirituality, Religion, and Morality in Contemporary Psychoanalysis,* edited by C. Spezzano and G. J. Gargiulo. Hillsdale, NJ: Analytic Press.

Kohon, Gregorio. (1999). *No Lost Certainties to Be Discovered.* London: Karnac.

Kohut, Heinz. (1984). *How Does Analysis Cure?* Chicago: University of Chicago Press.

Koppel, Moshe. (1997). *Metahalakhah.* Northvale, NJ: Aronson.

Korman, Avraham. (2002) "*Vidduy Maaser.*" Pp. 103–4 in *Musagim be-Halakhah.* Tel-Aviv: Re' emim.

Koyeve, Alexandre. (1939). *Introduction to the Work of Hegel.* Trans. J. Nichols, Jr. New York: Basic Books, 1969.

Kriegman, Daniel. (1998). "Interpretation, the Unconscious, and Analytic Authority." Pp. 187-272 in *Empirical Perspectives on the Psychoanalytic Unconscious,* edited by

R. F. Bornstein and J. B. Masling. Washington, DC: American Psychological Association.

Kristeva, Julia. (1977). "*Stabat Mater.*" Pp. 160–85 in *The Kristeva Reader*, translated and edited by T. Moi. London: Blackwell.

———. (1981). *Language: The Unknown.* Trans. A. Menke. Columbia University Press, 1989.

———. (1983). "Freud and Love: Trauma and Its Discontents." Pp. 238–71 in *The Kristeva Reader*, edited by T. Moi, translated by L. S. Roudiez. London: Blackwell, 1986.

———. (1987). *Black Sun: Depression and Melancholia.* Trans. L. S. Roudiez. Cambridge, UK: Cambridge University Press, 1989.

———. (1994). *Time and Sense: Proust and the Experience of Literature.* Trans. R. Guberman, New York: Columbia University Press, 1996.

Kubler, George. (1962). *The Shape of Time.* New Haven, Conn.: Yale University Press.

Kuk, Abraham Issac ha-Cohen. (1924). *Orot ha-Teshuvah.* Ed. Y. Halevi Filber. Jerusalem: Gal-Or, 1987.

Lacan, Jacques. (1946). "*Le temps logique et l'assertion de certitude anticipee.*" Pp. 197–213 in *Ecrits.* Paris: du Seuil, 1966. ["Logical time and the assertion of unanticipated certainty." Trans. M. Silver and B. Fink. *Newsletter of the Freudian Field* 2, no. 2 (1988): 4–22.]

———. (1953–1954). *Seminar I: Freud's Papers on Technique: 1953–1954.* Trans. S. Tomaselli. Ed. J.-A. Miller. New York: W.W. Norton, 1988.

———. (1954–1955). *Seminar II: The Ego in Freud's Theory and in the Technique of Psychoanalysis: 1954–1955.* Trans. S. Tomaselli. Ed. J.-A. Miller. New York: W.W. Norton, 1988.

———. (1955–1956). *Seminar III: The Psychoses.* Trans. R. Grigg. New York: W. W. Norton, 1993.

———. (1962–1963). Seminare X: L'Angoisse. Ed. J.-A. Miller. Unpublished.

———. (1964). *Seminar XI: The Four Fundamental Concepts of Psycho-Analysis.* Trans. A. Sheridan. Ed. J.-A. Miller. London: Routledge, 1977.

———. (1972–1973 [1975]). *Seminar XX: Encore.* Trans. B. Fink. New York: W. W. Norton, 1998.

LaCapra, Dominick. (1994). *Representing the Holocaust.* Ithaca, NY: Cornell University Press.

Lamm, Norman. (1990). *Torah Umadda.* Northvale, NJ: Aronson.

Langer, Susanne. (1942). *Philosophy in a New Key.* Cambridge, MA: Cambridge University Press.

Langs, Robert. (1978). *Technique in Transition.* New York: Aronson.

Laplanche, Jean. (1992a). "Time and the Other." Pp. 234–59 in *Essays on Otherness*, translated by L. Thurston. London: Routledge, 1999.

———. (1992b). "Transference: Its Provocation by the Analyst." Pp. 214–33 in *Essays on Otherness*, translated by L. Thurston. London: Routledge, 1999.

Laplanche, Jean, and Pontalis, Jean-Bertrand. (1967). *The Language of Psycho-Analysis.* Trans. D. Nicholson-Smith. New York: Norton, 1973.

Leary, Kimberlyn. (1994). "Psychoanalytic 'Problems' and Postmodern 'Solutions.'" *Psychoanal. Quart.* 68:433–65.

Leavy, Stanley A. (1988). *In the Image of God*. New Haven, CT: Yale University Press.

Lefort, Rosine, and Lefort, Robert. (1980). *Birth of the Other*. Trans. M. Du Ry, L. Watson, and L. Rodriguez. Chicago: University of Illinois Press, 1994.

Leiber, Jerome. (1996). "Helen Keller as Cognitive Scientist." *Phil. Psychol.* 9:419–40.

Lewis, C. S. (1961). *The Screwtape Letters*. New York: Simon and Schuster.

Lichtenstein, Aharon. (1999). "The Duties of the Heart and Response to Suffering." Pp. 21–61 in *Jewish Perspectives on the Experience of Suffering*, edited by S. Carmy. Northvale, NJ: Aronson.

Loewald, Hans W. (1973). "Some Considerations on Repetition and the Repetition-Compulsion." Pp. 87–101 in *Papers on Psychoanalysis*. New Haven, CT: Yale University Press, 1980.

———. (1988). *Sublimation: Inquiries into Theoretical Psychoanalysis*. New Haven, CT: Yale University Press.

Lyotard, Jean-Francois. (1988). *Heidegger and "the jews."* Trans. A. Michel and M. Roberts. Minneapolis: University of Minnesota Press, 1995.

Matte-Blanco, Ignacio. (1975). *The Unconscious as Infinite Sets*. London: Duckworth.

———. (1988). *Thinking, Feeling, and Being*. London: Routledge.

McDargh, John. (1986). *Psychoanalytic Object Relations Theory and the Study of Religion*. New York: University Press of America.

Meares, Russell. (2000). *Intimacy and Alienation: Memory, Trauma and Personal Being*. London: Routledge.

Meltzer, Donald. (1983). *Dream-Life*. Perthshire: Clunie.

Mester, Roberto, and Klein, Hillel. (1981). "The Young Jewish Revivalist." *Brit. J. Med. Psychol.* 54:299–306.

Metzger, Alter, B. Z. (1968). *Rabbi Kuk's Philosophy of Repentance*. New York: Yeshiva University Press.

Milner, Marion. (1956). "The Sense in Nonsense (Freud and Blake's *Job*)." Pp. 169–91 in *The Suppressed Madness of Sane Men: Forty-Four Years of Psychoanalysis*. London: Routledge, 1996.

Mitchell, Stephen. (1988). *Relational Concepts in Psychoanalysis: An Integration*. Cambridge, MA: Harvard University Press.

Mitrani, Judith L. (1995). "Toward an Understanding of Unmentalized Experience." Pp. 205–47 in *A Framework for the Imaginary*. Northvale, NJ: Aronson, 1996.

Montefiore, Charles, G. (1904). "Rabbinic Conception of Repentance." *Jew. Quart Rev.* 16:209–57.

Muller, John P. (1996). *Beyond the Psychoanalytic Dyad*. London: Routledge.

Muller, John P., and Richardson, William J., eds. (1988). *The Purloined Poe*. Baltimore: Johns Hopkins University Press.

Nachmani, Gil. (1995). "Trauma and Ignorance." *Contemp. Psychoanal.* 33:423–51.

Neuberger, Yaakov. (1992). "Halakhah and Scientific Method." *Torah Umaada J.* 3:82–89.

Newberg, Andrew B., d'Aquili, Eugene, Newberg, Stephanie K., and deMarici, Verushka. (2000). "The Neuropsychological Correlates of Forgiveness." Pp. 91–110 in *Forgiveness: Theory, Research, and Practice*, edited by M. McCullough, K. Pargament, and C. E. Thoresen. New York: Guilford.

Nobus, Dany. (2000). *Jacques Lacan and the Freudian Practice of Psychoanalysis.* London: Routledge.

Ogden, Thomas H. (1986). *Matrix of the Mind.* Northvale, NJ: Aronson.

———. (1989). "On the Structure of Experience." Pp. 69–96 in *Master Clinicians on the Treatment of the Regressed Patient,* edited by L. B. Boyer and P. Giovachinni Northvale, NJ: Aronson.

———. (1992). *The Primitive Edge of Experience.* London: Karnac.

———. (1994). *Subjects of Analysis.* Northvale, NJ: Aronson.

———. (1997). *Reverie and Interpretation.* Northvale, NJ: Aronson.

Ostow, Mortimer. (1995a). "The Sources and Signs of Jewish Identity." *Israel J. Psychiatry* 32:229–36.

———. (1995b). "Comment on the Sources and Signs of Jewish identity." *Israel J. Psychiatry* 32:248–51.

Pachter, Mordechai. (1985). "Repentance in the Thought of Rav Yisroel Salanter and the Musar Movement." Pp. 322–48 in *The World of Rav Kuk's Thought,* edited by B. Ish Shalom and S. Rosenberg, translated by S. Carmy and B. Casper. Jerusalem: Avi Chai.

Parsons, Michael. (1999). "Psychic Reality, Negation, and the Analytic Setting." Pp. 59–75 in *The Dead Mother: The Work of Andre Green,* edited by G. Kohon. London: Routledge.

Peirce, Charles S. (1953). *Collected Works.* Ed. C. Hartshorne and P. Weiss. Cambridge, MA: Harvard University Press.

Peraldi, Francois. (1988). "A Note on Time in 'The Purloined Letter.'" Pp. 330–42 in *The Purloined Poe,* edited by J. P. Muller and W. J. Richardson. Baltimore: Johns Hopkins University Press.

Proust, Marcel. (1913). "Swann's Way." Pp. 3–415 in *Remembrance of Things Past,* translated by C. K. Scott Moncrieff and T. Kilmartin. Vol. I. London: Penguin (1913) 1981.

Rabinowitz, Aaron. (1999). *Judaism and Psychology: Meeting Points.* Northvale, NJ: Aronson.

Radden, Jennifer. (1987). "Melancholy and Melancholia." Pp. 209–34 in *Pathologies of the Modern Self,* edited by D. M. Levine. New York: New York University Press.

Ragland, Ellie. (1996). "An Overview of the Real, With Examples from Seminar I." Pp. 192–215 in *Reading Seminars I and II: Lacan's Return to Freud,* edited by R. Feldman, B. Fink, and M. Jaanus. Albany, NY: SUNY Press.

Randour, Mary Lou, ed. (1993). *Exploring Religious Landscapes: Religious and Spiritual Experiences in Psychotherapy.* New York: Basic Books.

Rayner, Eric. (1995). *Unconscious Logic.* London: Routledge.

Rice, Emanuel. (1999). "Religion and the Adolescent." *Psychoanal. Psychol.* 16:58–75.

Riesenberg-Malcom, Rosine. (1981). "Expiation as a Defense." *Int. J. Psychoanal. Psychother.* 8:549–70.

Rizzuto, Ana-Maria. (1979). *The Birth of the Living God.* New Haven, CT: Yale University Press.

Rose, Gilbert. (1986). *The Power of Form.* New York: International University Press.

Rosenberg, Shalom. (1976). "'Continuing Revelation'—Three Directions." [Hebrew]. Pp. 131–43 in *Revelation, Belief, and Understanding*, edited by M. Hallamish and M. Schwartz. Ramat-Gan, Israel: Bar-Ilan University Press.

———. (2000). "ha-Ani u-Teshuvato." Pp. 108-19 in *Bi-heyoto Karov: Asufat Maamarim la-Yamim ha-Noraim*, edited by E. Ganzel, E. Plas, O. Gartner, and D. Yegal-Fiegelstock. Jerusalem: Or Etzyon.

Rothenberg, Daniel J. (1997). "Formulation, Psychic Space, and Time: New Dimensions in Psychoanalysis and Jewish Spirituality." Pp. 57–77 in *Soul on the Couch*, edited by C. Spezzano and G. Gargiulo. Hillsdale, NJ: Analytic Press.

Roudinesco, Elisabeth. (1993). *Jacques Lacan*. Trans. B. Bay. New York: Columbia University Press.

Rucker, Nancy, and Lombardi, Karen. (1998). *Subject Analysis*. London: Routledge.

Sacks, Oliver. (1984). *A Leg to Stand On*. New York: Touchstone, 1993.

Sandler, Joseph, and Fonagy, Peter, eds. (1997). *Recovered Memories of Abuse*. London: Karnac.

Sartre, Jean-Paul. (1943). *Being and Nothingness*. Trans. H. Barnes, New York: Philosophical Library, 1956.

Schachter, Daniel J. (1996). *Searching for Memory*. New York: Basic Books.

Schafer, Roy. (1960). "The Loving and Beloved Superego in Freud's Structural Theory." Pp. 163–88 in *The Psychoanalytic Study of the Child*, vol. XV, edited by A. Freud, M. Kris, and R. Loewenstein. New York: International University Press.

———. (1970). "The Psychoanalytic Vision of Reality." *Int. J. Psycho.-Anal.* 51:279–97.

Schecter, Solomon. (1909). *Aspects of Rabbinic Theology*. New York: Schocken, 1961.

Schiffer, Irving. (1978). *The Trauma of Time*. New York: International University Press.

Schimmel, H. Chaim, and Carmell, Aryeh, eds. (1989). *Encounter*. Jerusalem: Feldheim.

Schweid, Eliezer. (1985). "Repentance in Twentieth-Century Jewish Thought." Pp. 349–72 in *The World of Rav Kuk's Thought*, edited by B. Ish Shalom and S. Rosenberg, translated by S. Carmy and B. Casper. Jerusalem: Avi Chai, 1985.

Sebald, W. G. (2001). *Austerlitz*. Trans. A. Bell. New York: Random House.

Segal, Hanna. (1957). "Notes on Symbol Formation." Pp. 49–65 in *The Work of Hanna Segal*. London: Karnac, 1981.

Shaffir, William. (1991). "Conversion Experiences: Newcomers to and Defectors from Orthodox Judaism." Pp. 173–202 in *Tradition, Innovation, and Conflict: Jewishness and Judaism in Contemporary Israel*, edited by Z. Sobel and B. Bet-Hallahmi. New York: SUNY Press.

Shanon, Benny. (1993). *The Representational and the Presentational*. London: Harvester.

Shokek, Shimon. (1985). "Status of *Baale Teshuvah*." Pp. 305–21 in *The World of Rav Kuk's Thought*, edited by B. Ish Shalom and S. Rosenberg, translated by S. Carmy and B. Casper. Jerusalem: Avi Chai.

Siggins, Lorraine D. (1983) "Psychoanalysis and Ritual." *Psychiatry* 46:2–15.

Snelling, David. (2000). "Subject, Object, World: Some Reflections on the Kleinian Origins of the Mind." Pp. 101–18 in *The Analytic Freud*, edited by M. P. Levine. London: Routledge.

Sokol, Moshe Z., ed. (1992). *Rabbinic Authority and Personal Autonomy.* Northvale, NJ: Aronson.

———. (1997). *Engaging Modernity.* Northvale, N.J.: Aronson.

Soloveitchik, Joseph Dov. (1944). *The Halakhic Mind.* New York: Seth Press, 1986.

———. (1980 [1974]). *Al ha-Teshuvah: Devarim she-be-Al Peh.* Ed. and trans. P. Peli. Jerusalem: Torah Education Department-World Zionist Organization, 1984.

———. (1978). "Redemption, Prayer, Talmud Torah." *Tradition* 17:55–72.

Spero, Moshe H. (1977). "Religious Anxiety and the Experience of God." *Judaism* 26:206–16.

———. (1982). *Judaism and Psychology.* New York: Ktav/Yeshiva University Press.

———. (1986a). "Aspects of Identity Development among Nouveaux Religious Patients." Pp. 379–416 in *The Psychoanalytic Study of the Child,* vol. XLI, edited by A. J. Solnit, P. B. Neubauer, S Abrams, and A. S. Dowling. New Haven, CT: Yale University Press.

———. (1986b). *Handbook of Psychotherapy and Jewish Ethics.* New York: Feldheim.

———. (1992). *Religious Objects as Psychological Phenomena.* Chicago: University of Chicago Press.

———. (1993). "The Temporal Framework and Lacan's Concept of the Unfixed Psychoanalytic Hour." Pp. 115–42 in *The Psychoanalytic Study of the Child,* vol. XLVIII, edited by A. J. Solnit, P. B. Neubauer, S. Abrams, and A. S. Dowling. New Haven, CT: Yale University Press.

———. (1994). "The Talmudic Perception of Narcissisus: The Subversion of Mirroring by Symbolizing Death." *Psychoanal. and Contemp. Thought* 17:137–69.

———. (1997). "Modes of Knowing the Self in the Clinical Encounter: Toward the Recognition of Subjectivity in the Postmodern Psychoanalytic Hour." *J. Soc. Wk. and Policy in Israel* 9–10:139–82.

———. (1998a). "'Brilliantly Lit Streets': Oedipal and Preoedipal Influences on the Transformation of God Representations in a Schizoid Male." *Psychoanal. and Contemp. Thought* 21:521–91.

———. (1998b). *Psychotherapy of the Religious Patient.* Northvale, NJ: Aronson.

———. (1998c). "Divine Space: Contemplating the Object behind a Religious Patient's Spatial Metaphor." *J. Amer. Acad. Psychoanal.* 26:457–85.

———. (1999). "Metapsychological Dimensions of Religious Suffering: Common Ground between Halakhic Judaism and Psychoanalysis." Pp. 213–76 in *Jewish Perspectives on the Experience of Suffering,* edited by S. Carmy. New York: Aronson.

———. (2000) "'A Garland for Roses': Regarding the Diagnosis of Religious Rituality in Jay F. Schachter's 'Diagnosis and Treatment of a Psychotic Depressive.'" *Torah Umadda J.* 9:152–68.

Spero, Shubert. (1986). "Towards a Philosophy of Modern Orthodoxy." *Mod. Judaism* 6:72–89.

———. (1996). "Rabbi Joseph Dov Soloveitchik and the Philosophy of Halakha." *Tradition* 30:41–64.

———. (2004). "What is Self-Theory, and Does Judaism Need One?" *Torah U-Madda Journal* 12:130–157.

Spezzano, Charles. (1993). *Affect in Psychoanalysis.* Hillsdale, NJ: Analytic Press.

Stein, Samuel M., ed. (1999). *Beyond Belief: Psychotherapy and Religion.* London: Karnac.

Steiner, Robert. (1992). "The Paranoid-Schizoid and Depressive Positions." Pp. 25–39 in *Psychic Retreats: Pathological Organizations in Psychotic, Neurotic, and Borderline Patients.* London: Routledge, 1993.

———. (1993). "Revenge, Resentment, Remorse, and Reparation." Pp. 74–87 in *Psychic Retreats.* London: Routledge.

Steinsaltz, Adin. (1988). *The Strife of the Spirit.* New York: Aronson.

Stern, Daniel N. (1985) *The Interpersonal World of the Infant.* New York: Basic Books.

Stolorow, Robert D., and Atwood, George. (1992). *The Contexts of Being: The Intersubjective Foundations of Psychological Life.* Hillsdale, NJ: The Analytic Press.

Stolorow, Robert D., Orange, Donna M., and Atwood, George. (2001). "Cartesian and Post-Cartesian Trends in Relational Psychoanalysis." *Psychoanal. Psychol.* 18:468–84.

Ullman, Chana. (1989). *The Transformed Self.* New York: Plenum.

Urbach, Ephraim. E. (1950). "*Teshuvatan Shel Nineveh ve-ha-Vikkuah ha-Yehudi-Notzri.*" In *Sefer ha-Yovel li-Kevod Epstein.* Jerusalem: Hebrew University Press, 1966.

———. (1971). *The Sages: Their Concepts and Beliefs.* Jerusalem: Hebrew University/Magnes Press.

Ury, Zalman. (1970). *The Musar Movement.* Ed. Leon Stitskin. New York: Yeshiva University Press.

Van Buren, Jane. (2002). "The Emergence of the Ineffable." Pp. 141–58 in *Primitive Mental States,* vol. II, edited by S. Alhanati. New York: Karnac.

Van Pelt, Tamise. (2000). *The Other Side of Desire.* Albany, NY: SUNY Press.

Viderman, Serge. (1979). "The Analytic Space." *Psychoanal. Quart.* 48:257–91.

Winnicott, Claire. (1991). "D.W.W.: A Reflection." Pp. 3–20 in *Tactics and Techniques in Psychoanalytic Therapy,* vol. III, edited by P. L. Giovacchini. Northvale, NJ: Aronson.

Winnicott, Donald W. (1953) "Transitional Objects and Transitional Phenomena." Pp. 1–25 in *Playing and Reality.* London: Tavistock, 1971.

Wright, Kenneth. (1991). *Vision and Separation: Between Mother and Baby.* London: Free Association.

Zahn-Waxler, Carrolyn, and Robinson, Joann. (1995). "Empathy and Guilt: Early Origins of Feelings of Responsibility." Pp. 143–73 in *Self-Conscious Emotions,* edited by J. P. Tangney and K. Fischer. New York: Guilford.

Zevin, Shmuel Yosef. (1944). *Ha-Moadim ba-Halakhah.* Jerusalem: Bet Hillel.

Index

R. Abraham ben David (Ravad), 71, 85, 89
Adam, 72–73, 229
Agushewitz, R. Reuven, 104, 147n12, 148n23
Ainslie, George, 111
akrasia, 87
Albo, R. Yosef, 132, 151n58
anxiety, 112–13, 239, 256
Aragno, Anne, 236
Aristotle, 83
astrology, 71–72
Aulaugnier, Piera, 236

Baal Halakhot Gedolot, 157
Becker, Ernest, 260
Berlin, Isaiah, 156
Berlin, R. Naftali Tzvi Yehudah (Netziv), 157
Beruriah, 100n81, 163
Bion, W. R., 217
Bok, Hilary, 133
Boswell, James, 104
Buridan's ass, 124–25

causality, 15, 27, 29–30, 66, 69, 129; agent causation, 66, 77, 120; downward causation, 35–36, 38, 39; universal causality, 111, 122
Chajes, R. Zvi Hirsh, 145
chaos, 16, 31–32
character, 82–84, 109, 120, 124, 125, 138
child abuse, 7, 114, 140, 160–61, 167n20, 169–71
choice, 80–81, 104, 114–15, 125, 135
choice-diminished behavior, 6, 159–65
Christianity, 119, 156
Churchland, Paul, 60
Cohen, R. David, 170, 171n3
communal responsibility, 6, 159–66, 169–71
compatibilism, 5, 76–79, 84, 85, 87, 100n83, 105. *See also* free will
compulsion, 84, 85, 87
Crick, Francis, 57

David, King of Israel, 236–37, 240
Dennett, Daniel, 39–40, 57, 99n76, 122
Descartes, René, 40
Dessler, R. Eliyahu, 5, 68, 73–75, 84, 89–90, 98n65, 107, 126–31, 138, 144, 206, 237

determinism, 13, 40, 65, 87;
attractiveness of, 5, 106, 112–19; and educational techniques, 73, 83, 145; intuition about, 81–82, 112, 119–20; and language, 133–37; physical determinism, 14–18, 31–32, 38–39, 88, 89, 90, 106; psychic determinism, 81–83, 88, 90, 101n95; scientific vs. theological determinism, 68–71, 79, 96n48, 100n82; soft vs. hard determinism, 105, 112, 113, 144; universal vs. specific determinism 67–68. *See also* free will, indeterminism
Dostoevsky, Feodor, 82, 139
Dreiser, Theodore, 134–35
dualism, 56, 91n6, 148n18

Eccles, John, 28–29, 32, 34, 40
Eddington, Arthur, 28
education, 74, 114, 118, 128, 138, 164–65
Einstein, Albert, 18, 40
R. Eliezer ben Durdaya, 244, 246–47
Elisha ben Abuyah, 245–47
emergentism, 29, 33, 35, 38
evil eye, 146
evolution, 55, 60, 63, 90n4, 95n41

Fink, Bruce, 215
Fischer, John Martin, 86
Forking Paths principle. *See* Principle of Alternative Possibilities
Frankfurt, Harry, 80–81, 85, 88
free will: and agency, 77, 86, 107–8, 120, 124, 126, 144; arguments against, 110–20; arguments for, 107–10; compatibilist version of, 3, 5, 28, 81, 85, 88; and creativity, 5, 109–10, 122–26, 129, 139, 143; devalued, 72–74; difficulty defining, 88, 107, 119, 120; and divine foreknowledge, 68–70, 79, 94n29, 155; effort model of, 5, 71–72, 122–24, 126, 128, 129; and genetics, 1, 3, 45–53, 67, 72, 106, 117, 150n55; and good acts, 73, 98n66, 127–28; and Hasidic thought, 64, 68, 70–71, 75, 96n56; imagination of, 120–24; incompatibilist version of, 3, 5, 28, 37, 39, 40, 76, 81, 88; intuition about, 1, 2, 59–60, 65, 79–80, 85, 105–108, 118, 129, 143; Jewish view of, 2, 6, 48–49, 53, 74, 75, 89, 105, 137, 142, 155–56; limitations of, 6, 67–68, 74, 115, 130, 136, 137, 143, 145, 159; and materialism, 64–67; metaphysical free will, 27–28, 32, 34, 37, 39, 40; and neuropsychology, 1, 3, 34–38, 44n45, 46–47, 72, 106, 140–41; and reason, 85–88, 143, 151n60; and theodicy, 73
Freud, Sigmund, 63, 86, 95n39, 189, 197–98, 211, 212, 213, 220, 224–28, 239, 242

genetics, 57; and free will, 1, 3, 45–53, 67, 72, 106, 117, 150n55; genetic screening, 50–53; genetic vs. social inheritance, 47–49, 53
Gersonides, R. Levi (Ralbag), 63, 72, 94n29
God, 40, 51, 57, 61, 93n24, 66–67, 110, 129, 131, 143, 242, 262n1; fear of, 64–65, 178; freedom of, 126, 129, 131–33; love of, 64, 178, 236
man's relationship with, 116, 126, 177, 179, 186, 192, 209, 251–52, 261;
Green, Andre, 227–28, 254
Gruenwald, Itamar, 194
guilt, 113

Heisenberg's Uncertainty Principle, 18, 28, 111, 120, 121
heliocentrism, 54, 63, 90nn1–2
Hirsch, R. Samson Raphael, 270–1 n32
homosexuality, 67, 114, 117, 132, 162–63
Horowitz, R. Yosef (Sage of Navaradok), 196, 238
Hume, David, 63, 81, 86
Hutner, R. Yitzhak, 104

indeterminism: and free will, 77, 112, 120; and neurophysiology, 31–33, 37, 44n45; and quantum mechanics, 18–19, 28–30, 32, 77, 111–12. *See also* determinism, free will

James, Henry, 135–36
James, William, 110, 111, 124, 131, 143
Job, 251–53, 257, 278–79n67
Johnson, Samuel, 104, 129
Joseph, 68, 69, 96n52, 146

Kafka, Franz, 104, 120
Kane, Robert, 83, 99n75
Kant, Immanuel, 5, 63, 93n21, 102n104, 108, 128, 129
Kaplan, Mordechai, 62, 93n24
Kaufman, Yehezkel, 69
Keller, Helen, 189, 247–50, 253
Kierkegaard, Soren, 112, 121, 125, 132, 222–24, 238, 239, 262n2, 265n12
Kleinian school, 202, 203–5, 254, 255
Kook, R. Abraham Isaac: on death, 232; on the evil eye, 146; on free will, 5–6, 107, 130–33, 143, 144, 149nn42–44; on history 64–65; on modernity, 54, 55, 89; on *teshuvah*, 137, 200, 232, 243
Kramer, Peter D., 58–59, 61, 62
Kranzler, Chaim, 170
Kristeva, Julia, 216–17, 236, 267n22, 272–73n38
Kuhn, Thomas, 57

Lacan, Jacques, 198, 211–12, 219–21, 226–28, 239, 267n21
language, 134–37, 205, 212, 218–22, 226–27, 235, 239, 241, 248–49, 251, 260–61, 279–80n73
Laplace, Pierre Simon de, 14–15, 17, 18, 93n25, 95n47, 112
learning, 46–48
Lewis, C. S., 274n42
libertarianism, 37, 43n45, 76, 82–85, 109, 112, 114, 118, 141
Libet, Benjamin, 2, 26
Lichtenstein, R. Aharon, 61, 92n20
Lincoln, Abraham, 148n17
Locke, John, 79

macrostates vs. microstates, 16–17, 19, 22, 30, 32, 33–37, 43n41
Maimonides, R. Moses (Rambam), 54; on astrology, 71, 72; on behavior, 171; on choice, 151n60; on divine law, 166n10, 168n28; on fear of God 64; on free will 2, 62, 63, 71, 72, 85, 86, 89, 97nn62–63, 102n102, 103n107, 108, 151n58, 155; on inheritance of Torah, 47–48; on punishment, 99n78, 167n17; on *teshuvah*, 195, 201, 202, 205, 229, 230,232–33, 242–43, 253; on the *yetzer ha-ra*, 157
man, nature of, 121–22
Manekin, Charles H., 71, 72, 85, 94n29
materialism, 4, 56–58, 60, 66–67; cosmic materialism, 62; non-reductive materialism, 30; religious approach to, 55–56, 58, 60–65, 88; sin as result of, 62–63, 72
medicine, 50–53
R. Meir Simhah of Dvinsk, 196
memory, 213–14, 217–18, 241, 245
mind-body interaction, 28–30, 40
moral luck, 109, 127, 131–32
moral responsibility, 2, 5–6, 45, 67, 75–76, 77–79, 84, 87, 100n81, 108–9, 113, 130–31, 139–40, 144, 156, 158, 160. *See also* reactive attitudes
mourning, 202, 216, 218, 221, 235, 244

Nahmanides, R. Moses (Ramban), 72–73, 89
nekkudat ha-behirah, 74, 126, 128, 131
neuroscience, 19–22, 31, 34, 38–39, 56, 91n7, 180; of decision-making, 22–27, 34; and free will 1, 3, 34–38, 44n45, 46–47, 72, 106, 140–41
Newcombe, Don, 107, 115, 118
Newton, Isaac, 14, 16, 67, 93n24

Nietzsche, Friedrich, 108, 147n5
Nozick, Robert, 88, 95n44, 98n66

Ogden, Thomas H., 254

PAP. *See* Principle of Alternative Possibilities
Penrose, Roger, 29
physicalism. *See* materialism
predictability, 15–18
Principle of Alternative Possibilities, 27, 42n25, 79–88
privacy, 146, 161–62
probability, 17, 31, 32
Proust, Marcel, 217, 263n8
psychotherapy, 58–59, 113, 116–17, 140, 176–78, 191, 197–200, 222–25, 254; case studies in, 180–86; and retrieval of loss, 260–270 n31 and *teshuvah*, 7, 176–80, 182, 183–84, 186–88, 226
punishment, 77–78, 99n78, 144–45, 161, 167n17
Putnam, Hilary, 29

quantum mechanics, 18–19, 28–30, 41n5

Rabinowitz, Aron, 207
racism, 47, 50, 53n1
Ragland, Ellie, 211
reactive attitudes, 67, 76, 78–79, 108–9, 126
Resh Lakish, 244, 263n9
repentance. *See teshuvah*
reward and punishment, 67, 74, 108, 126, 129, 130

Salanter, R. Israel, 207
Sartre, Jean Paul, 112, 211, 212
Schechter, Solomon, 190
Schimmel, Solomon, 74, 83, 87
Schnabel, Freya, 52
science, 50, 61, 111, 138
scientism, 61, 138
Searle, John 29, 30, 36, 39
Seidel, Moshe, 130, 132
self, 8, 58, 85–86, 92n16, 105, 120–22, 133, 140–41, 190, 193, 204, 211–13, 217, 221, 226, 242, 262n2, 263n4. *See also teshuvah* and the self
self-esteem, 138–39
R. Shnayer Zalman of Lyadi, 70, 97nn57–58
sin, 73, 100n81, 163, 193, 200, 228, 240–41, 253
Skinner, B. F., 87, 103n111
Solomon, Andrew, 141
Soloveitchik, R. Joseph B., 57, 58, 60, 64; on free will, 82, 83, 143; on heroic man, 51; on man's drives, 63, 71–72; on man's relationship with God, 116, 232; on physicality of man, 62; on the self, 211; on *teshuvah*, 126, 137, 158, 200, 202, 206–7, 210, 222, 228, 230, 236–40, 253, 255, 263n10, 273n39
Sorotzkin, Benzion, 89
soul, 56, 61, 93n25, 120–21
Sperry, Roger, 24, 29–30, 36
Spinoza, Benedict, 113
split-brain studies, 24–25
statistical physics, 16–18, 41n4
Stern, Josef, 85
Strawson, Peter, 95n45, 108–9
suicide, 45, 160, 167n15, 170
symbolization, 269n30

teshuvah, 7–8, 53, 126, 133–33, 137, 158, 175–76, 182, 189–92, 218; attainment of, 165, 194, 200–203; *be-shaah ahat*, 195, 199, 230–31, 243–44, 247, 249, 251; case studies in, 243–50; and death, 229–32, 245; defining, 190, 192–94, 201, 255; and free choice, 82; impossibility of, 194, 195–99, 210, 230, 258; *mei-ahavah* vs. *mi-yirah*, 194, 205, 207, 229, 253–55; and *nehamah*, 195, 201–5, 252, 263–265n10; process of, 175–76; and psychoanalysis, 191–92, 197–200, 203–5, 208–9, 240–41, 258–59; and

psychotherapy, 7, 176–80, 182, 183–84, 186–88, 192; rabbi's role in, 179, 187–88; and repetition, 222–30, 258; and the self, 190–91, 193, 201, 240, 253, 256–58; and time, 233, 237–38, 243, 259, 276–77n53
time, 214–218, 221, 226, 228, 230–31, 232–43
Torah: as antidote to *yetzer ha-ra*, 156–57
Crown of, 48
transference, 7, 178–79, 185, 186, 198, 223–26, 239, 259

van Inwagen, Peter, 2, 124–25

Watson, James D., 50–51
Weigel, George, 140
Winnicott, Donald, 249–51, 253, 266n19
Wolbe, R. Shlomo, 83

yetzer ha-ra, 63, 82, 86, 127, 128, 156–59

About the Series Editor

Rabbi Robert S. Hirt is the senior advisor to the President of Yeshiva University and Vice President Emeritus of its affiliated Rabbi Isaac Elchanan Theological Seminary. In 1991, he was named inaugural occupant of the Rabbi Sidney Shoham Chair in Rabbinic and Community Leadership at RIETS.

Since 1985, he has overseen the "Torah U-Madda Project," the purpose of which is to explore the interaction of Torah and *Madda* (secular studies) within the institution and raise the level of discussion about the complexities and challenges that "Torah U-Madda" poses to the community at large.

In 1987, Rabbi Hirt—who formerly directed the University's array of Holocaust studies programs—coedited Shimon Huberband's critically acclaimed book on the Holocaust, *Kiddush Hashem: Jewish Religious and Cultural Life in Poland during the Holocaust.* He serves as the series editor of the Orthodox Forum.

About the Contributors

Yitzhak Berger, Ph.D., is assistant professor of biblical studies at Hunter College of the City University of New York. He received his doctoral degree and rabbinic ordination from Yeshiva University. He has taught at several Jewish institutions and has served as educational director of the Torah-Umadda Project at Yeshiva University. He is the author of a number of forthcoming studies on medieval biblical interpretation.

Rivkah Teitz Blau, Ph.D., is the author of essays in popular and academic journals, and of a biography of her father, *Learn Torah, Love Torah, Live Torah—Harav Mordechai Pinchas Teitz, the Quintessential Rabbi*. She wrote her doctoral thesis in English and Comparative Literature at Columbia University on Psalms in seventeenth-century English poetry. She has taught Jewish Studies and English in high school and college, served as principal of two yeshiva high schools, and is a member of the editorial board of *Tradition*.

Shalom Carmy teaches Bible, Jewish thought, and philosophy at Yeshiva University and is editor of *Tradition*. He has published extensively and is the editor of two volumes in the Orthodox Forum series, most recently *Jewish Perspectives on the Experience of Suffering*.

Basil Herring, Ph.D., is currently the executive vice president of the Rabbinical Council of America. Prior to assuming this position, he served a variety of pulpits over a thirty-year career. He has also taught on the university level, and is currently an adjunct professor of Jewish studies at Stern College for Women of Yeshiva University. Rabbi Herring has authored a number of volumes in Jewish law, Bible, and philosophy, including the two-volume *Jewish Ethics and Halachah for Our Time.*

Michelle Friedman, M.D., is assistant professor of Clinical Psychiatry at the Mount Sinai School of Medicine and director of Pastoral Counseling at Yeshivat Chovevei Torah. She also practices psychiatry and psychoanalysis in Manhattan. Her interests include the interface of psychiatry and obstetrics/gynecology and the interface of religion and psychiatry.

Robert Pollack, Ph.D., is professor of Biological Sciences, adjunct professor of Religion, lecturer in psychiatry at the Center for Psychoanalytic Training and Research, and director of the Earth Institute's Center for the Study of Science and Religion, all at Columbia University; and adjunct professor of Science and Religion at Union Theological Seminary. He is the author of more than a hundred research papers on the oncogenic phenotype of mammalian cells in culture, and has written and edited many books and reviews on aspects of molecular biology.

David Shatz, Ph.D., is professor of philosophy at Yeshiva University, editor of *The Torah u-Madda Journal*, and editor of the series *MeOtzar HoRav: Selected Writings of Rabbi Joseph B. Soloveitchik.* He has published ten books and over fifty articles and reviews, dealing with both general and Jewish philosophy. His work in general philosophy focuses on the theory of knowledge, free will, ethics, and the philosophy of religion, while his work in Jewish philosophy focuses on Jewish ethics, Maimonides, Torah and science, and twentieth-century rabbinic figures. His most recent books include *Philosophy and Faith* and *Peer Review: A Critical Inquiry.* He was ordained by the Rabbi Isaac Elchanan Theological Seminary.

Moshe Halevi Spero, Ph.D., is full professor and director of the postgraduate program for Psychoanalytic Psychotherapy, School of Social Work, Bar-Ilan University; senior clinical psychologist, Department of Psychiatry, Sarah Herzog Memorial Hospital; senior clinical psychologist, Weinstock Hematology-Oncology Day Hospital, Shaare Zedek Medical Center, Jerusalem; and scientific associate, American Academy of Psychoanalysis.

Haim Sompolinsky, Ph.D., is professor of physics at Hebrew University in Jerusalem and a founding member of its Interdisciplinary Center for Neural Computation, a worldwide leading research institute in computational neuroscience. His current research interests include cortical dynamics and sensory processing, learning and memory, and action initiation and control. He has published extensively in the areas of computational neuroscience, dynamical systems, and condensed matter physics. His collaborations include research groups at Harvard University, New York University, MIT, Max-Plank Institute, Munich, and CNRS, Paris. He received advanced Talmudic education at the Rabbi Amiel Yeshiva, Tel-Aviv, and Ponevez Yeshiva, Bnei-Brak.

Rachel Yehuda, Ph.D., is professor of psychiatry and director of the Division of Traumatic Stress Studies at the Mount Sinai School of Medicine and Bronx Veterans Affairs in New York City. She has published extensively in the field of traumatic stress and has been the recipient of numerous grants and awards from the National Institute of Mental Health and other agencies.

www.ingramcontent.com/pod-product-compliance
Lightning Source LLC
LaVergne TN
LVHW101445200926
848483LV00001B/50

* 9 7 8 0 7 4 2 5 4 5 9 5 3 *